Circuits, Systems and Signal Processing

Suhash Chandra Dutta Roy

Circuits, Systems and Signal Processing

A Tutorials Approach

Suhash Chandra Dutta Roy
Department of Electrical Engineering
Indian Institute of Technology Delhi
New Delhi, Delhi
India

ISBN 978-981-13-3901-1   ISBN 978-981-10-6919-2 (eBook)
https://doi.org/10.1007/978-981-10-6919-2

Printed on acid-free paper

This Springer imprint is published by Springer Nature
The registered company is Springer Nature Singapore Pte Ltd.
The registered company address is: 152 Beach Road, #21-01/04 Gateway East, Singapore 189721, Singapore

Dedicated
to
My Parents
Shri Suresh Chandra Dutta Roy
who told me 'high quality steel cannot be made without burning iron'
and said, 'Suhash you will surely shine'
and
Shrimati Suruchi Bala Dutta Roy
who must have given me the genes to keep the fire on
and says, 'Son, I am with you, even though you lost me at nine'

Preface

Starting from 1962, I have written and published a large number of articles in IETE Journals. At various points of time, starting from the early 80s, I have been requested by students, as well as teachers and researchers, to publish a book of collected reprints, appropriately edited and sequenced. Of late, this request has intensified, and the demand for reprints of some of the tutorial papers I wrote, has increased considerably, not only from India, but also from abroad, because of my five video courses related to Circuits, Systems and Signal Processing (CSSP) successfully uploaded by NPTEL on the YouTube. I thought it would be a good idea to venture into such a project at this time. This is a book for you, students, teachers and researchers in the subjects related to CSSP.

As you would notice, I have written this book in a conversational style to make you feel at ease while reading it. I have also injected some wit and humour at appropriate places to make it enjoyable.

This book is divided into four parts, dealing with Signals and Systems, Passive Circuits, Active Circuits and Digital Signal Processing. An appendix has also been added to give simple derivations of mathematics used throughout this book.

In each chapter, I have added some examples so that the students may appreciate the fundamentals treated in the chapter and apply them to practical cases.

This book contains chapters based on only articles of tutorial nature and those containing educational innovations. Purely research papers are excluded.

The details of the parts are as follows:

Part I on *Signals and Systems* comprises six chapters on basic concepts in signals and systems, state variable characterization, some fundamental issues involving the impulse function and partial fraction expansion of rational functions in s and z, having repeated poles.

Part II on *Analysis of Passive Circuits* consists of 16 chapters on circuit analysis without transforms, transient response of RLC networks, circuits which are deceptive in appearance, resonance, many faces of the single tuned circuit, analysis and design of the parallel-T RC network, perfect transformer, capacitor charging through a lamp, difference equations and resistive networks, a third-order driving point synthesis problem, an example of LC driving point synthesis, low-order Butterworth filters, band-pass/band-stop filter design by frequency transformation and optimum passive differentiators.

Part III on *Active Circuits* contains eight chapters on BJT biasing, analysis of a high-frequency transistor stage, transistor Wien bridge oscillator, analysis of sinusoidal oscillators, triangular to sine wave converter and the Wilson current mirror.

Part IV *Digital Signal Processing* comprises eight chapters on the ABCD's of digital signal processing, second-order band-pass and band-stop filters, all-pass digital filters, FIR lattice and fast Fourier transform.

In the *Appendix*, simplified treatment of some apparently difficult topics is presented. These are roots of a polynomial, Euler's relation, approximation of the square root of the sum of two squares, solution to cubic and quartic equations, solving second-order linear differential equations and Chebyshev polynomials.

I hope students and teachers will benefit from these chapters. It is only then that I shall feel adequately rewarded. For any mistakes/confusions/clarifications, please feel free to contact me on email at s.c.dutta.roy@gmail.com. I take such mail on top priority, I assure you.

Happy learning!

New Delhi, India Suhash Chandra Dutta Roy

Acknowledgements

I owe so much to so many that it is not possible to acknowledge all of them in this limited space. To the ones whose credit I have not been able to mention, I ask for forgiveness.

To the successive three Presidents, Shri R. K. Gupta, Shrimati Smriti Dagur and Lt Gen (Dr.) AKS Chandele, PVSM, AVSM (retd), I am indebted for their enthusiasm and encouragement;

To the successive Publication Committee Chairpersons and Members of the Governing Council, I would like to say a big thank you for supporting my proposal of writing this book;

To Shrimati Sreelatha Menon, Former Managing Editor of IETE, for her constant advice and suggestions for taking the project forward;

To Shrimati Sandeep Kaur Mangat, I would like to express my deep gratitude for successfully leading this project to a conclusion, with her immense patience, and love and passion for work, much beyond her duty, and for carrying out all the necessary hard work;

To all the workers of the Publications Section and the Secretariat in general, I have no words to convey my heartfelt appreciation for their diligent work;

To my students—real and virtual, I would like to express my love and best wishes for their raising interesting questions in the class as well as outside, and through innumerable emails, which led to the polishing and refreshing of the contents;

To my wife and lifelong companion, Shrimati Sudipta Dutta Roy, I cannot say enough to express my love and adoration for being with me in all weathers—in rain and sunshine—and in all the moments of happiness and depression, particularly during the years that I devoted to compiling and editing this book;

To my two sons, Sumantra and Shoubhik, and their families, I would like to say 'I have been and shall always be with you, even though many a times I could not pay due attention to family duties and responsibilities';

Finally, to my grandson Soham, is due a special word. He stood by my table patiently and asked me to hurry up the work so that he gets his share of time to play with me. It is because of this that this book has seen the light of the day in such a short time.

Suhash Chandra Dutta Roy

About the Book

This book, I claim, is unique in character. Such a book has never been written. This book is unique, because it is innovative all throughout. It is innovative in Titles of Chapters, Abstracts, Headings of Articles, Subheadings, References and Problems. I have injected wit and humour, so as to retain the interest of the readers and to ignite their imagination. I shall not write more about this book. Read and you will know.

I do not wish to receive any compliments, whatsoever. What I wish to receive is your frank and blunt criticism, pointing out the deficiencies. I promise I shall take them seriously and make my best efforts to take care of them in the next edition.

Happy reading!

Contents

Part IV Digital Signal Processing

About the Author

Suhash Chandra Dutta Roy I was fortunate enough to be educated at the Calcutta University, where great professors and researchers like C. V. Raman, S. N. Bose, M. N. Saha and S. Radhakrishnan taught and made a name, globally, not only for them, but also for the university and the country. Immediately after my master's, I had to take up a job in a West Bengal Government Research Institute. I soon discovered that many things are done at the Institute, but not research (like an American Drug Store!). I therefore quit and shifted to the newly established University of Kalyani as a lecturer. There I spent a few years, and after my Ph.D., I left the country for taking up an Assistant Professorship at the University of Minnesota. A few years passed by, but I felt increasingly guilty that I was not doing anything for my motherland. I therefore returned to the country as an Associate Professor at IIT Delhi where I served for more than four decades as professor, head of the department and dean. The last two positions were imposed upon me, and I suffered because they cut down the time available for research and interaction with my dear students. I formally retired at the age of 60, but continued to teach and carry out research, as an Emeritus Fellow, followed by INSA Senior Scientist and INSA Honorary Scientist. Since my term as an Emeritus Fellow was over, I did not take any money from IIT Delhi and served voluntarily, simply for the love of teaching and research. I finally quit IIT Delhi 12 years after retirement. I am now settled in a DDA Flat at Hauz Khas, where I live happily with my wife, but still continue to do research. We both are reasonably healthy, because of strict diet and exercise, including pranayama. I have been extraordinarily lucky to have had gems of Ph.D. students, 30 of them, who have done exemplary work in research and innovation. In fact, I consider myself as shining from reflected glory. I have also been lucky to have received high recognition through Fellowship of IEEE, Distinguished Fellowship of IETE and Fellowship of all the relevant national academies. I have been awarded some prestigious national awards, including the Shanti Swarup Bhatnagar Prize. Over and above all the awards and recognition, however, what I value most is the love and affection of my students. I now spend time giving professional lectures and also delivering sermons on innovations in teaching and research, in general, and also on how to improve their standards in the country. I have the hobby of listening to Hindustani classical music and researching on its masters. I also love spending quality time with my grandson, Soham. Reading political history and detective stories are my other hobbies. I also read poetry and compose some poems and short stories, for my own pleasure. In short, I have lived a complete life, with nothing to complain about.

Part I
Signals and Systems

This part contains six chapters based on the same number of articles. Although they may appear to be disjointed, in reality, they are not. For example, the second chapter is about an impulse function, which often appears as a signal and in input–output characterization of systems. State variable characterization, dealt with in the following two parts, relates to a special kind of system, viz. linear system. State variable was a hot topic when the corresponding articles were written. They are still used, mostly in control systems. The fifth part relates to a rational function in s which is the Laplace transform of the output of a system. A simple method is presented, in contrast to usual textbook methods, for partial fraction expansion of a function with repeated poles. The last part relates to the same topic in Digital Signal Processing where s is replaced by the variable z.

All throughout here, as well as everywhere in the book, my effort has been to simplify life so that you can sail in gentle waters and do not have to shed tears. Tears are costly and should not be allowed to flow just like that! Reserve them for practical life where you would have ample opportunities to shed them.

All the examples here, at the end of each chapter, have some twists and turns; you have to unwind them to be able to figure out how you should proceed. Do not think along difficult lines because difficult problems always have simple solutions. This is true not only here but in life in general. Do not give up. If a path does not lead you to the goal, return and choose an alternative path. Sometimes, when you go some way in the latter, it appears that the first path gives you the solution! There are no straightforward rules; you have to find your own path. All the time, do not allow your mind to be polluted by complicated rules and procedures. All that is complicated is useless. This is also true for life in general.

But enough of this philosophical discourse. Coming down to the problem at hand, you should learn the fundamentals carefully and completely. Once you make them your own, no application will appear difficult. I have done this throughout my life, in teaching as well as research, and I have received excellent rewards, well beyond my expectations.

Help your classmates whenever they are in difficulty in solving problems, or otherwise. This helps in strengthening the bonds and also in clearing your own understanding at places where they were fuzzy.

The problems I have set have a different character as compared to the usual textbook ones. Some have intentionally designed wrong results. You will have to correct them before going ahead. Do workout all the problems. Intentionally, I have not given the correct answers. Consult your classmates; if a majority gets the same result, then you can be confident that your solution is correct. A good teacher is born and not made. Identify one or two such good teachers and consult them in case of difficulties. Don't give up if they turn you down on one pretext or other. After all, your teacher should realize that teaching–learning is an interactive process, and unless there is a strong and loving relationship among the three components, viz. teacher, student and subject, teaching–learning becomes routine, monotonous and repelling.

Consult the video courses, fully downloadable from the YouTube, I have created five of them, but there are many more from IITs through the NPTEL programme. Choose the best few and follow them. It is not that these are all faultless, but faults are few and far between. If you find a mistake, confirm with an email to the concerned teacher and clarify. I have been doing this every day since my first video course appeared in the YouTube.

Finally, I wish you happy reading and happy learning. Write to me at the email address given earlier, if you have a question. I enjoy interacting through emails as most of the teachers also do.

1 Basic Concepts in Signals and Systems

In this chapter, I shall tell you all you wanted to know about signals and systems but were afraid to ask your teacher. Starting with the definition of linear systems, some elementary signals are introduced, which is followed by the notion of time-invariant systems. Signals and systems are then coupled through impulse response, convolution and response to exponential signals. This naturally leads to Fourier series representation of periodic signals and consequently, signal representation in the frequency domain in terms of amplitude and phase spectra. Does this sound difficult? I shall simplify it to the extent possible, do not worry. Linear system response to periodic signals, discussed next, is then easy to understand. To handle non-periodic signals, Fourier transform is introduced by viewing a non-periodic function as the limiting case of a periodic one, and its application to linear system analysis is illustrated. The concepts of energy and power signals and the corresponding spectral densities are then introduced.

Keywords

Linear systems · Signals · Amplitude and phase spectra · Response to periodic signals · Non-periodic signals · Fourier series and transforms · Energy and power signals Spectral density

Linear System

The concept of linear systems plays an important role in the analysis and synthesis of most practical systems, be it communication, control or instrumentation. Consider a system S which produces an output y when an input x is applied to it (both y and x are usually functions of time). We shall denote this symbolically as

$$x \underset{\rightarrow}{S} y$$

Then S is said to be linear if it obeys two principles, viz. principle of superposition and principle of homogeneity. The former implies that if $x_1 \rightarrow y_1$ (note that we have omitted S above the arrow: this is implied) and $x_2 \rightarrow y_2$, then $x_1 + x_2 \rightarrow y_1 + y_2$. The principle of homogeneity implies that if $x \rightarrow y$, then $\alpha x \rightarrow \alpha y$ where α is an arbitrary constant. Note, in passing, that α could be zero, i.e. zero input should lead to zero output in a linear system. Combining the two

Source: S. C. Dutta Roy, "Basic Concepts in Signals and Systems," *IETE Journal of Education*, vol. 40, pp. 3–11, January–June 1999.

Lectures delivered at Trieste, Italy, in the URSI/ICTP Basic Course on Telecommunication Science, January 1989.

S. C. Dutta Roy, *Circuits, Systems and Signal Processing*,
https://doi.org/10.1007/978-981-10-6919-2_1

principles, we can now formally define a linear system as one in which

$$x_{1,2} \rightarrow y_{1,2} \Rightarrow \alpha x_1 + \beta x_2 \rightarrow \alpha y_1 + \beta y_2, \quad (1.1)$$

where the notation $\Rightarrow$ is used to mean 'implies'.

As an example, consider the system described by the well-known equation of a straight line

$$y = mx + c \quad (1.2)$$

It may seem surprising but Eq. 1.2 does not describe a linear system unless $c = 0$, simply because zero input does not lead to zero output. Another way of demonstrating this is to apply αx as input; then the output is

$$y' = m\alpha x + c \neq \alpha y = m\alpha x + c\alpha \quad (1.3)$$

By the same token, the dynamic system described by

$$\frac{\mathrm{d}y}{\mathrm{d}t} + 5y = 5x + 6 \quad (1.4)$$

is not linear, because ($x = 0$, $y = 0$) does not satisfy the equation.

Another, and a bit more subtle example, is shown in Fig. 1.1. Is this system linear? Obviously $x = 0$ leads to $z = 0$ but then this is only a necessary condition for a linear system. Is it sufficient? To test this, apply $x = x_0$; then $z = z_0$. Now apply $x = -x_0$; the output is still z_0 instead of $-z_0$. The obvious conclusion is that the system is nonlinear.

Almost all practical systems are nonlinear, which are usually much more difficult to handle than linear systems. Hence, we make our life comfortable by approximating (or idealizing?) a nonlinear system by a linear one. This should not surprise you, because life, in general, is nonlinear and we make it simple by approximating it by a linear one. Otherwise, life would have been so complicated that it would not be worth living. Enjoyment would have been out of question! Also, in many situations, a nonlinear system is 'incrementally' linear, i.e. the system is linear if an increment Δx in x is considered as the input and the corresponding increment Δy in y is considered as the output. Both Eqs. 1.2 and 1.4 are descriptions of such incrementally linear systems. A transistor amplifier is a highly nonlinear system, but it behaves as a linear one if the input is an AC signal superimposed on a much larger DC bias.

Elementary Signals

A signal, in the context of electrical engineering, is a time-varying current or voltage. An arbitrary signal can be decomposed into some elementary or 'basic' signals, which, by themselves also occur frequently in nature. These are (i) the exponential signal $e^{\alpha t}$ where α may be real, imaginary or complex, (ii) the unit step function $u(t)$ and (iii) the unit impulse function, $\delta(t)$. When α is purely imaginary in $e^{\alpha t}$, we get a particularly important situation, because if $\alpha = j\omega$ and ω is real, then

$$e^{j\omega t} = \cos \omega t + j \sin \omega t \quad (1.5)$$

Thus sinusoidal signals, $\cos \omega t$ and $\sin \omega t$, which are so important in the study of communications, are special cases of the exponential signal. The quantity ω, as is well known, is the frequency in radians/s, while $f = \omega/(2\pi)$ is the frequency in cycles/s or Hz.

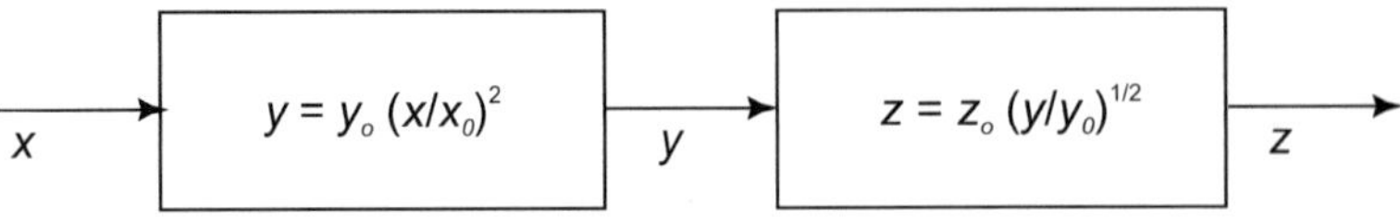

Fig. 1.1 A subtle example of a nonlinear system

The unit step function, shown in Fig. 1.2, is defined by

$$u(t) = \begin{cases} 0 & t<0 \\ 1 & t>0 \end{cases} \tag{1.6}$$

Note that it is discontinuous at $t = 0$. The unit impulse function $\delta(t)$ is related to $u(t)$ through

$$u(t) = \int_{-\infty}^{t} \delta(\tau)\mathrm{d}\tau \tag{1.7}$$

or

$$\delta(t) = \frac{\mathrm{d}u(t)}{\mathrm{d}t} \tag{1.8}$$

Obviously, it exists only at $t = 0$, and the value there is infinitely large, but

$$\int_{-\infty}^{\infty} \delta(\tau)\mathrm{d}\tau = \int_{0^-}^{0^+} \delta(\tau)\mathrm{d}\tau = 1, \tag{1.9}$$

i.e. the area under the plot of $\delta(t)$ versus t is unity. This is called the strength of the impulse; for example, the strength of the impulse $K\delta(t)$ is K. Obviously, there is some formal difficulty with regard to the definition of $\delta(t)$, but we shall not enter into this debate here. $\delta(t)$ can be viewed as the limit of the rectangular pulse shown in Fig. 1.3 as Δ tends to 0; Fig. 1.3 also shows the representation of $\delta(t)$. Two important properties of $\delta(t)$ are that

$$x(t)\delta(t-t_o) = x(t_o)\delta(t-t_o) \tag{1.10}$$

and

$$\int_{-\infty}^{\infty} x(\tau)\delta(t-\tau)\mathrm{d}\tau = x(t) \tag{1.11}$$

Equation 1.11 easily follows from Eqs. 1.9 and 1.10, and represents the 'Sifting' or 'Sampling' property of the impulse function.

Time Invariance

At this point, we need to introduce another concept, viz. that of time invariance of a system. A system S is time invariant if a time shift in the input signal causes the same time shift in the output signal, i.e. if $x(t) \to y(t)$ implies $x(t - t_0) \to y(t - t_0)$.

Both Eqs. 1.2 and 1.4 are descriptions of time-invariant systems. On the other hand, $y(t) = tx(t)$ represents a time-varying system. Most of the practical systems we encounter are time-invariant systems.

Systems which are linear and time invariant (LTI) are particularly simple to analyze in terms of their impulse responsc or frequency response function, as will be demonstrated in what follows.

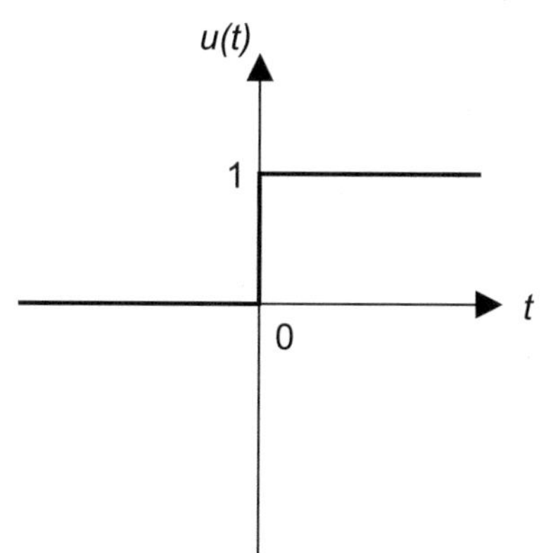

Fig. 1.2 The unit step function

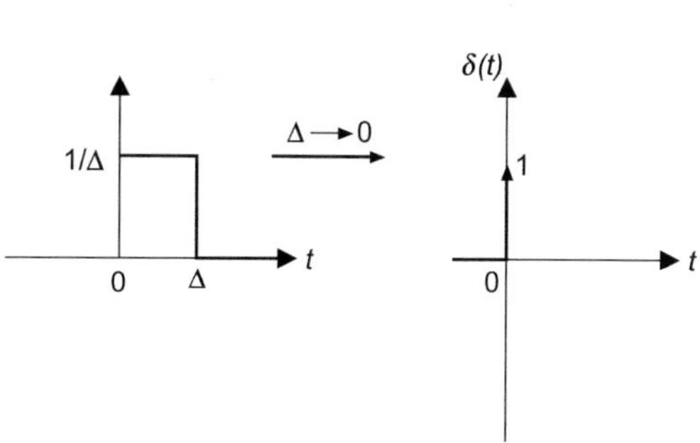

Fig. 1.3 A limiting view of δ(t)

Impulse Response and Convolution

Consider an LTI system whose response to a unit impulse function is $h(t)$, i.e.

$$\delta(t) \rightarrow h(t) \tag{1.12}$$

By time invariance, therefore

$$\delta(t-\tau) \rightarrow h(t-\tau) \tag{1.13}$$

By homogeneity, if we multiply the left-hand side of Eq. 1.13 by $x(\tau)\mathrm{d}\tau$, the right-hand side should also get multiplied by $x(\tau)\mathrm{d}\tau$, i.e.

$$x(\tau)\delta(t-\tau)\mathrm{d}\tau \rightarrow x(\tau)h(t-\tau)\mathrm{d}\tau \tag{1.14}$$

By superposition, if we integrate the left-hand side of Eq. 1.14, we should do the same for the right-hand side, i.e.

$$\int_{-\infty}^{\infty} x(\tau)\delta(t-\tau)\mathrm{d}\tau \rightarrow \int_{-\infty}^{\infty} x(\tau)h(t-\tau)\mathrm{d}\tau \tag{1.15}$$

But, by Eq. 1.11, the left-hand side of Eq. 1.15 is simply $x(t)$, so the right-hand side should be $y(t)$. Thus, if the unit impulse response $h(t)$ of an LTI system is known, then one can find the output of the system due to an arbitrary excitation $x(t)$ as

$$y(t) = \int_{-\infty}^{\infty} x(\tau)h(t-\tau)\mathrm{d}\tau \tag{1.16}$$

$$= \int_{-\infty}^{\infty} x(t-\tau)h(\tau)\mathrm{d}\tau, \tag{1.17}$$

where the second form follows simply, through a change of variable. The integral Eqs. 1.16 or 1.17 is called the convolution integral and the operation of convolution is symbolically denoted as

$$y(t) = x(t) * h(t)$$

It is a simple matter to prove that convolution operation is commutative (i.e. $x(t) * h(t) = h(t) * x(t)$); in fact, this is what equivalence of Eqs. 1.16 and 1.17 implies), associative (i.e. $x(t) * [h_1(t) * h_2(t)] = [x(t) * h_1(t)] * h_2(t)$); this is useful in the analysis of cascade connection of systems) and distributive (i.e. $x(t) * [h_1(t) + h_2(t)] = x(t) * h_1(t) + x(t) * h_2(t)$); this is useful in the analysis of parallel systems).

As an example of application of the convolution integral, consider the *RC* network shown in Fig. 1.4, where both $x(t)$ and $y(t)$ are voltages, and the capacitor is uncharged before application of $x(t)$ (an alternate way of expressing this is to say that C is initially relaxed). When $x(t) = \delta(t)$, the current in the circuit is $i(t) = \delta(t)/R$. This impulse of current charges the capacitor to a voltage

$$\frac{1}{C}\int_{0^-}^{0^+} \frac{\delta(\tau)}{R}\mathrm{d}\tau = \frac{1}{RC} \tag{1.18}$$

at $t = 0^+$. For $t > 0^+$, $\delta(t) = 0$; hence the capacitor charge decays exponentially; so does the voltage across it, according to

$$y(t) = \frac{1}{RC}\mathrm{e}^{-t/(RC)} \tag{1.19}$$

Thus, the impulse response of the *RC* network is

$$h(t) = \frac{1}{T}\mathrm{e}^{-t/T}u(t), \tag{1.20}$$

where $T = RC$ is called the time constant of the network. Now, suppose the input is changed to a unit step voltage, i.e. $x(t) = u(t)$. Then the response is, by Eq. 1.17,

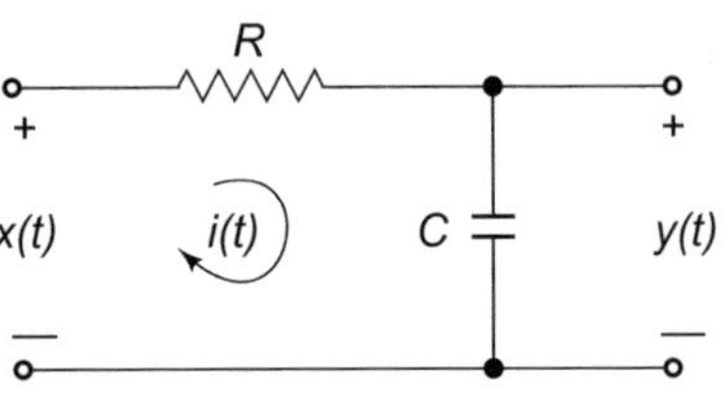

Fig. 1.4 An RC network

$$y(t) = \int_{-\infty}^{\infty} \frac{1}{T} e^{-\tau/T} u(\tau) u(t-\tau) d\tau \quad (1.21)$$

$$= \frac{1}{T} \int_{0}^{t} e^{-\tau/T} d\tau = \left(1 - e^{-t/T}\right) u(t), \quad (1.22)$$

where the lower limit arises due to the factor $u(\tau)$ and the upper limit arises as a consequence of the factor $u(t - \tau)$ in the integrand.

LTI System Response to Exponential Signals

Let $x(t) = e^{st}$ where s, as you will see later, is the complex frequency $\sigma + j\ \omega$, be applied to a system with impulse response $h(t)$; then by Eq. 1.17, the response is

$$y(t) = \int_{-\infty}^{\infty} h(\tau)\, e^{s(t-\tau)} d\tau \quad (1.23)$$

$$= e^{st} \int_{-\infty}^{\infty} h(\tau)\, e^{-s\tau} d\tau \quad (1.24)$$

$$= H(s) e^{st}, \quad (1.25)$$

where

$$H(s) = \int_{-\infty}^{\infty} h(\tau)\, e^{-s\tau} d\tau \quad (1.26)$$

is called the system function or transfer function of the system and is a function of s only. A signal for which the output differs from the input only by a scaling factor (perhaps complex) is called the eigenfunction of the system, and the scaling factor is called the eigenvalue of the system. Obviously, e^{st} is an eigen function of an LTI system, and $H(s)$ is its eigenvalue.

When $s = j\omega$, H represents the frequency response of the system, i.e. if $x(t) = e^{j\omega t}$ or its real part (cost ωt) or imaginary part (sin ωt), then the output will be $H(j\omega)\, e^{j\omega t}$ or Re $[H(j\omega)\, e^{j\omega t}]$ or Im $[H(j\omega) e^{j\omega t}]$, respectively. For example, if $H(j\omega) = |H(j\omega)|\, e^{j\angle H(j\omega)}$ and the input is cos ωt, then the output shall be $|H(j\omega)|$ cos $(\omega t + \angle H(j\omega))$. $H(j\omega)$ varies with frequency, and the plots of $|H(j\omega)|$ and $\angle H(j\omega)$ versus ω are known as magnitude and phase responses, respectively.

Since the principle of superposition holds in a linear system, the response to a linear combination of exponential signals, $\sum_i a_i e^{s_i t}$, will be of the form $\sum_i a_i H(s_i) e^{s_i t}$. It is precisely this fact which motivated Fourier to explore if an arbitrary signal could be represented as a superposition of exponential signals. As is now well known, this can indeed be done by a Fourier series for a periodic signal and by the Fourier transform for a general, not necessarily periodic, signal.

The Fourier Series

Consider a linear combination of the exponential signal $e^{j\omega_0 t}$ with its harmonically related exponential signals $e^{jk\omega_0 t}$ $k = 0, \pm 1, \pm 2, \ldots$:

$$x(t) = \sum_{k=-\infty}^{\infty} a_k e^{jk\omega_0 t} \quad (1.27)$$

In this, $k = 0$ gives a constant term or dc, in electrical engineering language; $e^{j\omega_0 t}$ is the smallest frequency term, with a frequency ω_0 and period $T = 2\pi/\omega_0$, and is called the fundamental frequency. The term $e^{j2\omega_0 t}$ has a frequency $2\omega_0$, while $e^{-j2\omega_0 t}$ has a frequency $-2\omega_0$; the period of either term is $T/2$, and both the terms represent what is known as the second harmonic. A similar interpretation holds for the general term $e^{jk\omega_0 t}$, which has a period $T/|k|$. Note that we take the frequency as positive or negative, but the period is taken as positive. Obviously, the summation Eq. 1.27 is periodic with a period equal to T, in which there are $|k|$ periods of the general term $e^{jk\omega_0 t}$ but only one period of the fundamental.

What about a given periodic function $x(t)$ with a period T, i.e. $x(t + mT) = x(t)$, $m = 0, \pm 1, \pm 2, \ldots$? Can it be decomposed into the form

Eq. 1.27? It turns out that under certain conditions which are satisfied by all but a few exceptional cases, one can indeed do so. To determine a_k's, multiply both sides of Eq. 1.27 by $e^{-jn\omega_0 t}$ and integrate over the interval 0 to *T*. Obviously, this results in an integral $\int_0^T e^{j(k-n)\omega_0 t} dt$ on the right-hand side, which is zero if $k \neq n$, and *T* if $k = n$. Thus, $a_n = (1/T) \int_0^T x(t)\, e^{-jn\omega_0 t} dt$ or

$$a_k = \frac{1}{T} \int_0^T x(t)\, e^{-jk\omega_0 t} dt \tag{1.28}$$

a_k represents the weight of the *k*th harmonic and is called the spectral coefficient of *x*(*t*). a_k is, in general, complex. A plot of $|a_k|$ versus *k* will consist of discrete lines at $k = 0, \pm 1, \pm 2, \ldots$, it resembles a spectrum as observed on a spectroscope and is called the amplitude spectrum. Similarly, one can draw a phase spectrum.

It is obvious from Eq. 1.27 that *x*(*t*) could be written as the summation of a sine and a cosine series, and that the corresponding coefficients could be found from $\int_0^T x(t) \cos k\omega t\, dt$ and $\int_0^T x(t) \sin k\omega t\, dt$. It is, however, much more convenient to handle the exponential form of the Fourier series as given in Eq. 1.27.

As an example of application of the Fourier series, consider the pulse stream shown in Fig. 1.5.

At this point, note that in Eq. 1.28, the lower and the upper limit of integration are not important so long as their difference is *T*. This is so because $\int_{t_0}^{t_0+T} e^{j(k-n)\omega_0 t}$ is independent of t_0. In the example under consideration, it is obviously convenient to choose the interval $-\frac{T}{2} \leq t \leq T/2$ which virtually becomes $-\tau/2 \leq t \leq +\tau/2$, because $x(t) = 0$ at other values of *t* within the chosen interval. Hence

$$\begin{aligned} a_k &= \frac{A}{T} \int_{-\tau/2}^{\tau/2} e^{-jk\omega_0 t} dt \\ &= \frac{2A}{k\omega_0} \sin \frac{k\omega_0 \tau}{2} \\ &= \tau A \frac{\sin \frac{k\omega_0 \tau}{2}}{\frac{k\omega_0 \tau}{2}} \end{aligned} \tag{1.29}$$

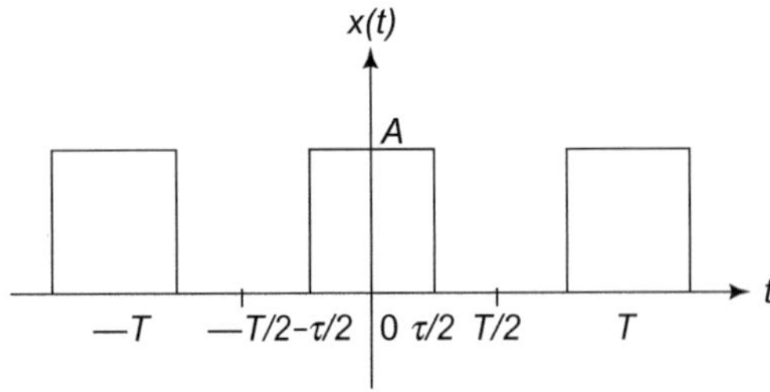

Fig. 1.5 A rectangular pulse stream

This is of the form $\tau A \sin x/x$, where $x = k\omega_0 \tau/2$. Note that a_k is real, and can be positive, zero or negative. Hence, separate amplitude and phase spectrum plots are not necessary; a single diagram suffices and is shown in Fig. 1.6. Note that a_k has a maximum value at DC, i.e. $k = 0$, the value being τA (this checks with direct calculation from Fig. 1.5). The envelope of the spectrum is of the form sin *x*/*x* and exhibits damped oscillations with zeros at $x = \pi$ (i.e. $k\omega_0 = 2\pi/\tau$), 2π (i.e. $k\omega_0 = \pi/\tau$), … Further, the sketch is symmetrical about $x = 0$, because sin *x*/*x* is an even function. The spectrum consists of discrete lines, two adjacent lines being separated by $\frac{2\pi}{T} = \omega_0$ rad/s.

Some important points emerge from the sketch of Fig. 1.6. As *T* increases, the lines get closer and ultimately when $T \to \infty$, corresponding to a single pulse, the spectrum becomes continuous and will be characterized by the function $\tau A \frac{\sin \omega\tau/2}{\omega\tau/2}$, where $k\omega_0$ has become the continuous variable ω. This, as we shall see, is the Fourier transform of the single pulse.

Second, since the lines concentrated in the lower frequency range are of higher amplitude, most of the energy of the periodic wave of Fig. 1.5 must be confined to lower frequencies. Third, as τ decreases, the spectrum spreads out, i.e. there is an inverse relationship between pulse width and frequency spread.

Since the energy of the periodic wave is mostly confined to the lower frequency range, a convenient measure of bandwidth of the signal is from zero frequency to the frequency of the first

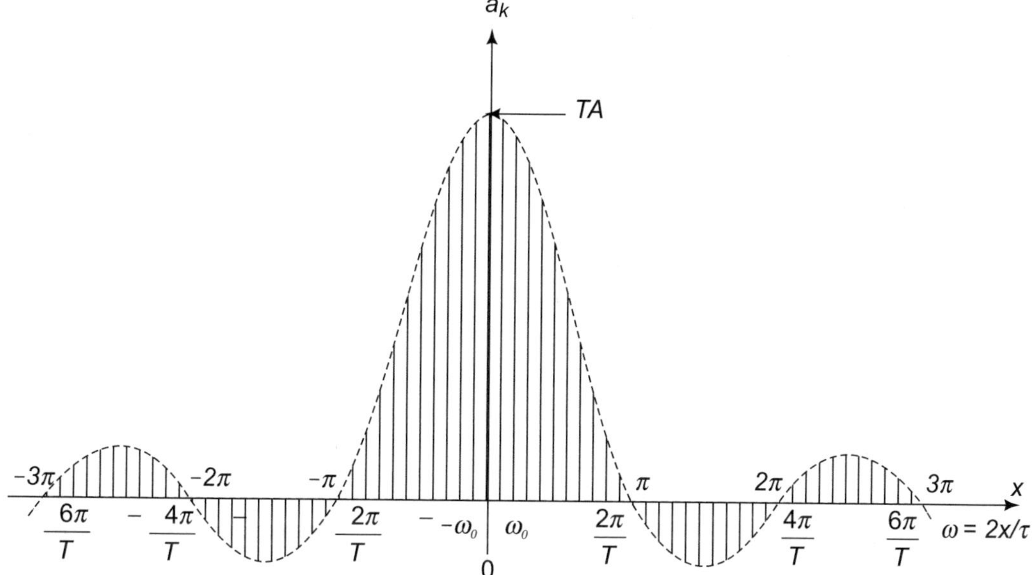

Fig. 1.6 Spectrum of $x(t)$ of Fig. 1.5

zero crossing, i.e. the bandwidth in Hz, B, can be taken as $1/\tau$.

If $x(t)$ of Eq. 1.27 is the voltage across or the current through a one-ohm resistor, then the average power dissipated is $\frac{1}{T}\int_0^T |x(t)|^2 \mathrm{d}t$. If one writes $|x(t)|^2 = x(t)\ \bar{x}(t)$, where bar denotes complex conjugate, and substitutes for $x(t)$ and $\bar{x}(t)$ from Eq. 1.27, there results the following:

$$|x(t)|^2 = \sum_{k=-\infty}^{\infty} \sum_{n=-\infty}^{\infty} a_k \bar{a}_n \mathrm{e}^{j(k-n)\omega_0 t} \qquad (1.30)$$

As we have already seen, $\int_0^T \mathrm{e}^{j(k-n)\omega_0 t} \mathrm{d}t$ is zero if $k \neq n$ and equals T when $k = n$. Thus, the average power becomes

$$\frac{1}{T}\int_0^T |x(t)|^2 \mathrm{d}t = \sum_{k=-\infty}^{\infty} |a_k|^2 \qquad (1.31)$$

This is known as Parseval's theorem.

A periodic signal that is of great importance in digital communication is the impulse train

$$x(t) = \sum_{k=-\infty}^{\infty} \delta(t - kT) \qquad (1.32)$$

as shown in Fig. 1.7. If this is expanded in Fourier series

$$x(t) = \sum_{k=-\infty}^{\infty} a_k \mathrm{e}^{jk\omega_0 t}, \qquad (1.33)$$

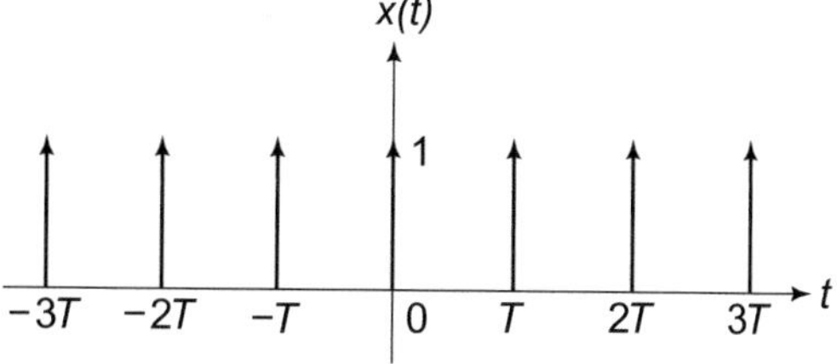

Fig. 1.7 A periodic impulse train

where $\omega_o = 2\pi/T$, then

$$a_k = \frac{1}{T}\int_{-T/2}^{T/2} \delta(t)e^{-jk\omega_0 t}dt = \frac{1}{T} \quad (1.34)$$

The spectrum is sketched in Fig. 1.8

What is the bandwidth of this signal? The amplitude is a constant at all frequencies, unlike the spectrum of Fig. 1.6. Hence, the bandwidth is infinite. This agrees with our observation about bandwidth and pulse duration, because Fig. 1.7 is the degenerate form of Fig. 1.5 with $\tau \to 0$ and $A \to \infty$.

Linear System Response to Periodic Excitation

From the discussion on LTI system response to exponential signals, it follows that a linear system, excited by the periodic signal of Eq. 1.27, will produce an output signal which is also periodic with the same period, and is given by

$$y(t) = \sum_{k=-\infty}^{\infty} a_k H(jk\omega_0)e^{jk\omega_0 t}, \quad (1.35)$$

where

$$H(j\omega) = \int_{-\infty}^{\infty} h(t)e^{-j\omega t}dt \quad (1.36)$$

and $h(t)$ is the unit impulse response.

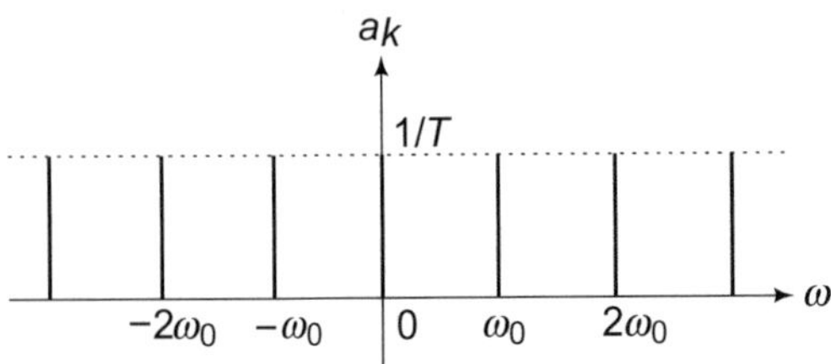

Fig. 1.8 Spectrum of the impulse train of Fig. 1.7

As a simple example, consider the RC network of Fig. 1.4; we have already derived its impulse response as

$$h(t) = \frac{1}{RC}e^{-t/(RC)}u(t) \quad (1.37)$$

so that

$$H(j\omega) = \int_{0}^{\infty} \frac{1}{RC}e^{-t\left(j\omega + \frac{1}{RC}\right)}dt \quad (1.38)$$

$$= \frac{1}{j\omega RC + 1} \quad (1.39)$$

When excited by the periodic impulse train of Fig. 1.7, the response $y(t)$ can be found in two ways. First, since

$$\delta(t) \to h(t),$$

it follows that

$$\sum_{k=-\infty}^{\infty} \delta(t - kT) \to \sum_{k=-\infty}^{\infty} h(t - kT)$$

so that

$$y(t) = \frac{1}{RC}\sum_{k=-\infty}^{\infty} e^{-(t-kT)/(RC)}u(t - kT) \quad (1.40)$$

Should you try to sketch this waveform, you would realize how messy it looks; also not much information about the effect of the RC network will be obvious from this sketch. On the other hand, the Fourier series method gives, from Eqs. 1.35, 1.34 and 1.39.

$$y(t) = \sum_{k=-\infty}^{\infty} \frac{1/T}{jk\omega_0 RC + 1}e^{jk\omega_0 t} \quad (1.41)$$

Let this be written as $\sum_{k=-\infty}^{\infty} b_k e^{jk\omega_0 t}$; then the sketch of $|b_k|$ versus $\omega = k\omega_0$ looks like that shown in Fig. 1.9. Comparing this with Fig. 1.8, we note that the RC network attenuates higher frequencies as compared to lower ones and hence acts as a low-pass filter. The bandwidth of the

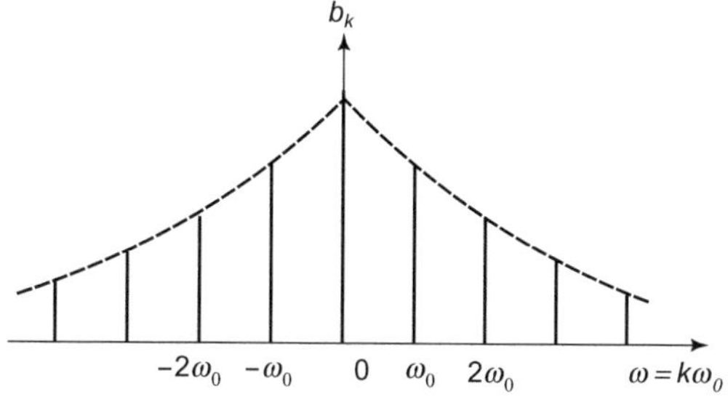

Fig. 1.9 Spectrum of output $y(t)$ given in Eq. 1.41

filter B_f is defined as the frequency at which the $|H(j\omega)|$ falls down by 3 dB as compared to its dc value, i.e.

$$|H(j2\pi B_f)| = H(j0)/\sqrt{2} \tag{1.42}$$

Combining this with Eq. 1.39 gives $B_f = 1/(2\pi RC)$.

What would be the response of the RC filter to the rectangular pulse stream of Fig. 1.5? This will of course depend on the relative values of T, τ and B_f.

First let us confine ourselves to the time domain. If the product RC is comparable to T, then the output will consist of overlapping pulses, and will retain very little similarity to the input. Let, therefore, $RC \ll T$; then depending on τ, the response during one period will be of the form shown in Fig. 1.10. It is obvious that for fidelity, i.e. if the output is to closely resemble the input, we require $RC \ll \tau < T$.

Now, turn to the frequency domain. If the signal bandwidth is taken as $B = 1/\tau$ Hz, then obviously for fidelity, the RC filter must pass all frequencies up to $1/\tau$ Hz with as little attenuation as possible. Thus, B_f must be at least equal to $B = 1/\tau$. Since the attenuation is 3 dB instead of zero at B_f and the input spectrum is not limited to B, the pulse shape will be distorted. For reduced distortion, we need to increase B_f and we expect good results if $B_f \gg B$ i.e. $RC \ll \tau$.

The Fourier Transform

Now consider a non-periodic function $x(t)$ which exists in the range $-T/2 \leq t \leq T/2$, and is zero outside this range. Consider a periodic extension $x_p(t)$ of $x(t)$, as shown in Fig. 1.11. $x_p(t)$ can be expanded in Fourier series as

$$x_p(t) = \sum_{k=-\infty}^{\infty} a_k e^{jk\omega_0 t}, \tag{1.43}$$

where $\omega_0 = 2\pi/T$ and

$$a_k = \frac{1}{T}\int_{-T/2}^{T/2} x_p(t)e^{-jk\omega_0 t}dt \tag{1.44}$$

$$= \frac{1}{T}\int_{-T/2}^{T/2} x(t)e^{-jk\omega_0 t}dt \tag{1.45}$$

Fig. 1.10 Response of RC network in one period of Fig. 1.5

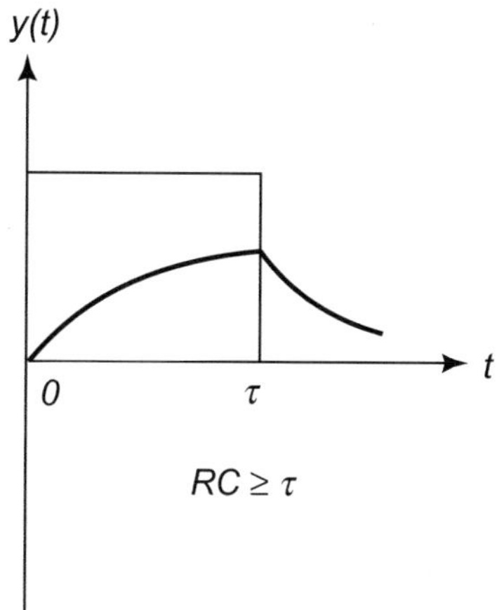

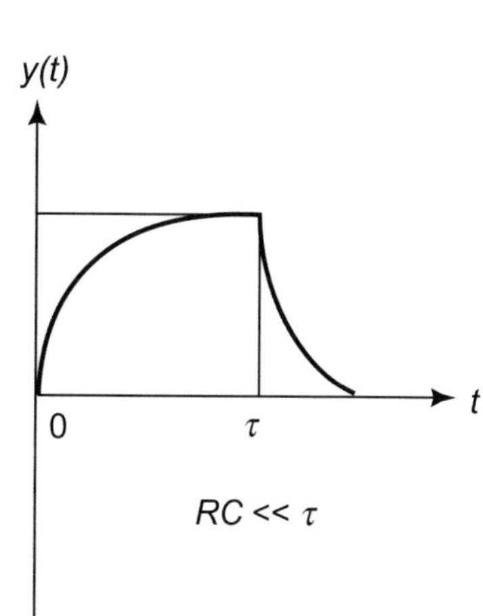

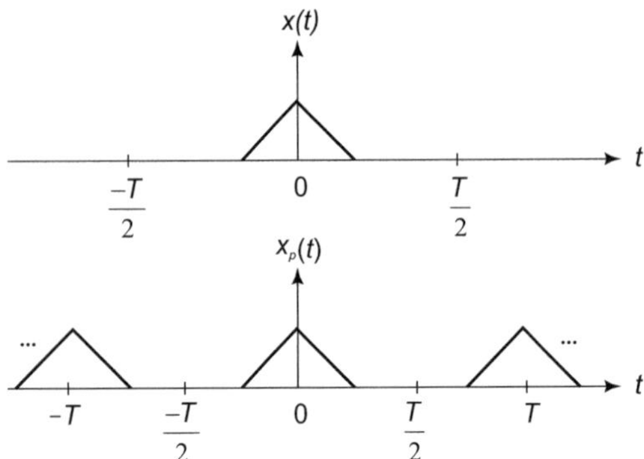

Fig. 1.11 A non-periodic function $x(t)$ and its periodic extension $x_p(t)$

because for $|t| \leq T/2$, $x_p(t) = x(t)$. Also, since $x(t) = 0$ for $|t| > T/2$, we can write

$$a_k = \frac{1}{T} \int_{-\infty}^{\infty} x(t) e^{-jk\omega_0 t} dt \tag{1.46}$$

If we define

$$X(j\omega) = \int_{-\infty}^{\infty} x(t) e^{-j\omega t} dt \tag{1.47}$$

then from Eq. 1.46, we get

$$a_k = \frac{1}{T} X(jk\omega_0) = \frac{1}{2\pi} X(jk\omega_0)\omega_0 \tag{1.48}$$

Thus, Eq. 1.43 can be written as

$$x_p(t) = \frac{1}{2\pi} \sum_{k=-\infty}^{\infty} X(jk\omega_0)\omega_0 e^{jk\omega_0 t} \tag{1.49}$$

Now let T tend to infinity; then $x_p(t)$ tends to $x(t)$, $k\omega_0$ tends to ω, a continuous variable, ω_0 tends to $d\omega$, and the summation becomes an integral. Thus, Eq. 1.49 becomes

$$x(t) = \frac{1}{2\pi} \int_{-\infty}^{\infty} X(j\omega) e^{j\omega t} d\omega \tag{1.50}$$

Combining Eqs. 1.47 and 1.48, we now formally define the Fourier transform of $x(t)$ as

$$X(j\omega) = \mathcal{F}[x(t)] = \int_{-\infty}^{\infty} x(t) e^{-j\omega t} dt \tag{1.51}$$

and the inverse Fourier transform as

$$x(t) = \mathcal{F}^{-1}[X(j\omega)] = \frac{1}{2\pi} \int_{-\infty}^{\infty} X(j\omega) e^{j\omega t} d\omega \tag{1.52}$$

Without entering into the question of existence, we simply state below the conditions, named after Dirichlet, under which $x(t)$ is Fourier transformable. These are

(1) $\int_{-\infty}^{\infty} |x(t)| dt < \infty$
(2) finite number of maxima and minima within any finite interval, and
(3) finite number of finite discontinuities within any finite interval.

Referring to Eqs. 1.26 or 1.36, it should be obvious that the impulse response $h(t)$ and the frequency response $H(j\omega)$ are Fourier transform pairs. Explicitly

$$H(jw) = \mathcal{F}[h(t)] \tag{1.53}$$

As an example of Fourier transformation, consider the rectangular pulse shown in Fig. 1.12. Notice that this is the limiting form of

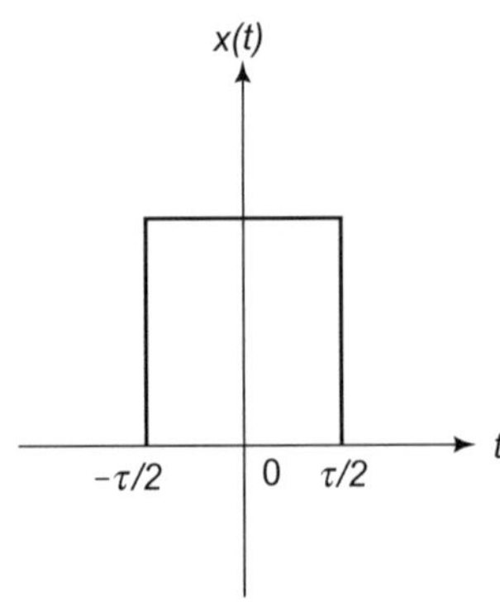

Fig. 1.12 A rectangular pulse

the periodic function of Fig. 1.5 with $T \to \infty$. Applying Eq. 1.51, we get

$$X(j\omega) = A \int_{-\tau/2}^{\tau/2} e^{-j\omega t} dt = \tau A \frac{\sin(\omega\tau/2)}{(\omega\tau/2)} \quad (1.54)$$

This, as will be easily recognized, is the limiting form of Fig. 1.6, and is the envelope of the same figure. This verifies the observation made in the discussion on Fourier series.

The Fourier transform has many important properties, the most important in the context of analysis of linear systems being that it converts a convolution in the time domain to a multiplication in the frequency domain, i.e. if

$$y(t) = x(t) * h(t) = \int_{-\infty}^{\infty} x(\tau)h(t-\tau)d\tau \quad (1.55)$$

then, assuming that $y(t)$, $x(t)$ and $h(t)$ are Fourier transformable, and $\mathcal{F}[y(t)] = Y(j\omega)$, we get

$$Y(j\omega) = X(j\omega)H(j\omega) \quad (1.56)$$

The proof of Eq. 1.56 is simple and proceeds as follows:

$$\begin{aligned} Y(j\omega) &= \mathcal{F}[y(t)] \\ &= \int_{-\infty}^{\infty} \left[\int_{-\infty}^{\infty} x(\tau)h(t-\tau)d\tau \right] e^{-j\omega t}dt \end{aligned} \quad (1.57)$$

Interchange the order of integration and notice that $x(\tau)$ does not depend on t; the result is

$$Y(j\omega) = \int_{-\infty}^{\infty} x(\tau) \left[\int_{-\infty}^{\infty} h(t-\tau)e^{-j\omega t}dt \right] d\tau \quad (1.58)$$

Let $t - \tau = \xi$; then the integral inside the bracket becomes $e^{-j\omega\tau} H(j\omega)$, so that

$$Y(j\omega) = \int_{-\infty}^{\infty} H(j\omega)x(\tau)e^{-j\omega t}d\tau, \quad (1.59)$$

i.e.

$$Y(j\omega) = H(j\omega)X(j\omega) \quad (1.60)$$

In words, this amounts to saying that the spectrum of the output of a linear system is simply the product of the spectrum of the input signal and the frequency response of the system. The output in the time domain, $y(t)$ can be simply found by taking the inverse Fourier transform of $Y(j\omega)$.

To illustrate the application of Eq. 1.60, consider a linear system having the impulse response

$$h(t) = -e^{-\alpha t}u(t), \ \alpha > 0 \quad (1.61)$$

which is excited by an input signal

$$x(t) = e^{-\beta t}u(t), \ \beta > 0 \quad (1.62)$$

By direct integration, it is easily shown that

$$H(j\omega) = \frac{1}{\alpha + j\omega} \text{ and } X(j\omega) = \frac{1}{\beta + j\omega} \quad (1.63)$$

Thus

$$Y(j\omega) = \frac{1}{(\alpha + j\omega)(\beta + j\omega)} \quad (1.64)$$

To determine $y(t)$, one may write

$$Y(j\omega) = \frac{A}{(\alpha + j\omega)} + \frac{B}{(\beta + j\omega)} \quad (1.65)$$

and find A and B as

$$A = -B = \frac{1}{\beta - \alpha} \tag{1.66}$$

so that

$$y(t) = \mathcal{F}^{-1}\left[\frac{1}{\beta - \alpha}\left\{\frac{1}{\alpha + j\omega} - \frac{1}{\beta + j\omega}\right\}\right] \tag{1.67}$$

$$= \frac{1}{\beta - \alpha}\left[e^{-\alpha t} - e^{-\beta t}\right] u(t) \tag{1.68}$$

Things are of course, different if $\alpha = \beta$; then one goes back to Eq. 1.64 and uses the property that if $\mathcal{F}\,[x(t)] = X(j\omega)$ then $\mathcal{F}[tx(t)] = j\,\mathrm{d}X(j\omega)/\mathrm{d}\omega$. Accordingly if $\alpha = \beta$, then

$$y(t) = t\,e^{-\alpha t}\,u(t) \tag{1.69}$$

Spectral Density

In using Fourier transform to calculate the energy or power of a signal, the notion of spectral density is an important one. The total energy and average power of a signal $x(t)$ are defined as

$$E = \lim_{T \to \infty} \int_{-T}^{T} |x(t)|^2 \mathrm{d}t = \int_{-\infty}^{\infty} |x(t)|^2 \mathrm{d}t \tag{1.70}$$

and

$$P = \lim_{T \to \infty} \frac{1}{2T} \int_{-T}^{T} |x(t)|^2 \mathrm{d}t, \tag{1.71}$$

respectively. A signal $x(t)$ is called an energy signal if $0 < E < \infty$ and a power signal if $0 < P < \infty$. A given signal $x(t)$ can be either an energy signal or a power signal but not both. A periodic signal (e.g. the one of Fig. 1.5) is usually a power signal, while a non-periodic signal (e.g. the one of Fig. 1.12) is usually an energy signal. Power and energy signals are mutually exclusive because the former has infinite energy while the latter has zero average power. Depending on the nature of the signal, the spectral density is also to be qualified as power or energy.

Consider an energy signal $x(t)$. Using the facts that $|x(t)|^2 = x(t)\overline{x}(t)$ and $\mathcal{F}[\overline{x}(t)] = \overline{X}(-j\omega)$ and combining with the inversion integral Eq. 1.52, it is not difficult to show that

$$E = \int_{-\infty}^{\infty} |x(t)|^2 \mathrm{d}t = \frac{1}{2\pi} \int_{-\infty}^{\infty} |X(j\omega)|^2 \mathrm{d}\omega \tag{1.72}$$

The rightmost expression in Eq. 1.72 shows that $|X(j\omega)|^2/(2\pi)$ has the dimension of energy per unit radian frequency, i.e. $|X(j\omega)|^2$ has the dimension of energy per unit Hz. For this reason, $|X(j\omega)|^2$ is referred to as the energy density spectrum of the signal $x(t)$. Incidentally, Eq. 1.72 is known as the Parseval's relation (cf. Eq. 1.31).

For a periodic signal, which is a power signal, we have already seen in Eq. 1.31 that the average power P is given by $\sum_{k=-\infty}^{\infty} |a_k|^2$, where $|a_k|$ is the amplitude of the kth harmonic. If we define, in similarity with Eq. 1.72,

$$P = \int_{-\infty}^{\infty} S_x(f)\mathrm{d}f \tag{1.73}$$

then $S_x(f)$ qualifies as the power per unit Hz and is called the power spectral density. In terms of $|a_k|$, it is easily seen that

$$S_x(f) = \sum_{k=-\infty}^{\infty} |a_k|^2 \delta(f - kf_o), \tag{1.74}$$

where $f_0 = \omega_0/(2\pi)$ is the fundamental frequency.

Concluding Comments

For more on the impulse function, see the next chapter. In this chapter, we have talked about the fundamentals of signals and systems and their relationship. Fourier series and Fourier transform and the concepts of spectra—amplitude as well as phase—and the concepts of energy, power and their density functions have also been introduced.

Problems

You have to think carefully. These are designed also carefully, as the problems are designed to test your grasp of the fundamentals and the ease with which you can find a clue.

P.1 Sketch (−1), (−½), (0), (½), (1), (1 − t)

P.2 The impulse response of a system is $h(t - t_0)$. Find an expression for the output $y(t)$, if the input is $t^2x(t^{1/2})$.

P.3 Determine the Fourier transform of the function $tu(t)$.

P.4 Determine and sketch the spectrum of the following function:

$$x(t) = \begin{cases} 0, & t<0 \\ 1, & 0<t<\frac{T_0}{2} \\ \delta(t-T), & t>0 \end{cases}$$

P.5 Determine the impulse response of a system consisting of a cascade of two systems characterized by the impulse response

$$f(t) = \mathrm{e}^{-\alpha t}u(t)$$

and

$$g(t) = \mathrm{e}^{-\beta t}u(t),$$

where $\alpha > 0$, $\beta > 0$ and (i) $\alpha \neq \beta$ (ii) $\alpha = \beta$.

Bibliography

1. A.V. Oppenheim, A.S. Willsky, I.T. Young, Signals and systems. Prentice Hall (1983)

The Mysterious Impulse Function and its Mysteries

2

Some fundamental issues relating to this mysterious but fascinating impulse function in continuous as well as discrete time domain are discussed first. These are: definition and relation to the unit step function, dimension of impulse response, solution of differential and difference equations involving the impulse function and Fourier transform of the unit step function. Conceptual understanding is emphasized at every point. This is very important, particularly for beginners like you. So read the chapter carefully and grasp the contents. This will help you to understand the later course on signals and systems, control, DSP, etc.

Keywords
Impulse function · Unit step function
Impulse response · Differential equation
Difference equation · Fourier transform
Signals and systems · Digital signal processing

Introduction

The issues considered in this chapter are usually treated in a routine manner in courses on circuit theory, signals and systems and digital signal processing. That problems may arise without a thorough understanding of the concepts involved are usually not clearly brought out, or avoided. These problems are usually related to the occurrence of the impulse function $\delta(t)$ in the continuous time domain and $\delta(n)$ in the discrete time domain in various situations. The chapter makes an attempt to clarify the concepts involved in a few of such problem cases and the manner in which they can be solved. In particular, starting with the definitions of $\delta(t)$ and $\delta(n)$ and their relationship with the unit step functions $u(t)$ and $u(n)$, respectively, we discuss the following issues: dimension of impulse response, solution of differential and difference equations involving the impulse function, and Fourier transforms of $u(t)$ and $u(n)$. The last discussion is a concluding one.

The chapter is written in a tutorial style so as to be appealing to students and teachers alike. The chapter also asks some questions, the answers to which are left as open problems.

Source: S. C. Dutta Roy, "Some Fundamental Issues Related to the Impulse Function," *IETE Journal of Education*, Vol 57, pp 2–8, January–June 2016.

S. C. Dutta Roy, *Circuits, Systems and Signal Processing*,
https://doi.org/10.1007/978-981-10-6919-2_2

Definitions

The unit impulse function $\delta(t)$ in the continuous time domain is defined as

$$\delta(t) = 0, t \neq 0 \text{ and } \delta(t) \to \infty, t = 0 \quad (2.1)$$

such that

$$\int_a^b \delta(t)\, \mathrm{d}t = 1, \quad (2.2)$$

where $a \leq 0-$ and $b \geq 0+$. The adjective 'unit' refers to Eq. 2.2 and the area under the function, which is also known as the strength of the impulse. An impulse $A\ \delta(t)$ will have an area A under the integral and will be called an impulse of strength A. The definition must include Eqs. 2.1 and 2.2 together. Often Eq. 2.2 is not paid attention to, but it must be understood that infinity is incomprehensible in a finite world. If Eq. 2.2 was not coupled to Eq. 2.1, then $\delta(t)$ would not be admissible in the realm of functions, particularly for engineering analysis and design. Even otherwise, the impulse function is a mathematical anomaly. From a purely mathematical point of view, $\delta(t)$ is not strictly a function because any extended real function that is zero everywhere except at a single point must have a total integral equal to zero. There is a fascinating history behind the acceptance of $\delta(t)$, not as a function, but as a distribution or a generalized function by mathematicians. However, it would be too much of a diversion to go into the details in this chapter. Instead, we refer to an excellent tutorial article by Balakrishnan [1] on the subject. For our purposes here, Eqs. 2.1 and 2.2 as the definition of $\delta(t)$ will suffice.

It must be understood that $\delta(t)$ is only a mathematical concept and is introduced to facilitate analysis and design. It cannot be generated in the laboratory. Textbooks usually give a visual aid of a rectangular pulse of duration τ and height $1/\tau$ or a triangular pulse of width 2τ and height $1/\tau$. By allowing $\tau \to 0$, one can visualize $\delta(t)$. However, this limiting approach may fail to give correct results in system analysis, because if the starting time is $t = 0$, then the value of the pulse, rectangular or triangular, is 0 at $t = 0-$, 0, as well as 0+, which is not the case with $\delta(t)$.

The impulse function may also be looked upon as the limiting value of some distributions, *e.g.* $(a\sqrt{\pi})^{-1} \exp(-t^2/a^2)$ or $a^{-1}\text{sinc}\,(x/a)$, the integral of which from $-\infty$ to $+\infty$ is unity, as $a \to 0$. However, this point of view is not relevant in the context of this chapter, and will not be pursued further.

The function $\delta(t)$ is best visualized by relating it to the unit step function $u(t)$, defined by

$$u(t) = 1, t \geq 0+ \text{ and } u(t) = 0, t \leq 0- \quad (2.3)$$

Clearly, the differentiation of $u(t)$ will give $\delta(t)$, satisfying both Eqs. 2.1 and 2.2. Hence, the relationship between the two functions can be formally written as

$$\delta(t) = \mathrm{d}u(t)/\mathrm{d}t \text{ and } u(t) = \int_a^t \delta(t)\, \mathrm{d}t, \quad (2.4)$$

where $a \leq 0-$. Unlike $\delta(t)$, $u(t)$ can be generated in the laboratory by a switch in series with a voltage source.

On the other hand, in the discrete time domain, there is no uncertainty or confusion about the unit impulse function $\delta(n)$, which is defined as

$$\delta(n) = 1, n = 0 \text{ and } \delta(n) = 0, n \neq 0 \quad (2.5)$$

The unit step function $u(n)$ is defined as

$$u(n) = 1, n \geq 0 \text{ and } u(n) = 0, n < 0 \quad (2.6)$$

Thus, the relationships between $\delta(n)$ and $u(n)$ are

$$\delta(n) = u(n) - u(n-1) \text{ and } u(n) = \sum_{k=0}^{n} \delta(n-k) \quad (2.7)$$

In order to distinguish between the continuous and discrete time domains, $\delta(n)$ should more appropriately be called the 'unit sample function'. However, in conformity with the common usage, we shall continue to call $\delta(n)$ as the unit impulse function, it being implied that the argument of δ will make it clear which domain we are referring to.

Impulse Response

In the continuous time domain, the impulse response *h*(*t*) of a system is generally interpreted as the response of the system to an excitation $\delta(t)$. What is the dimension of *h*(*t*)? To answer this question, look at the first relation in Eq. 2.4; if *u*(*t*) is a voltage, then $\delta(t)$ has the dimension of volts/second, which we can neither generate nor apply to a system. However, for a linear system, differentiation of the unit step response will give *h*(*t*). This is how we can measure *h*(*t*) in the laboratory. Obviously, *h*(*t*) will have the dimension of $(\text{second})^{-1}$.

Now look at the convolution relation:

$$y(t) = \int_{-\infty}^{+\infty} x(\tau)h(t-\tau)\mathrm{d}\tau, \qquad (2.8)$$

where *y*(*t*) is the response of a system to an excitation *x*(*t*). Clearly, if both *x*(*t*) and *y*(*t*) are voltages or currents, then *h*(*t*) will have the dimension of $(\text{second})^{-1}$. However, if *x*(*t*) is a voltage and *y*(*t*) is a current, then *h*(*t*) will have the dimension of $(\text{ohm-second})^{-1}$. Similarly, if *x* (*t*) is a current and *y*(*t*) is a voltage, then the dimension of *h*(*t*) will be ohm/second.

In the discrete time domain, *h*(*n*) is dimensionless, as are all signals, because the processing is concerned only with numbers. The convolution relation

$$y(n) = x(n) * h(n) = \sum_{k=-\alpha}^{\infty} x(k)h(n-k) \qquad (2.9)$$

does not involve any dimensions.

In the context of the continuous time domain, the transfer function is defined as

$$H(s) = L[y(t)/L[x(t)]|_{\text{zero initial conditions}}, \qquad (2.10)$$

where *L* stands for the Laplace transform and the initial conditions are on *y*(*t*) and its derivative(s), if the order of the system is more than one. It is known that the *h*(*t*) and *H*(*s*) are related to each other by

$$H(s) = L[h(t) \text{ evaluated under zero initial conditions}] \qquad (2.11)$$

Textbooks usually omit the condition under which *h*(*t*) is to be evaluated. It must be emphasized that *h*(*t*) is the impulse response under zero-state condition.

Similarly, for the discrete time domain, the transfer function is defined as

$$H(z) = Z\,[y(n)/Z\,[x(n)]|_{\text{zero initial conditions}} \qquad (2.12)$$

and

$$H(z) = Z\,[h(n) \text{ evaluated under zero initial conditions}], \qquad (2.13)$$

where *Z* stands for the *z*-transform.

How Do You Solve Differential Equations Involving an Impulse Function?

We shall illustrate the kind of difficulties that arise in solving differential equations involving an impulse function, with a typical problem. Consider the following differential equation:

$$y'' + 2y' + 2y = \delta(t), \qquad (2.14)$$

where prime denotes differentiation with respect to *t*. Let the given initial conditions be

$$y(0-) = 1 \text{ and } y'(0-) = -2 \qquad (2.15)$$

Let us try to solve Eq. 2.14 by using Laplace transforms, as is usually done. In order to take account of $\delta(t)$ on the right-hand side of Eq. 2.14, we take the Laplace transform of $y(t)$ as

$$Y(s) = \int_{0-}^{\infty} y(t)\exp(-st)\mathrm{d}t \qquad (2.16)$$

Then, the Laplace transformation of Eq. 2.14 gives

$$s^2Y(s) - sy(0-) - y'(0-) + 2[sY(s) - y(0-)] + 2Y(s) = 1 \qquad (2.17)$$

Combining Eqs. 2.17 and 2.15 and simplifying, we get

$$Y(s) = (s+1)/(s^2+2s+2) \qquad (2.18)$$

Using the standard table of Laplace transforms, inversion of Eq. 2.18 gives

$$y(t) = [\exp(-t)\cos t]\,u(t) \qquad (2.19)$$

Differentiation of Eq. 2.19 gives

$$y'(t) = [\exp(-t)\cos t]\delta(t) - \exp(-t).[\cos t + \sin t]\,u(t) \qquad (2.20)$$

The solutions Eqs. 2.19 and 2.20 are valid for $t \geq 0+$. In particular, at $t = 0+$,

$$y(0+) = 1 \text{ and } y'(0+) = -1 \qquad (2.21)$$

Note that while $y(0+) = y(0-)$, $y'(0+) \neq y'(0-)$. This discontinuity in initial condition is typical, whenever an impulse function figures in the right-hand side of a differential equation. Also note that if Eqs. 2.19 and 2.20 are applied at $t = 0-$, the values are: $y(0-) = 0$ and $y'(0-) = 0$, because both $u(t)$ and $\delta(t)$ are zero at $t = 0-$.

Can we get a solution which is valid at $t = 0-$ also? The answer is yes, if we apply the superposition of 'zero-input' and 'zero-state' solutions. This method is seldom emphasized in circuit theory courses. Instead, one uses the superposition of the complementary function and the particular solution, and then evaluates the constants in the former from the initial conditions. The method gives correct results if there is no impulse function in the excitation function, but not in the present case. Why? The question is left to you as an open problem.

Applying the first method, we first find the zero-input solution, i.e. the solution of Eq. 2.14 with the right-hand side equal to zero. It can be easily verified that the solution is of the form

$$y_{zi}(t) = A\cos(t+\theta), \qquad (2.22)$$

where A and θ are constants to be evaluated from the initial conditions given by Eq. 2.15. Carrying out the steps, we get

$$A = \sqrt{5} \text{ and } \theta = \arctan 2 \qquad (2.23)$$

Now consider the zero-state solution where the initial conditions are to be put equal to zero. Obviously, this can be done by the Laplace transform method by putting $y(0-) = 0$ and $y'(0-) = 0$ in Eq. 2.17. Then, Eq. 2.17 gives

$$Y_{zs}(s) = 1/(s^2+2s+2) \qquad (2.24)$$

On inversion, Eq. 2.24 gives

$$y_{zs}(t) = [\exp(-t)\sin t]\,u(t) \qquad (2.25)$$

Combining Eqs. 2.22, 2.23 and 2.25, we get the total solution as

$$y(t) = \sqrt{5}\cos(t + \arctan 2) + [\exp(-t)\sin t]\,u(t) \qquad (2.26)$$

Note that the first term does not involve $u(t)$ because the zero-input solution is valid for all t. Differentiating Eq. 2.26, we get

$$y'(t) = -\sqrt{5}\sin(t + \arctan 2) + [\exp(-t)\sin t]\,\delta(t) - [\exp(-t)(\sin t + \cos t)]u(t) \qquad (2.27)$$

Putting $t = 0-$ in Eqs. 2.26 and 2.27 and noting that both $u(t)$ and $\delta(t)$ are zero at $t = 0-$, we get the same values as in Eq. 2.15.

On the other hand, putting t = 0+ in Eqs. 2.26 and 2.27 and noting that $u(0+) = 1$ and $\delta(0+) = 0$, we get the same values as in Eq. 2.21.

How Do You Solve Difference Equations Involving an Impulse Function?

Do the types of difficulties discussed in the previous section arise in the discrete time domain also? The answer is yes, if z-transform method is applied blindly because the resulting solution is strictly valid for $n \geq 0$. However, the situation can be corrected by adding the necessary terms representing the initial conditions for $n < 0$. As in the continuous time domain, the complementary function and particular solution method does not work here. Why? The answer to this question is again left to you as an open problem. On the other hand, as in the continuous time case, if the difference equation is solved in the n- domain by the zero-input and zero-state method, correct solution is obtained, with initial conditions taken into account. As in the previous case, we shall illustrate these facts with an example.

Let it be required to solve the difference equation

$$y(n) + y(n-1) - 6y(n-2) = \delta(n) \qquad (2.28)$$

subject to the initial conditions

$$y(-1) = 1 \text{ and } y(-2) = -1 \qquad (2.29)$$

If we use the z-transform method blindly and take the z-transform of $y(n)$ as

$$Y(z) = \sum_{n=0}^{\infty} y(n) z^{-n} \qquad (2.30)$$

then the z-transform of Eq. 2.28 gives

$$\begin{aligned} Y(z) + z^{-1}Y(z) + y(-1) - 6\big[z^{-2}Y(z) \\ + z^{-1}y(-1) + y(-2)\big] = 1 \end{aligned} \qquad (2.31)$$

Combining Eqs. 2.31 with 2.29 and simplifying gives

$$Y(z) = 6(z^{-1} - 1)/(1 + z^{-1} - 6z^{-2}) \qquad (2.32)$$

Expanding Eq. 2.32 in partial fractions and taking the inverse transform gives

$$y(n) = -\,[4.8(-3)^n + 1.2(2)^n]\,u(n) \qquad (2.33)$$

This solution is strictly valid for $n \geq 0$. However, it can be easily modified to take care of $n < 0$ by adding the terms $\delta(n+1)$ and $-\delta(n+2)$ to the right-hand side of Eq. 2.33.

Consider now the zero-input, zero-state method of solution. The zero-input solution is of the form

$$y_{zi}(n) = K_1(-3)^n + K_2(2)^n, \qquad (2.34)$$

where K_1 and K_2 are to be evaluated from Eq. 2.29. The result is

$$y_{zi}(n) = -\,[5.4(-3)^n + 1.6\,(2)^n] \qquad (2.35)$$

This part of the solution is valid for all n.

The zero-state solution can be obtained by the z-transform technique by putting $y(-1) = y(-2) = 0$ in Eq. 2.31. Inverting the resulting $Y(z)$, we get

$$y_{zs}(n) = [0.6\,(-3)^n + 0.4\,(2)^n]\,u(n) \qquad (2.36)$$

which is valid for $n \geq 0$. The total solution is the sum of Eqs. 2.35 and 2.36, i.e.

$$y(n) = -\,[5.4(-3)^n + 1.6\,(2)^n] + [0.6\,(-3)^n + 0.4\,(2)^n]\,u(n) \qquad (2.37)$$

Clearly, Eq. 2.37 will give the correct values for $y(-1)$ and $y(-2)$. Also, for $n \geq 0$, Eq. 2.37 gives the same result as Eq. 2.33, as expected.

Fourier Transform of the Unit Step Function

For ready reference, recall that the Fourier transform of a continuous time function $y(t)$ is defined as

$$Y(j\omega) = \int_{-\infty}^{+\infty} y(t)\exp(-j\omega t)\mathrm{d}t \quad (2.38)$$

while that of the discrete time function $y(n)$ is defined as

$$Y[\exp(j\omega)] = \sum_{n=-\infty}^{\infty} y(n)\exp(-j\omega n) \quad (2.39)$$

Note that in either case, the transform is a continuous function of ω, although $y(n)$ is a discrete variable. In Eq. 2.38, $\omega = 2\pi f$, where f is the frequency in Hz, while in Eq. 2.39, $\omega = 2\pi f/f_s$, where f_s is the sampling frequency in Hz. Consequently, in Eq. 2.38, ω has the dimension of (second)$^{-1}$, while in Eq. 2.39, ω is dimensionless.

In order to be useful, any transformation must be reversible, i.e. one must be able to recover the original signal from the transformed one in a unique way. The Fourier transform is no exception for which the inverse Fourier transforms is

$$y(t) = [1/(2\pi)]\int_{-\infty}^{+\infty} Y(j\omega)\exp(j\omega t)\mathrm{d}\omega \quad (2.40)$$

corresponding to Eq. 2.38 and

$$y(n) = [1(2\pi)]\int_{-\infty}^{\infty} Y[\exp(j\omega)]\exp(j\omega n)\mathrm{d}\omega \quad (2.41)$$

corresponding to Eq. 2.40. The limits of the integral in Eq. 2.41 cover a range of 2π because $Y[\exp(j\omega)]$ is periodic with a period of 2π. That both Eqs. 2.38 and 2.40 or Eqs. 2.39 and 2.41 cannot be definitions is often not appreciated by students.

Now, we turn to the main item of discussion, i.e. the Fourier transforms of $u(t)$ and $u(n)$. Applying the definition, we get

$$\begin{aligned} \mathrm{F}[u(t)] &= \int_{-\infty}^{+\infty} u(t)\exp(-j\omega t)\mathrm{d}t, \\ &= \int_{0}^{+\infty} \exp(-j\omega t)\mathrm{d}t \\ &= [\exp(-j\omega t)/(-j\omega)]\big|_0^{\infty} \end{aligned} \quad (2.42)$$

where F stands for the Fourier transform. Since $\exp(-j\infty)$ is not known, we evaluate Eq. 2.42 indirectly. It is easy to show that

$$\mathrm{F}[\exp(-at)\,u(t)] = 1/(a+j\omega) \quad (2.43)$$

It is tempting to put $a = 0$ in Eq. 2.43 and conclude that $F[u(t)] = 1/(j\omega)$, but this is part of the story. Hidden here is the fact that Eq. 2.43 has a real part as well as an imaginary part, i.e.

$$1/(a+j\omega) = [a/(a^2+\omega^2)] - [j\omega/(a^2+\omega^2)] \quad (2.44)$$

While $1/(j\omega)$ takes care of the imaginary part when $a = 0$, the real part becomes $1/\omega^2$ when $a = 0$, and it tends to infinity when $\omega \to 0$. Hence, there is an impulse function at $\omega = 0$. (It would be wrong to say that the real part becomes of the form 0/0 when both a and ω are zero, because the denominator becomes 0^2 so that the real part becomes 1/0, i.e. infinite). To determine the strength of the impulse, we find the area under $a/(a^2+\omega^2)$. This is given by

$$\int_{-\infty}^{+\infty} [a/(a^2+\omega^2)]\mathrm{d}\omega = [\arctan(\omega/a)]\big|_{-\infty}^{\infty} = \pi \quad (2.45)$$

Thus, the real part in Eq. 2.44 becomes $\pi\delta(\omega)$ when both a and ω tend to zero. Finally, therefore,

$$\mathrm{F}[u(t)] = \pi\delta(\omega) + [1/j\omega)] \quad (2.46)$$

For finding the Fourier transform of $u(n)$, if we apply the definition Eq. 2.39 blindly, then it is tempting to conclude that the required transform is $1/[1-\exp(-j\omega)]$. However, as in the case of $u(t)$, this is a part of the story; hidden here is the fact that $u(n)$ has an average value of ½. To show that $1/[1-\exp(-j\omega)]$ is not $F[u(n)]$, consider the sequence

$$\mathrm{sgn}\,(n) = -1/2 \text{ for } n<0 \text{ and } +1/2 \text{ for } n \geq 0 \tag{2.47}$$

It is not difficult to show, by applying (2.39) that $F\,[\mathrm{sgn}(n)] = 1/[1-\exp(-j\omega)]$. Clearly,

$$u(n) = \mathrm{sgn}(n) + (1/2) \tag{2.48}$$

Thus, $F[u(n)]$ will have an additional term corresponding to the average value of ½. To find this term, consider the following inverse z-transform:

$$F^{-1}[\pi \sum_{-\infty}^{+\infty} \delta(\omega + 2\pi k)] = [1(2\pi)] \int_{-\pi}^{+\pi} \pi \cdot \sum_{-\infty}^{+\infty} \delta(\omega + 2\pi k) \exp(j\omega n) d\omega \tag{2.49}$$

In the range of integration, only the k = 0 term will have an effect, which gives the right-hand side of Eq. 2.49 as ½. Since the Fourier transform gives a one-to-one correspondence between the n and ω domains, we conclude that

$$F[u(n)] = 1/[1-\exp(-j\omega)] + \pi \sum_{-\infty}^{\infty} \delta(\omega + 2\pi k) \tag{2.50}$$

Conclusion

In this chapter, some fundamental and conceptual issues, relating to the unit impulse function, in both continuous and discrete time domains, are presented and some possible misconceptions and confusions have been clarified. That the complementary function, and particular solution method does not work if the excitation contains an impulse function does not appear to have been recorded earlier. The reason why it does not work has been left as an open problem for you.

The reason why misconceptions and confusions persist in the minds of students and teachers lies in the kind of rote learning that is emphasized and practiced in the class. Also, even if the teacher is knowledgeable and wishes to explain the concepts and subtle points, he/she finds no time to do this because of the pressure to 'cover' the syllabus, which invariably is overloaded.

Problems

P.1 Solve the equation

$$y''' + 3y'' + 3y' + y = \delta(t),$$

given

$$y(0) = 0, y'(0) = 0, y''(0)$$

P.2 Solve the equation

$$y(n-4) + y(n-3) + y(n-1) = \delta(n-1)$$

given

$$y(0) = y(-1) = y(-2) = y(-3) = 1.$$

P.3 Solve the equation

$$\frac{dy(t)}{dt} + q_0 y(t) = Ae^{-\alpha t} u(t)$$

given that

$$y(0) = y_0$$

P.4 Solve the equation

$$y' + q_0 y = q_0 x' + x$$

given that

$$y(0) = 0$$

P.5 Solve the equation

$$y(n) + q_0 y(n-1) = x(n),$$

given that

$$y(-1) = y_0.$$

Acknowledgement The author acknowledges the help received from Professor Yashwant V Joshi in the preparation of this chapter.

Reference

1. V. Balakrishnan, All about the Dirac delta function(?). Resonance **8**(8), 48–58 (2003)

State Variables—Part I

3

Here, we introduce state variables, which were the hot topics in the middle of 1960s. Later, they have seeped into and made deep impact on signals and systems, circuit theory, controls, etc. You better get familiar with them and make friends with them as early as possible.

Keywords
State variables · Standard forms · Choice of state variables · Solution of state equations

Why State Variables?

The state variable approach provides an extremely powerful technique for the study of system theory and has led to many results of far-reaching importance. This chapter is intended to serve as an introduction to the basic concepts and techniques involved in the state variable characterization of systems. The discussion is restricted to linear and time-invariant systems only. Linearity implies that if inputs $x_1(t)$ and $x_2(t)$ produce outputs $y_1(t)$ and $y_2(t)$, respectively, then an input $ax_1(t) + bx_2(t)$, where a and b are arbitrary constants, should lead to the output $ay_1(t) + by_2(t)$. This involves the two principles of homogeneity and superposition. Time invariance implies that if an input $x(t)$ produces the output $y(t)$, then the delayed input $x(t - \tau)$ should produce an output $y(t - \tau)$, delayed by the same amount τ. For state variable characterization of systems which do not obey linearity and/or time invariance, the reader is referred to the literature listed under references.

This discussion on state variables is organized in two parts. In Part I, we introduce the concept of state; clarify definitions and symbols; present the standard form of linear state equations; discuss how state variables are chosen in a given physical system, or a differential equation description of the same and elaborate on the methods of solving the state equations. In Part II of this presentation, in the next chapter, we shall deal with the properties of the fundamental matrix and procedures for evaluation of the same, and dwell upon the state transition flow graph method in considerable details. The references for the whole chapter will be given in Part II, which will also include an appendix on the essentials of matrix algebra.

Source: S. C. Dutta Roy, "State Variables—Part I," *IETE Journal of Education*, vol. 38, pp. 11–18, January–March 1997.

This chapter and the next one are based on some notes prepared for some students at the University of Minnesota in USA as early as 1965.

S. C. Dutta Roy, *Circuits, Systems and Signal Processing*,
https://doi.org/10.1007/978-981-10-6919-2_3

What is a State?

Consider an electrical network containing resistances, capacitances and inductances, whose response to an excitation $e(t)$ is $r(t)$. Since the network is known, a complete knowledge of $e(t)$ over the time interval $-\infty$ to t is sufficient to determine the output $r(t)$ over the same time interval. However, if $e(t)$ is known over the time interval t_0 to t, as is usually the case (with t_0 taken as equal to zero), then the currents through the inductors and the voltages across the capacitors in the network must be known at some time t_1, $t_0 \le t_1 \le t$ (usually, $t_1 = t_0$, hence the name 'initial conditions'), in order to determine $r(t)$ over the interval t_0 to t. These currents and voltages constitute the 'state' of the network at time t_1. In this sense, the state of the network is related to its memory; for a purely resistive network (zero memory), only the present input is required to determine the present output.

Next, consider a general type of linear system, which is described by a set of linear differential equations for $t \ge t_0$. The complete solution of these equations will involve arbitrary constants which can be determined from 'initial', 'given' or 'boundary' conditions at time $t = t_0$ (or at $t = t_1$ where $t_0 \le t_1 \le t$). The boundary conditions can thus be termed as the 'state' of the system at $t = t_0$. Heuristically, the state of a system separates the future from the past, so that the state contains all relevant information concerning the past history of the system, which is required to determine the response for any input. The evolution of an excited system from one state to another may be visualized as a process of state transition.

Definitions and Symbols

The state of a dynamical system may now be formally defined as the *minimal* amount of information necessary at any time such that this information, together with the input or the excitation function and the equations describing the dynamics of the system, characterize completely the future state and output of the system. The state is described by a set of numbers, called state variables, which contains sufficient information regarding the past history of the system so that the future behaviour can be computed. For example, in an electrical network, the inductor currents and capacitor voltages constitute the state variables.

To represent a system of more than two or three state variables, it is convenient to use the vector space notation of modern algebra. A system defined by n independent states $x_i(t)$, $i = 1, 2, \ldots n$, (corresponding to an nth order system), can be represented by the state vector

$$\mathbf{x}(t) = \begin{bmatrix} x_1(t) \\ x_2(t) \\ \vdots \\ x_n(t) \end{bmatrix} \tag{3.1}$$

Consider the multiple-input, multiple-output (or multivariable) system shown in Fig. 3.1. It can be described by a set of simultaneous differential equations relating the state vector $\mathbf{x}(t)$ given by Eq. 3.1, the excitation vector[1]

$$\mathbf{u}(t) = \begin{bmatrix} u_1(t) \\ u_2(t) \\ \vdots \\ u_m(t) \end{bmatrix}, \tag{3.2}$$

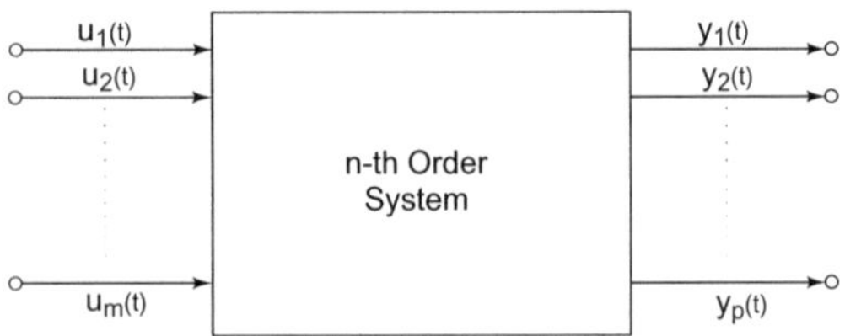

Fig. 3.1 Schematic representation of a multivariable system

[1]Note that the symbol $u(t)$ is usually reserved for the unit step function. We have used $u(t)$ to denote the excitation function for conformity with the control literature. We shall use the symbol $\mu(t)$ for the unit step function when the occasion arises.

where m is the number of inputs, and the response vector

$$\mathbf{y}(t) = \begin{bmatrix} y_1(t) \\ y_2(t) \\ \vdots \\ y_p(t) \end{bmatrix}, \quad (3.3)$$

where p is the number of outputs, through the system parameters, as discussed next.

Standard Form of Linear State Equations

In general, the equations of a dynamical system can be written in the following functional forms

$$\mathbf{x}(t) = \mathbf{f}[\mathbf{x}(t_0), \mathbf{u}(t_0, t)] \quad (3.4)$$

$$\mathbf{y}(t) = \mathbf{g}[\mathbf{x}(t_0), \mathbf{u}(t_0, t)] \quad (3.5)$$

for $t \geq t_0$. If the system can be described by a set of ordinary linear differential equations, then the state equations can be written as

$$\dot{\mathbf{x}}(t) = \mathbf{A}(t)\mathbf{x}(t) + \mathbf{B}(t)\mathbf{u}(t) \quad (3.6)$$

$$\mathbf{y}(t) = \mathbf{C}(t)\mathbf{x}(t) + \mathbf{D}(t)\mathbf{u}(t), \quad (3.7)$$

where the dot above the symbol $\mathbf{x}$ denotes differentiation with respect to time and $\mathbf{A}(t)$, $\mathbf{B}(t)$, $\mathbf{C}(t)$ and $\mathbf{D}(t)$ are, in general, time-varying real matrices. Equations 3.6 and 3.7 are the standard forms of linear state equations, as will be illustrated in the examples to follow. If the system is time invariant, then the matrices $\mathbf{A}(t)$, $\mathbf{B}(t)$, $\mathbf{C}(t)$ and $\mathbf{D}(t)$ are constants. A general block diagram representation for the state equations is given in Fig. 3.2.

Referring to Eqs. 3.1 to 3.7, we observe that the dimensions of the matrices **A**, **B**, **C** and **D** are $n \times n$, $n \times m$, $p \times n$ and $p \times m$, respectively.

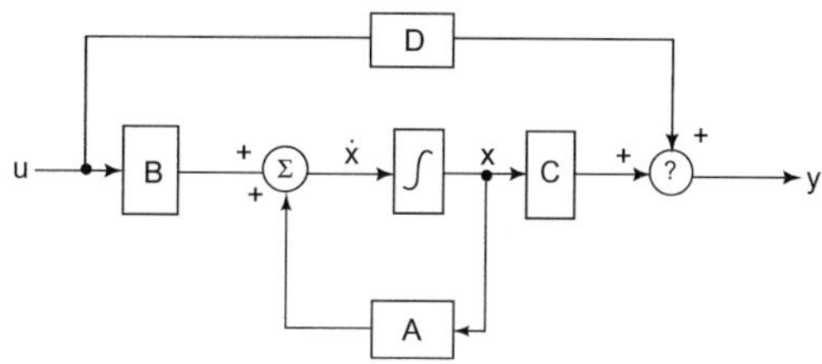

Fig. 3.2 Block diagram representation of the state equations

Example 1 Consider the network shown in Fig. 3.3. We identify the inductor current i_1 and the capacitor voltage v_2 as the state variables. There are two excitations, $e_1(t)$ and $e_2(t)$. Let the desired responses be i_1 and i_2. Thus, for this system, we identify

$$\mathbf{x} = \begin{bmatrix} i_1 \\ v_2 \end{bmatrix}, \mathbf{u} = \begin{bmatrix} e_1 \\ e_2 \end{bmatrix} \text{ and } \mathbf{y} = \begin{bmatrix} i_1 \\ i_2 \end{bmatrix} \quad (3.8)$$

A system of two first-order differential equations can be written for this electric circuit as follows:

$$L(\mathrm{d}i_1/\mathrm{d}t) = e_1 - (v_2 + e_2) = e_1 - e_2 - v_2 \quad (3.9)$$

$$i_2 = C(\mathrm{d}v_2/\mathrm{d}t) = i_1 - (v_2 + e_2)/R \quad (3.10)$$

Equations 3.9 and 3.10 can be rearranged as follows:

$$(\mathrm{d}i_1/\mathrm{d}t) = (-1/L)v_2 + (l/L)e_1 + (-1/L)e_2 \quad (3.11)$$

$$(\mathrm{d}v_2/\mathrm{d}t) = (1/C)i_1 + [-1/(RC)]v_2 + [-1/(RC)]e_2 \quad (3.12)$$

These two equations can be combined into the following single matrix equation:

$$\frac{\mathrm{d}}{\mathrm{d}t}\begin{bmatrix} i_1 \\ v_2 \end{bmatrix} = \begin{bmatrix} 0 & -1/L \\ 1/C & -1/(RC) \end{bmatrix}\begin{bmatrix} i_1 \\ v_2 \end{bmatrix} + \begin{bmatrix} 1/L & -1/L \\ 0 & -1/(RC) \end{bmatrix}\begin{bmatrix} e_1 \\ e_2 \end{bmatrix} \quad (3.13)$$

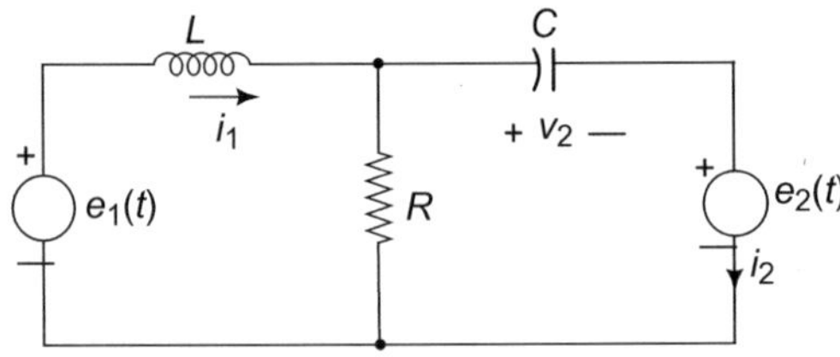

Fig. 3.3 Circuit for Example 1

A comparison of Eq. 3.13 with Eq. 3.6 yields the identification

$$\mathbf{A} = \begin{bmatrix} 0 & -1/L \\ 1/C & -1/(RC) \end{bmatrix} \text{ and } \quad \mathbf{B} = \begin{bmatrix} 1/L & -1/L \\ 0 & -1/(RC) \end{bmatrix} \tag{3.14}$$

The equation for the output vector **y** is derived by recognizing the fact that

$$i_1 = i_1 \tag{3.15}$$

and

$$i_2 = i_1 - (v_2/R) - (e_2/R) \tag{3.16}$$

Equations 3.15 and 3.16 may be combined into the following single matrix equation:

$$\begin{bmatrix} i_1 \\ i_2 \end{bmatrix} = \begin{bmatrix} 1 & 0 \\ 1 & -1/R \end{bmatrix} \begin{bmatrix} i_1 \\ v_2 \end{bmatrix} + \begin{bmatrix} 0 & 0 \\ 0 & -1/R \end{bmatrix} \begin{bmatrix} e_1 \\ e_2 \end{bmatrix} \tag{3.17}$$

Equation 3.17 is of the form Eq. 3.7 with

$$\mathbf{C} = \begin{bmatrix} 1 & 0 \\ 1 & -1/R \end{bmatrix} \text{ and } \mathbf{D} = \begin{bmatrix} 0 & 0 \\ 0 & -1/R \end{bmatrix} \tag{3.18}$$

In this chapter, we shall restrict the term 'state equation' to denote the set of first-order differential equations given by Eq. 3.6 only, because the other equation, viz. Eq. 3.7, relating the output to the input and the state vectors, can be easily obtained by algebraic means.

How Do You Choose State Variables in a Physical System?

The choice of state variables in Example 1 was very simple. We chose the capacitor voltage and the inductor current as the two state variables. This should not lead one to believe that, in general, an electrical network with l inductors and c capacitors would require $l + c$ state variables. This is because not all capacitor voltages can be specified independently when there are capacitor loops in the network. Similarly, not all inductor currents can be specified independently when there are inductance cut-sets[2] in the network. An allowable set of state variables is the set of all capacitor voltages, diminished by a subset equal to the number of capacitor loops, plus the set of inductor currents, diminished by a subset equal to the number of inductance cut-sets. Thus to find the state equations, we must eliminate, from the set of equations governing the network, all non-state variables, that is all resistance voltages and currents, and those capacitor voltages and inductor currents which correspond to certain members of capacitance loops and inductance cut-sets.

These topological circuit constraints, which reduce the total number of state variables from the total number of storage elements, apply to other systems also. However, the state variables of a non-electrical system can perhaps be detected more easily by drawing the analogous electrical system and by finding the all capacitor loops and all inductance cut-sets in this analogous representation.

Example 2 Consider the network shown in Fig. 3.4 in which there is one all capacitor loop and one all inductance node. Thus, although there are six energy storage elements, we shall require only $6 - 1 - 1 = 4$ state variables. This reduction occurs because we can assign arbitrary initial voltages to two capacitors only (the initial voltage on the remaining capacitor is determined by Kirchoff's voltage law); similarly, we can assign arbitrary initial currents in two inductors

[2]A cut-set is a set of branches, which, when removed, splits the network into two unconnected parts.

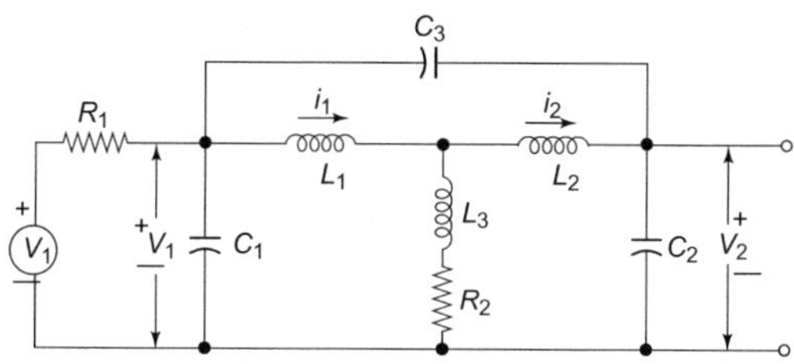

Fig. 3.4 Circuit for Example 2

only (the third inductor current is determined by Kirchoff's current law). Let us, therefore, choose v_1, v_2, i_1 and i_2 as the state variables. We can then write the following equations:

$$\left.\begin{aligned}&(v_i - v_1)/R_1 = C_1(\mathrm{d}v_1/\mathrm{d}t) + i_1 + C_3[\mathrm{d}(v_1 - v_2)/\mathrm{d}t]\\&C_2(\mathrm{d}v_2/\mathrm{d}t) = i_2 + C_3[\mathrm{d}(v_1 - v_2)/\mathrm{d}t]\\&v_1 - L_1(\mathrm{d}i_1/\mathrm{d}t) = v_2 + L_2(\mathrm{d}i_2/\mathrm{d}t)\\&\qquad = L_3[\mathrm{d}(i_1 - i_2)/\mathrm{d}t] + R_2(i_1 - i_2)\end{aligned}\right\} \tag{3.19}$$

By manipulating these equations, one can express the quantities ($\mathrm{d}v_1/\mathrm{d}t$), ($\mathrm{d}v_2/\mathrm{d}t$), ($\mathrm{d}i_1/\mathrm{d}t$) and ($\mathrm{d}i_2/\mathrm{d}t$) in terms of v_1, v_2, i_1, i_2 and v_i, and hence obtain the state formulation. This reduction to a canonical form is left to you as an exercise.

In the preceding example, it was easy to identify the all-capacitance loop and the all-inductance node. In a complicated network, however, difficulties may arise in counting the number of such subsets. To organize the counting in such a situation, first, open circuit all elements in the network excepting the capacitors and count the number of independent loops by the usual topological rule, namely $(N_l)_c = N_b - N_j + N_s$, where N_b is the number of branches, N_j is the number of nodes, and N_s is the number of separate parts. Next, short circuit all elements of the network excepting the inductors and find the number of independent nodes by the formula $(N_n)_l = N_j - N_s$. Then, the minimal number of state variables is $N = N_{l+c} - (N_l)_c - (N_n)_l$, where N_{l+c} is the total number of energy storage elements in the network. It should also become apparent now that the choice of state variables for a given physical system is not unique. For a more detailed exposition of this topic, see Kuh and Rohrer [1].

How Do You Choose State Variables When the System Differential Equation is Given?

When the physical system is given, the natural choice of state variables is the quantities associated with the energy storage elements. Suppose, however, that the only given information about the system is a single differential equation involving the output of the system, y, and its first k derivatives,[3] e.g.

$$F\left(y, \dot{y}, y^{(2)}, \ldots, y^{(k)}\right) + u(t) = 0 \tag{3.20}$$

If Eq. 3.20 can be rearranged to the form

$$y^{(k)} = f\left(y, \dot{y}, y^{(2)}, \ldots, y^{(k-1)}\right) + u(t) \tag{3.21}$$

then a natural, but quite arbitrary choice of state variables would be the following:

$$x_1 = y, x_2 = \dot{y}, x_3 = y^{(2)}, \ldots, x_k = y^{(k-1)} \tag{3.22}$$

It follows that the state variables obey the following differential equations:

$$\begin{aligned}\dot{x}_1 &= x_2\\\dot{x}_2 &= x_3\\&\cdots\\\dot{x}_{k-1} &= x_k\\\dot{x}_k &= f(x_1, x_2, \ldots, x_k) + u(t)\end{aligned} \tag{3.23}$$

For linear systems, $f(x_1, x_2, \ldots, x_k)$ will be of the form

$$f = a_1x_1 + a_2x_2 + \cdots + a_kx_k \tag{3.24}$$

[3]The symbol $y^{(k)}$ stands for the kth derivative of y with respect to t.

Hence, the state equation will be of the form Eq. 3.6 with

$$\mathbf{A} = \begin{bmatrix} 0 & 1 & 0 & 0 & 0 & \cdots & 0 \\ 0 & 0 & 1 & 0 & 0 & \cdots & 0 \\ 0 & 0 & 0 & 1 & 0 & \cdots & 0 \\ \cdots & \cdots & \cdots & \cdots & \cdots & \cdots & \cdots \\ 0 & 0 & 0 & 0 & 0 & \cdots & 1 \\ a_1 & a_2 & a_3 & a_4 & a_5 & \cdots & a_k \end{bmatrix} \tag{3.25}$$

which is of dimension $k \times k$, and

$$\mathbf{B} = \begin{bmatrix} 0 \\ 0 \\ 0 \\ \vdots \\ 1 \end{bmatrix} \tag{3.26}$$

which has the dimension $k \times 1$.

A more complicated situation arises when the differential equation also involves the derivatives of the input. For example, consider the equation

$$\begin{aligned} y(k) + a_1 y^{(k-1)} + \cdots + a_{k-1}\dot{y} + a_k y \\ = b_1 \dot{u}(t) + b_2 u(t) \end{aligned} \tag{3.27}$$

If we make the same choice of the state variables as in the previous case, i.e.

$$x_1 = y, x_2 = \dot{y}, \ldots, x_k = y^{(k-1)} \tag{3.28}$$

we would obtain the following dynamic equations:

$$\dot{x}_1 = x_2, \dot{x}_2 = x_3, \ldots, \dot{x}_{k-1} = x_k \tag{3.29}$$

and the remaining equation for $\dot{x}_k$ would be, from Eq. 3.27,

$$\dot{x}_k = -a_1 x_k - a_2 x_{k-l}, - \cdots - a_k x_1 + b_l \dot{u} + b_2 u \tag{3.30}$$

which is not in the normal form because of the presence of the $\dot{u}$ term.

To avoid this difficulty, let us choose

$$x_k = y^{(k-l)} - bu, \tag{3.31}$$

where b is a constant to be determined, and leave the other state variables unchanged. Then the dynamic equations become

$$\dot{x}_1 = x_2, \dot{x}_2 = x_3, \ldots, \dot{x}_{k-2} = x_{k-1}, \tag{3.32}$$

and

$$\dot{x}_{k-1} = y^{(k-1)} = x_k + bu$$

The last equation must satisfy Eq. 3.27, so that

$$\begin{aligned} \dot{x}_k = x_{k-1}^{(2)} - b\dot{u} = y^{(k)} - b\dot{u} \\ = -a_1(x_k + bu) - a_2 x_{k-1} - \cdots - a_k x_1 \\ + (b_1 - b)\dot{u} + b_2 u \end{aligned} \tag{3.33}$$

The term involving $\dot{u}$ disappears if we choose $b_1 = b$. Then

$$\dot{x}_k = -a_1 x_k - a_2 x_{k-l}, - \cdots - a_k x_l + (b_2 - a_1 b_1) u \tag{3.34}$$

and the state equation becomes of the form Eq. 3.6 with

$$\mathbf{A} = \begin{bmatrix} 0 & 1 & 0 & 0 & \cdots & 0 \\ 0 & 0 & 1 & 0 & \cdots & 0 \\ \cdots & \cdots & \cdots & \cdots & \cdots & \cdots \\ 0 & 0 & 0 & 0 & \cdots & 1 \\ -a_k & -a_{k-1} & \cdots & \cdots & \cdots & -a_1 \end{bmatrix} \tag{3.35}$$

which is of dimension $k \times k$, and

$$\mathbf{B} = \begin{bmatrix} 0 \\ 0 \\ \vdots \\ b \\ b_2 - a_1 b_1 \end{bmatrix} \tag{3.36}$$

which is of dimension $k \times 1$. For a more general procedure for determining the normal form of linear differential equations, you are referred to Schwartz and Friedland [2].

How Does One Solve Linear Time-Invariant State Equations?

Consider the familiar first-order differential equation

$$(\mathrm{d}x/\mathrm{d}t) = ax + bu, \qquad (3.37)$$

where $x(t)$ and $u(t)$ are scalar functions of time. The solution of this equation can be obtained by using either the integrating factor or the Laplace transform method. Using the latter method, we get, from Eq. 3.37,

$$sX(s) - x(0) = aX(s) + bU(s) \qquad (3.38)$$

or,

$$(s-a)X(s) = x(0) + bU(s) \qquad (3.39)$$

or,

$$X(s) = [x(0)/(s-a)] + [bU(s)/(s-a)] \qquad (3.40)$$

or,

$$x(t) = \mathcal{L}^{-1}X(s) = \mathrm{e}^{at}x(0) + \int_0^t \mathrm{e}^{a(t-\tau)}bu(\tau)\mathrm{d}\tau \qquad (3.41)$$

where the last term on the right-hand side is easily recognized as the familiar convolution integral. The same result could also have been obtained by using the integrating factor $\in^{-at}$.

Now, consider the vector differential equation

$$(\mathrm{d}\mathbf{x}/\mathrm{d}t) = \mathbf{Ax} + \mathbf{Bu} \qquad (3.42)$$

We expect the solution to be of the same form as Eq. 3.41, but involving matrix exponential functions.

Laplace Transform Method

Taking the Laplace transform of Eq. 3.42 gives

$$s\mathbf{X}(s) - \mathbf{x}(0) = \mathbf{AX}(s) + \mathbf{BU}(s) \qquad (3.43)$$

A block diagram representation of Eq. 3.43 is shown in Fig. 3.5. The matrix solution of this equation is given by

$$\mathbf{X}(s) = (s\mathbf{I}-\mathbf{A})^{-1}\mathbf{x}(0) + (s\mathbf{I}-\mathbf{A})^{-1}\mathbf{BU}(s), \qquad (3.44)$$

where **I** is the identity matrix

$$\mathbf{I} = \begin{bmatrix} 1 & 0 & 0 & \cdots & 0 \\ 0 & 1 & 0 & \cdots & 0 \\ \cdots & \cdots & \cdots & \cdots & \cdots \\ 0 & 0 & 0 & \cdots & 1 \end{bmatrix} \qquad (3.45)$$

of dimension $n \times n$ and $(\ )^{-1}$ indicates matrix inversion. Clearly, the solution is determined by the properties of the matrix ($s\mathbf{I} - \mathbf{A}$), aptly called the characteristic matrix of the system. If we write

$$\mathbf{L}(s) = s\mathbf{I} - \mathbf{A} \qquad (3.46)$$

then

$$\mathbf{L}^{-1}(s) = (s\mathbf{I}-\mathbf{A})^{-1} = \mathbf{L}_a(s)/\Delta(s), \qquad (3.47)$$

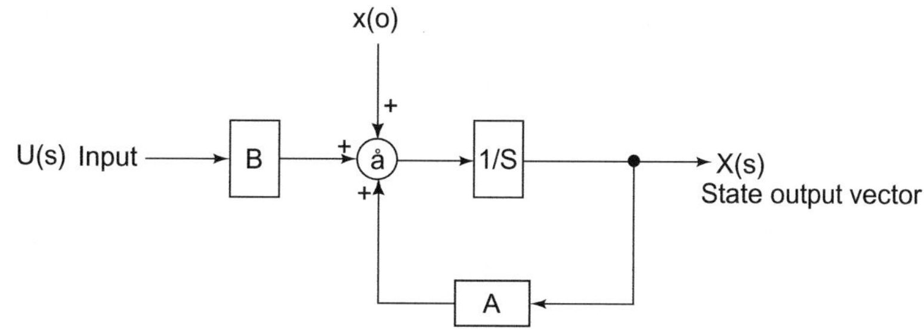

Fig. 3.5 Block diagram representation of (3.43)

where $\mathbf{L}_a(s)$ is the adjoint of the characteristic matrix and

$$\Delta(s) = \det(s\mathbf{I} - \mathbf{A}) \tag{3.48}$$

This determinant is a polynomial of degree n and is called the characteristic polynomial of the matrix $\mathbf{A}$ or of the system characterized by the matrix $\mathbf{A}$. The roots of this polynomial are called by various different names, some of them being eigenvalues, natural modes, natural frequencies and characteristic roots. As the elements in $\mathbf{L}_a(s)$ are the co-factors of $\mathbf{L}(s)$, these will be of order $(n - 1)$.

In the event that the entries in the adjoint matrix have a common factor, this will also appear in $\Delta(s)$. This will permit the cancellation of the common factor in $\mathbf{L}^{-1}(s)$. This cancellation should be done before evaluating the inverse Laplace transform of $\mathbf{L}^{-1}(s)$.

The time domain solution follows by taking the inverse Laplace transform of the expression

$$[1/\Delta(s)]\mathbf{L}_a(s)\mathbf{x}(0) + [1/\Delta(s)]\mathbf{L}_a(s)\mathbf{BU}(s) \tag{3.49}$$

As in the case of ordinary expressions in Laplace variable s, we expand each of the terms in (3.49) into a partial fraction and take the inverse Laplace transform. For example, for the force-free system, assuming that the roots of $\Delta(s) = 0$ are all distinct, we have

$$\begin{aligned} x(t) &= \mathcal{L}^{-l}\left[(s\mathbf{I} - \mathbf{A})^{-1}\right]x(0) \\ &= \mathcal{L}^{-1}\{[\mathbf{X}_1/(s - s_1)] + [\mathbf{X}_2/(s - s_2)] \\ &\quad + \cdots + [\mathbf{X}_n/(s - s_n)]\} \\ &= \mathbf{X}_1 e^{s_1 t} + \mathbf{X}_2 e^{s_2 t} + \cdots + \mathbf{X}_n e^{s_n t} \end{aligned} \tag{3.50}$$

where

$$\mathbf{X}_i = \lim_{s \to s_l}[(s - s_i)\mathbf{L}_a(s)/\Delta(s)]\mathbf{x}(0) \tag{3.51}$$

The case in which the roots are not all distinct can be handled by the same technique as in the ordinary expressions in the Laplace variable s.

The Laplace transform method of solution requires evaluation of the adjoint of a matrix. Also, in general, it is difficult to obtain the required inverse transformation.

Before we proceed to discuss a second method of solution, we digress a little in order to introduce the important concept of the transfer matrix. Taking the Laplace transform of Eq. 3.7 and substituting the value of $\mathbf{X}(s)$ from Eq. 3.44, we get

$$\mathbf{Y}(s) = \mathbf{C}(s\mathbf{I} - \mathbf{A})^{-1}\mathbf{x}(0) + [\mathbf{C}(s\mathbf{I} - \mathbf{A})^{-1}\mathbf{B} + \mathbf{D}]\mathbf{X}(s) \tag{3.52}$$

With zero initial conditions, this becomes

$$\mathbf{Y}(s) = [\mathbf{C}(s\mathbf{I} - \mathbf{A})^{-1}\mathbf{B} + \mathbf{D}]\mathbf{X}(s) = \mathbf{H}(s)\mathbf{X}(s), \tag{3.53}$$

where

$$\mathbf{H}(s) = [\mathbf{C}(s\mathbf{I} - \mathbf{A})^{-1}\mathbf{B} + \mathbf{D}] = \begin{bmatrix} H_{11}(s) & H_{12}(s) & \cdots & H_{1n}(s) \\ H_{21}(s) & H_{22}(s) & \cdots & H_{2n}(s) \\ \cdots & \cdots & \cdots & \cdots \\ H_{p1}(s) & H_{p2}(s) & \cdots & H_{pn}(s) \end{bmatrix} \tag{3.54}$$

Thus, the ith component $Y_i(s)$ of the transform $\mathbf{Y}(s)$ of the output vector may be written as

$$\begin{aligned} Y_i(s) =& H_{i1}(s)X_1(s) + H_{i2}(s)X_2(s) \\ &+ \cdots + H_{in}(s)X_n(s) \end{aligned} \tag{3.55}$$

Clearly, $H_{ij}(s)$, the (i, j)th element of $\mathbf{H}(s)$, is the transfer function between $X_j(s)$ and $Y_i(s)$ and

$$h_{ij}(t) = \mathcal{L}^{-1}[H_{ij}(s)] \tag{3.56}$$

is the response at the ith output terminal due to an unit impulse at the jth input terminal.

We have thus found a general expression for the transfer functions of a system in terms of the matrices appearing in the normal form characterization. The matrix $\mathbf{H}(s)$ is called the transfer matrix of the system.

Example 3 Consider again the circuit in Fig. 3.5 with L = (1/2) Henry, C = 1 Farad and R = (1/3) ohm. Then from Eqs. 3.14 and 3.18, **A**, **B**, **C** and **D** matrices become

$$\mathbf{A} = \begin{bmatrix} 0 & -2 \\ 1 & -3 \end{bmatrix} \; \mathbf{B} = \begin{bmatrix} 2 & -2 \\ 0 & -3 \end{bmatrix}$$
$$\mathbf{C} = \begin{bmatrix} 1 & 0 \\ 1 & -3 \end{bmatrix} \text{ and } \mathbf{D} = \begin{bmatrix} 0 & 0 \\ 0 & -3 \end{bmatrix} \tag{3.57}$$

Therefore

$$\mathbf{L}(s) = (s\mathbf{I} - \mathbf{A}) = \begin{bmatrix} s & 2 \\ -1 & s+3 \end{bmatrix} \tag{3.58}$$

and

$$\begin{aligned}(s\mathbf{I} - \mathbf{A})^{-1} &= \frac{1}{s^2 + 3s + 2} \begin{bmatrix} s+3 & -2 \\ 1 & s \end{bmatrix} \\ &= \begin{bmatrix} [-1/(s+2)] + [2/(s+1)][2/(s+2)] + [-2/(s+1)] \\ [-1/(s+2)] + [1/(s+1)][2/(s+2)] + [-1/(s+1)] \end{bmatrix}\end{aligned} \tag{3.59}$$

For the force-free case, the time domain solution is

$$\mathbf{x}(t) = \begin{bmatrix} -e^{-2t} + 2e^{-t} & 2e^{-2t} - 2e^{-t} \\ -e^{-2t} + e^{-t} & 2e^{-2t} - e^{-t} \end{bmatrix} \mathrm{x}(0) \tag{3.60}$$

$$\begin{aligned}\mathbf{H}(s) &= \mathbf{C}(s\mathbf{I} - \mathbf{A})^{-1}\mathbf{B} + \mathbf{D} \\ &= \frac{1}{s^2 + 3s + 2} \begin{bmatrix} 1 & 0 \\ 1 & -3 \end{bmatrix} \begin{bmatrix} s+3 & -2 \\ 1 & s \end{bmatrix} \begin{bmatrix} 2 & -2 \\ 0 & -3 \end{bmatrix} + \begin{bmatrix} 0 & 0 \\ 0 & -3 \end{bmatrix} \\ &= \frac{1}{s^2 + 3s + 2} \begin{bmatrix} 2(s+3) & -2s \\ 2s & -(3s^2 + 2s) \end{bmatrix}\end{aligned} \tag{3.61}$$

The transfer matrix is given by Eq. 3.61

Thus, the response $i_2(t)$ to the input $e_1(t) = \delta(t)$ is the inverse transform of $H_{21}(s)$, giving

$$h_{21}(t) = 4e^{-2t} - 2e^{-t} \tag{3.62}$$

which checks with the result of a direct calculation. It may also be noted that only $H_{22}(s)$ has a numerator of second degree, so that only $h_{22}(t)$ involves an impulsive term. This is exactly what we expect, since $e_2(t) = \delta(t)$ will result in an impulse of current through C.

By Series Expansion

We next consider a solution of the vector differential equation by series expansion. First, consider the homogeneous equation

$$\dot{\mathbf{x}} = \mathbf{A}\mathbf{x} \tag{3.63}$$

Performing a Taylor series expansion about the origin of the state space, we get

$$\mathbf{x}(t) = \mathbf{E}_0 + \mathbf{E}_1 t + \mathbf{E}_2 t^2 + \cdots \tag{3.64}$$

where $\mathbf{E}_i$'s are column vectors, whose elements are constants. In order to determine these vectors, we successively differentiate Eq. 3.64 and set $t = 0$; thus

$$\mathbf{x}(0) = \mathbf{E}_0, \dot{\mathbf{x}}(0) = \mathbf{E}_1, \mathbf{x}^{(2)}(0) = 2\mathbf{E}_2, \ldots \tag{3.65}$$

However, from Eq. 3.63, we note that

$$\dot{\mathbf{x}}(0) = \mathbf{A}\mathbf{x}(0), \mathbf{x}^{(2)}(0) = \mathbf{A}\,\dot{\mathbf{x}}(0) = \mathbf{A}[\mathbf{A}\mathbf{x}(0)] = \mathbf{A}^2\mathbf{x}(0), \ldots \tag{3.66}$$

where $\mathbf{A}^2$ implies the matrix multiplication $\mathbf{A} \times \mathbf{A}$. Continuing the process, we obtain the following series solution for $\mathbf{x}(t)$:

$$\mathbf{x}(t) = \left[\mathbf{I} + \mathbf{A}t + (\mathbf{A}^2t^2/2!) + (\mathbf{A}^3t^3/3!) + \cdots\right]\mathbf{x}(0) \tag{3.67}$$

This infinite series defines the matric exponential function $e^{\mathbf{A}t}$ which may be shown to be convergent for all square matrices $\mathbf{A}$. Therefore

$$\mathbf{x}(t) = e^{\mathbf{A}t}\mathbf{x}(0) \tag{3.68}$$

Comparing this result with Eq. 3.41, we note the equivalence of the unforced part of the solution. The function $e^{\mathbf{A}t}$ contains a great deal of information about the system behaviour and, as such, is called the fundamental or the transition matrix, denoted by $\boldsymbol{\phi}(t)$. It possesses all the important properties of the scalar exponential function. In particular, the derivative of the function yields the function itself pre-multiplied by a constant. Using this property and the method of variation of parameters, we now proceed to determine the forced part of the solution, i.e. the particular integral of the original equation given by Eq. 3.6.

By analogy with the particular solution of the scalar equation given by Eq. 3.37, we let the particular integral of the vector differential equation be

$$\mathbf{x}_p(t) = \boldsymbol{\phi}(t)\mathbf{q}(t) \tag{3.69}$$

Substituting this expression in Eq. 3.6, we obtain

$$(\mathrm{d}\boldsymbol{\phi}/\mathrm{d}t)\mathbf{q} + \boldsymbol{\phi}(\mathrm{d}\mathbf{q}/\mathrm{d}t) = \mathbf{A}\boldsymbol{\phi}\mathbf{q} + \mathbf{B}\,\mathbf{u} \tag{3.70}$$

But

$$\mathrm{d}\boldsymbol{\phi}/\mathrm{d}t = \mathrm{d}\,e^{\mathbf{A}t}/\mathrm{d}t = \mathbf{A}e^{\mathbf{A}t} = \mathbf{A}\boldsymbol{\phi} \tag{3.71}$$

so that

$$\boldsymbol{\phi}(\mathrm{d}\mathbf{q}/\mathrm{d}t) = \mathbf{B}\,\mathbf{u} \tag{3.72}$$

or

$$\mathrm{d}\mathbf{q}/\mathrm{d}t = \boldsymbol{\phi}^{-1}(t)\mathbf{B}\,\mathbf{u}(t) \tag{3.73}$$

or

$$\mathbf{q}(t) = \int_0^t \boldsymbol{\phi}^{-1}(\tau)\mathbf{B}\mathbf{u}(\tau)\,\mathrm{d}\tau \tag{3.74}$$

Thus, the particular integral is given by

$$\mathbf{x}_p(t) = \boldsymbol{\phi}(t)\int_0^t \boldsymbol{\phi}^{-1}(\tau)\mathbf{B}\mathbf{u}(\tau)\,\mathrm{d}\tau \tag{3.75}$$

and the complete solution is

$$\mathbf{x}(t) = \boldsymbol{\phi}(t)\mathbf{x}(0) + \int_0^t \boldsymbol{\phi}(t)\boldsymbol{\phi}^{-1}(\tau)\mathbf{B}\mathbf{u}(\tau)\,\mathrm{d}\tau \tag{3.76}$$

This expression is analogous to the familiar complete solution of a scalar first-order differential equation, given in Eq. 3.41. The properties of the fundamental matrix $\boldsymbol{\phi}(t)$ allow a further simplification of Eq. 3.76. These properties and methods for evaluation of $\boldsymbol{\phi}(t)$ will be discussed in details in Part II, which will also include a bibliography and an appendix on matrix algebra.

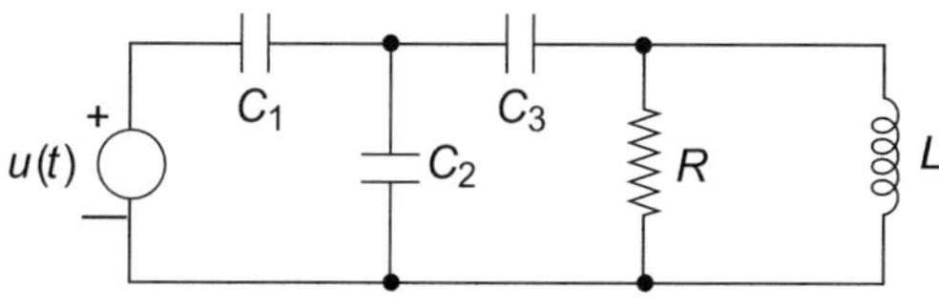

Fig. P.1 Circuit for Problem P.1

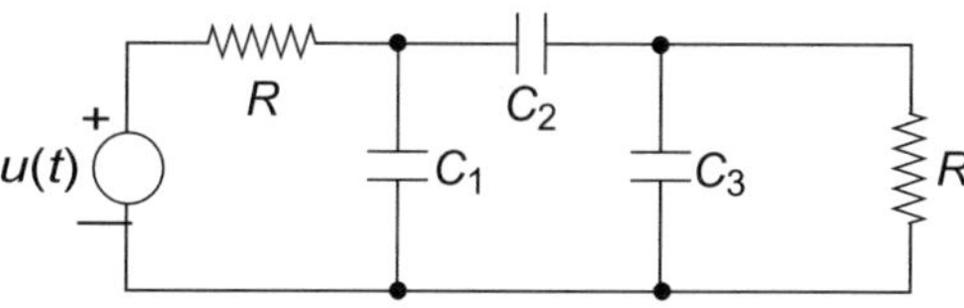

Fig. P.2 Circuit for Problem P.2

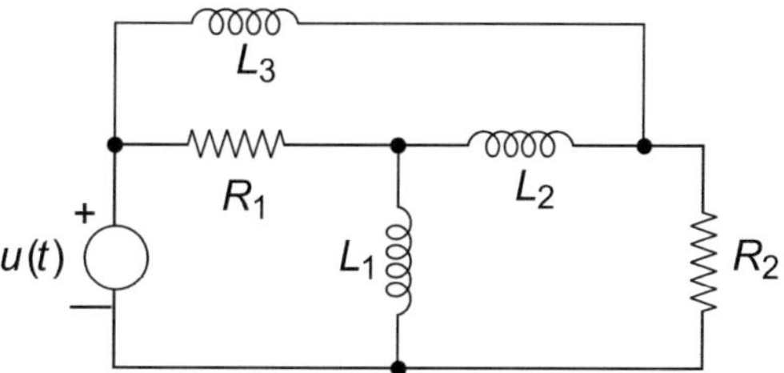

Fig. P.3 Circuit for Problem P.3

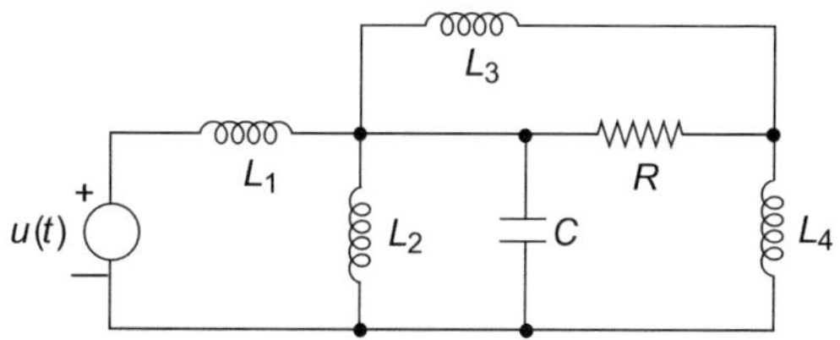

Fig. P.4 Circuit for Problem P.4

An Advice

The best way to comprehend a topic like this is to follow the examples given here as well as those in the next chapter and in addition, solve as many examples as possible, from the reference books, cited below.

Problems

I believe, as I mentioned on earlier occasion, that they are not difficult.

P.1. Write the state equations for the circuit in Fig. P.1.

P.2. Write the state equations for the circuit in Fig. P.2.

P.3. Write the state equations for the circuit shown in Fig. P.3.

P.4. Write the state equations for the circuit shown in Fig. P.4.

P.5. Solve the state equations for the circuit in Fig. P.4.

References

1. E.S. Kuh, R.A. Rohrer, The state variable approach to network theory, in *Proc IRE*, vol 53 (July 1965), pp. 672–686
2. R.J. Schwartz, B. Friedland, *Linear Systems* (McGraw Hill, 1965)

State Variables—Part II

4

In the first part of this discussion on state variables, which hopefully you have grasped, we presented the basic concepts of state variables and state equations, and some methods for solution of the latter. In this second and concluding part, we dwell upon the properties and evaluation of the fundamental matrix. An appendix on matrix algebra is also included. Several examples have been given to illustrate the techniques. This is the last part, be assured!

Keywords
Fundamental matrix · Fundamental state equation · Evaluating fundamental state matrix · State transition flow graphs
Review of matrix algebra

In Part I of this discussion, we introduced the basic concepts of state variables and state equations, and discussed several methods of solution of the latter. In this second and the concluding part, we deal with the properties of the fundamental matrix and procedures for evaluation of the same. In particular, we dwell upon the state transition flow graph method in considerable details. We provide a bibliography at the end and include an appendix on the essentials of matrix algebra.

Equations and figures occurring in Part I and referred to here are not reproduced, for brevity, and it is suggested that you make it convenient to have Part I at hand for ready reference.

Properties of the Fundamental Matrix

One important property of the fundamental matrix $\boldsymbol{\phi}(t)$, namely that

$$\mathrm{d}\boldsymbol{\phi}/\mathrm{d}t = A\boldsymbol{\phi}$$

has already been mentioned.

Consider now the solution to the homogeneous vector differential equation, given by Eq. 3.68 in Part I:

$$\mathbf{x}(t) = \boldsymbol{\phi}(t)\mathbf{x}(0) \tag{4.1}$$

Or, explicitly

Source: S. C. Dutta Roy, "State Variables—Part II," *IETE Journal of Education*, vol. 38, pp. 99–107, April–June 1997.

S. C. Dutta Roy, *Circuits, Systems and Signal Processing*,
https://doi.org/10.1007/978-981-10-6919-2_4

$$\begin{bmatrix} x_1(t) \\ x_2(t) \\ \vdots \\ x_n(t) \end{bmatrix} = \begin{bmatrix} \phi_{11}(t) & \phi_{12}(t) & \cdots & \phi_{1n}(t) \\ \phi_{21}(t) & \phi_{22}(t) & \cdots & \phi_{2n}(t) \\ \cdots & \cdots & \cdots & \cdots \\ \phi_{n1}(t) & \phi_{n2}(t) & \cdots & \phi_{nn}(t) \end{bmatrix} \begin{bmatrix} x_1(0) \\ x_2(0) \\ \vdots \\ x_n(0) \end{bmatrix} \tag{4.2}$$

We can determine the elements of the fundamental matrix by setting to zero all initial conditions except one, evaluating the output of the states, and repeating the procedure. For example, setting

$$x_1(0) = 1, x_2(0) = x_3(0) = \cdots = x_n(0) = 0 \tag{4.3}$$

we obtain

$$x_1(t) = \phi_{11}(t), x_2(t) = \phi_{21}(t), \ldots, x_n(t) = \phi_{nl}(t) \tag{4.4}$$

Thus, in general, $\phi_{ij}(t)$ is the transient response of the ith state due to unit initial condition of the jth state, when zero initial conditions apply to all other states. This property will be utilized later to determine the state equations of any system represented by an analog computer simulation.

So far, in solving the state equations, the initial conditions were taken at $t = 0$. If, instead, the initial time is taken as $t = t_0$, then the fundamental matrix becomes

$$\boldsymbol{\phi}(t - t_0) = \mathrm{e}^{\mathbf{A}(t-t_0)} \tag{4.5}$$

and the complementary solution changes to

$$\mathbf{x}(t) = \boldsymbol{\phi}(t-t_0)\mathbf{x}(t_0) \tag{4.6}$$

Furthermore, we have only considered time-invariant systems. For a time-varying system, we must use the fundamental matrix $\boldsymbol{\phi}(t, t_0)$ which depends on both the present time t and the initial time t_0, and not just on the time difference $t - t_0$. Also the matrices **A**, **B**, **C** and **D** become functions of time t.

Consider the evaluation of the transient response of a time-invariant system at various values of time t_1 and t_2 while the initial time was t_0. At $t = t_1$,

$$\mathbf{x}(t_1) = \boldsymbol{\phi}(t_1-t_0)\mathbf{x}(t_0) \tag{4.7}$$

At $t = t_2$, if the initial time is considered as t_1, then

$$\begin{aligned} \mathbf{x}(t_2) &= \boldsymbol{\phi}(t_2-t_1)\mathbf{x}(t_1) \\ &= \boldsymbol{\phi}(t_2-t_1)\boldsymbol{\phi}(t_1-t_0)x(t_0) \end{aligned} \tag{4.8}$$

On the other hand, considering the initial time as t_0, we get

$$\mathbf{x}(t_2) = \boldsymbol{\phi}(t_2-t_0)\mathbf{x}(t_0) \tag{4.9}$$

Comparing Eqs. 4.8 and 4.9, we get

$$\boldsymbol{\phi}(t_2-t_0) = \boldsymbol{\phi}(t_2-t_1)\boldsymbol{\phi}(t_1-t_0) \tag{4.10}$$

This important relationship justifies the name 'state transition matrix' for $\boldsymbol{\phi}(t)$. Clearly, $\boldsymbol{\phi}(t_2 - t_0)$ is a sequence of state transitions. Since the relation must hold at $t_2 = t_0$, we obtain

$$\begin{aligned} \boldsymbol{\phi}(t_2-t_0) &= \boldsymbol{\phi}(t_2-t_2) = \mathbf{I} \\ &= \boldsymbol{\phi}(t_0-t_1)\boldsymbol{\phi}(t_1-t_0) \end{aligned} \tag{4.11}$$

and

$$\boldsymbol{\phi}^{-1}(t_1-t_0) = \boldsymbol{\phi}(t_0-t_1) \tag{4.12}$$

Now setting the initial time of interest to zero, we get

$$\boldsymbol{\phi}^{-1}(t_1) = \boldsymbol{\phi}(-t_1) \tag{4.13}$$

Or, in general

$$\boldsymbol{\phi}^{-1}(t) = \boldsymbol{\phi}(-t) \tag{4.14}$$

This relation can be used to simplify the general solution, given by Eq. 3.76 of the state equation.

The Fundamental State Transition Equation

For an initial time $t = t_0$, the general solution of the vector differential equation given by Eq. 3.6, as found in Eq. 3.76, modifies as follows:

$$\mathbf{x}(t) = \phi(t-t_0)\,\mathbf{x}(t_0) + \int_{t_0}^{t} \phi(t)\phi^{-1}(\tau)\mathbf{B}\mathbf{u}(\tau)\mathrm{d}\tau \tag{4.15}$$

Using Eqs. 4.10 and 4.14, we may write the fundamental state transition equation of a time-invariant system as

$$\mathbf{x}(t) = \phi(t-t_0)\,\mathbf{x}(t_0) + \int_{t_0}^{t} \phi(t - \tau)\mathbf{B}\mathbf{u}(\tau)\mathrm{d}\tau \tag{4.16}$$

For a time-varying system, the more general form of the transition matrix $\phi(t, t_0)$ must be used and then the state transition equation becomes

$$\mathbf{x}(t) = \phi(t, t_0)\,\mathbf{x}(t_0) + \int_{t_0}^{t} \phi(t, \tau)\mathbf{B}(\tau)\mathbf{u}(\tau)\mathrm{d}\tau \tag{4.17}$$

The solution in either Eqs. 4.16 or 4.17 consists of the unforced natural response due to the initial conditions plus a matrix convolution integral containing the matrix of the inputs $\mathbf{u}$ (τ).

Procedures for Evaluating the Fundamental Matrix: Described in Steps

The fundamental matrix $\phi(t) = e^{\mathbf{A}t}$ can be evaluated by various methods which are now discussed through some examples.

By Exponential Series Expansion

Example 4 For the network in Fig. 3.3 with the specified element values, as in Example 3, the **A** matrix was shown to be

$$\mathbf{A} = \begin{bmatrix} 0 & -2 \\ 1 & -3 \end{bmatrix} \tag{4.18}$$

Thus

$$\mathbf{A}^2 = \begin{bmatrix} 0 & -2 \\ 1 & -3 \end{bmatrix}\begin{bmatrix} 0 & -2 \\ 1 & -3 \end{bmatrix} = \begin{bmatrix} -2 & 6 \\ -3 & 7 \end{bmatrix} \tag{4.19}$$

$$\mathbf{A}^3 = \begin{bmatrix} -2 & 6 \\ -3 & 7 \end{bmatrix}\begin{bmatrix} 0 & -2 \\ 1 & -3 \end{bmatrix} = \begin{bmatrix} 6 & -14 \\ 7 & -15 \end{bmatrix} \tag{4.20}$$

and so on. Hence

$$\begin{aligned}\phi(t) &= \begin{bmatrix} 1 & 0 \\ 0 & 1 \end{bmatrix} + \begin{bmatrix} 0 & -2 \\ 1 & -3 \end{bmatrix} t + \begin{bmatrix} -2 & 6 \\ -3 & 7 \end{bmatrix} \frac{t^2}{2!} \\ &\quad + \begin{bmatrix} 6 & -14 \\ 7 & -15 \end{bmatrix} \frac{t^3}{3!} + \cdots \\ &= \begin{bmatrix} 1 + 0.t - 2\frac{t^2}{2!} + 6\frac{t^3}{3!} - \cdots & 0 - 2t + 6\frac{t^2}{2!} - 14\frac{t^3}{3!} + \cdots \\ 0 + t - \frac{3t^2}{2!} + 7\frac{t^3}{3!} + \cdots & 1 - 3t + \frac{7t^2}{2!} - 15\frac{t^3}{3!} + \cdots \end{bmatrix}\end{aligned} \tag{4.21}$$

If the calculation is continued, each series in Eq. 4.21 turns out to be the expansion of the sum of two exponentials, and the fundamental matrix becomes

$$\phi(t) = \begin{bmatrix} -e^{-2t} + 2e^{-t} & 2e^{-2t} - 2e^{-t} \\ -e^{-2t} + e^{-t} & 2e^{-2t} - e^{-t} \end{bmatrix} \tag{4.22}$$

This method is effective only in very simple cases.

By Solution of the Homogeneous Differential Equations using Classical Methods

Example 5 Consider the network in Fig. 3.3 again. The homogeneous differential equation for this system is

$$\dot{\mathbf{x}} = \mathbf{A}x \tag{4.23}$$

or,

$$\begin{bmatrix} \dot{x}_1 \\ \dot{x}_2 \end{bmatrix} = \begin{bmatrix} 0 & -2 \\ 1 & -3 \end{bmatrix} \begin{bmatrix} x_1 \\ x_2 \end{bmatrix} \tag{4.24}$$

This matrix equation represents two simultaneous differential equations, namely

$$\left.\begin{aligned} \dot{x}_1 &= -2x_2 \\ \dot{x}_2 &= x_1 - 3x_2 \end{aligned}\right\} \tag{4.25}$$

Eliminating x_2, we obtain

$$\begin{aligned} x_1^{(2)} &= -2x_2 = -2x_1 + 6x_2 \\ &= -2x_1 + 6(-\dot{x}_1/2) = -2x_1 - 3\dot{x}_1 \end{aligned}$$

or

$$x_1^{(2)} + 3x_1 + 2x_1 = 0 \tag{4.26}$$

The general solution for this homogeneous second-order differential equation is

$$x_1(t) = C_1 \mathrm{e}^{-2t} + C_2 \mathrm{e}^{-t}, \tag{4.27}$$

where C_1 and C_2 are arbitrary constants. From Eqs. 4.25 and 4.27, we get

$$x_2(t) = C_1 \mathrm{e}^{-2t} + (C_2/2)\mathrm{e}^{-t} \tag{4.28}$$

We have to determine the matrix that multiplies the initial state vector $\mathbf{x}(0)$ to yield $\mathbf{x}(t)$. Thus, we must express C_1 and C_2 in terms of the components of $\mathbf{x}(0)$. From Eqs. 4.27 and 4.28, we find that

$$\begin{aligned} C_1 + C_2 &= x_1(0) \\ C_1 + (C_2/2) &= x_2(0) \end{aligned} \tag{4.29}$$

Solving Eq. 4.29, we get

$$\begin{aligned} C_1 &= -x_1(0) + 2x_2(0) \\ C_2 &= 2x_1(0) - 2x_2(0) \end{aligned} \tag{4.30}$$

Substitution of these expressions for the constants C_1 and C_2 into Eqs. 4.27 and 4.28 gives

$$\begin{bmatrix} x_1(t) \\ x_2(t) \end{bmatrix} = \begin{bmatrix} -\varepsilon^{-2t} + 2\varepsilon^{-t} & 2\varepsilon^{-2t} - 2\varepsilon^{-t} \\ -\varepsilon^{-2t} + \varepsilon^{-t} & 2\varepsilon^{-2t} - \varepsilon^{-t} \end{bmatrix} \begin{bmatrix} x_1(0) \\ x_2(0) \end{bmatrix} \tag{4.31}$$

Since $\mathbf{x}(t) = \phi(t)\ \mathbf{x}(0)$, the square matrix in Eq. 4.31 is $\phi(t)$. This result checks with the one obtained by the previous method.

This method also works out well for simple systems but, for higher order systems, it becomes laborious and time consuming.

By Evaluating the Inverse Laplace Transform of (sI–A)

Comparing the first equation in Eqs. 3.50 and 4.1, we obtain the important relationship

$$\phi(t) = \mathcal{L}^{-1}[s\mathbf{I} - \mathbf{A})^{-1}] \tag{4.32}$$

This relation can be used to determine the fundamental matrix, as illustrated in the following example.

Example 6 Consider the circuit in Fig. 3.3 once more. For this,

$$\begin{aligned} (s\mathbf{I} - \mathbf{A}) &= \begin{bmatrix} s & 0 \\ 0 & s \end{bmatrix} - \begin{bmatrix} 0 & -2 \\ 1 & -3 \end{bmatrix} \\ &= \begin{bmatrix} s & 2 \\ -1 & s+3 \end{bmatrix} \end{aligned} \tag{4.33}$$

The characteristic function of the system is

$$\Delta(s) = \det(s\mathbf{I} - \mathbf{A}) = (s+1)(s+2) \tag{4.34}$$

which has two real distinct roots. Thus

$$(s\mathbf{I} - \mathbf{A})^{-1} = \frac{\text{adjoint of } (s\mathbf{I} - \mathbf{A})}{(s+1)(s+2)} \tag{4.35}$$

$$= \frac{1}{(s+1)(s+2)} \begin{bmatrix} s+3 & -2 \\ 1 & s \end{bmatrix} \tag{4.36}$$

$$= \begin{bmatrix} \frac{2}{s+1} - \frac{1}{s+2} & -\frac{2}{s+1} + \frac{2}{s+2} \\ \frac{1}{s+1} - \frac{1}{s+2} & -\frac{1}{s+1} + \frac{2}{s+2} \end{bmatrix} \tag{4.37}$$

Evaluating the inverse transform of each of the terms, we find

$$\boldsymbol{\phi}(t) = \mathcal{L}^{-1}[(s\mathbf{I}-\mathbf{A})^{-1}] \quad (4.38)$$

$$= \begin{bmatrix} 2e^{-t} - e^{-2t} & -2e^{-t} + 2e^{-2t} \\ e^{-t} - e^{-2t} & -e^{-t} + 2e^{-2t} \end{bmatrix} \quad (4.39)$$

which agrees with Eq. 4.31.

Next, consider the system response to a unit step in place of $e_2(t)$, while $e_1(t) = 0$. To simplify matters, assume that the initial conditions are zero. Then

$$\mathbf{x}(t) = \int_{t_0}^{t} \begin{bmatrix} \phi_{11}(t-\tau) & \phi_{12}(t-\tau) \\ \phi_{21}(t-\tau) & \phi_{22}(t-\tau) \end{bmatrix} \begin{bmatrix} 0 \\ 1 \end{bmatrix} d\tau \quad (4.40)$$

since[1] $\mu(t - t_0) = 1$ for $t > t_0$. Thus

$$\mathbf{x}(t) = \begin{bmatrix} \int_{t_0}^{t} \phi_{12}(t-\tau)d\tau \\ \int_{t_0}^{t} \phi_{22}(t-\tau)d\tau \end{bmatrix} \quad (4.41)$$

Combining Eqs. 4.40 with 4.41, we get

$$x_1(t) = \int_{t_0}^{t} \left(e^{-(t-\tau)} - e^{-2(t-\tau)}\right) d\tau = \frac{1}{2} - e^{-(t-t_0)} + \frac{1}{2}e^{-2(t-t_0)} \quad (4.42)$$

$$x_2(t) = \int_{t_0}^{t} \left(-e^{\ (t-\tau)} + 2e^{-2(t-\tau)}\right) d\tau = e^{-(t-t_0)} - e^{-2(t-t_0)} \quad (4.43)$$

In this method, we must determine the adjoint of a matrix ($s\mathbf{I} - \mathbf{A}$), the roots of the characteristic function $\Delta(s)$ and the inverse transform of each term in ($s\mathbf{I} - \mathbf{A}$) matrix. This process will demand an increasingly tedious calculation for higher order systems.

In view of the shortcomings of the three methods illustrated in the preceding examples, it would be worthwhile to develop a method of determining the state transition matrix which simplifies the necessary steps. The state transition flow graph provides such a method; in addition, it intuitively illustrates the physical foundation of the vector state transition equation.

[1]Recall that $\mu(t)$ stands for the unit step function.

State Transition Flow Graphs

The state transition flow graphs differ from Mason's signal flow graphs in that, while the latter precludes consideration of initial conditions, the former does include all initial conditions.

Analog computers are considered obsolete nowadays, but for the purpose of developing state transition flow graphs, consider, for a moment, how systems are simulated on an analog computer. For linear systems, we require only the following computing abilities:

(i) Multiply a machine variable by a positive or a negative constant coefficient (In practice, this can be done by potentiometers and amplifiers).
(ii) Sum up two or more machine variables (This is accomplished by summing amplifiers).
(iii) Produce the time integral of a machine variable (This is done by an integrating amplifier).

The mathematical descriptions of these three basic analog computations are

$$\left.\begin{array}{ll} (1) & x_2(t) = ax_1(t) \\ (2) & x_3(t) = x_0(t)x_1(t) - x_2(t) \\ (3) & x_2(t) = \int x_1(t)dt = \int_0^t x_1(t)dt + x_2(0) \end{array}\right\}, \quad (4.44)$$

respectively, where a is a constant, positive or negative.

Taking the Laplace transform of each of these equations, we get

$$\left.\begin{array}{ll} (1) & X_2(s) = aX_1(s) \\ (2) & X_3(s) = X_0(s)X_1(s) - X_2(s) \\ (3) & X_2(s) = \frac{X_1(s)}{s} + \frac{x_2(0)}{s} \end{array}\right\} \quad (4.45)$$

The analog computer representations of Eq. 4.44 and the flow graph representations of Eq. 4.45 are shown in Fig. 4.1 on the left- and right-hand sides, respectively. Note that the representations of operations (1) and (2) are identical to the corresponding flow graph notation but the integral operation is different.

A state transition of flow graph is defined as a cause and effect representation of a set of system equations in normal form using the flow graphs of basic computer elements such as those shown in Fig. 4.1. The only dynamic (or storage) element in the computer is the integrator. The output of each integrator (or each node point in a signal flow graph in which a branch with transmittance s^{-1} ends) can therefore be considered as a state variable (or transform of the state variable). These constitute a set of state variables (or their transforms). Therefore, if we can present the system under investigation by an analog computer diagram or a state transition flow graph, we can easily determine the state vector $\mathbf{x}(t)$ or its transform $\mathbf{X}(s)$.

The state transition equation and its Laplace transform are given by

$$\mathbf{x}(t) = \phi(t-t_0)\mathbf{x}(t_0) + \int_{t_0}^{t} \phi(t-\tau)\mathbf{B}\mathbf{u}(\tau)\mathrm{d}\tau \tag{4.46}$$

$$\mathbf{X}(s) = (s\mathbf{I}-\mathbf{A})^{-1}\mathbf{x}(t_0) + (s\mathbf{I}-\mathbf{A})^{-1}\mathbf{B}U(s) \tag{4.47}$$

Now if we can determine Eq. 4.47 directly from the state transition flow graph, then we can avoid the matrix inversion, and by taking the inverse Laplace transform of each element, we can obtain Eq. 4.46 directly. The system equation is obtained in the form of Eq. 4.47 by applying Mason's gain formula to the state transition flow graph. This formula relates the gain $M_{xy}(s)$ between an independent node x and a dependent node y in the following manner:

$$M_{xy}(s) = \frac{X_y(s)}{X_x(s)} = \sum_k \frac{M_k(s)\Delta_k(s)}{\Delta(s)}, \tag{4.48}$$

where

$$M_k(s) = \text{gain of the } k\text{th forward path}, \tag{4.49a}$$

$$\begin{aligned}\Delta(s) &= \text{system determinant or characteristic function}\\ &= 1-(\text{sum of all individual loop gains})\\ &\quad + (\text{sum of gain products of all possible combinations of two non - touching loops, i.e., two loops having no common node})\\ &\quad - (\text{sum of gain products of all possible combinations of three non - touching loops}) + \cdots.\end{aligned} \tag{4.49b}$$

$$\Delta_k(s) = \text{the value of } \Delta(s) \text{ for that part of the graph not touching the } k\text{th forward path.} \tag{4.49c}$$

The method can best be illustrated by considering an example.

Example 7 Consider, for a change, another second-order system, described by the following state equations:

$$\dot{\mathbf{x}} = \begin{bmatrix} 0 & 1 \\ -2 & -3 \end{bmatrix}\mathbf{x} + \begin{bmatrix} 0 \\ 1 \end{bmatrix}u(t) \tag{4.50}$$

Taking the Laplace transform of both sides, we get

$$\begin{bmatrix} X_1(s) \\ X_2(s) \end{bmatrix} = \frac{1}{s}\left\{\begin{bmatrix} 0 & 1 \\ -2 & -3 \end{bmatrix}\begin{bmatrix} X_1(s) \\ X_2(s) \end{bmatrix} + \begin{bmatrix} 0 \\ 1 \end{bmatrix}U(s) + \begin{bmatrix} x_1(t_0) \\ x_2(t_0) \end{bmatrix}\right\} \tag{4.51}$$

The state transition flow graph diagram can be drawn as shown in Fig. 4.2, where the initial time has been assumed to be t_0. There are only two loops in the graph of gains $-2s^{-2}$ and $-3s^{-1}$ which touch at a node. Thus, the system characteristic function is

$$\Delta(s) = 1 + \frac{3}{s} + \frac{2}{s^2} = \frac{(s+1)(s+2)}{s^2} \tag{4.52}$$

Now, applying Mason's gain formula Eq. 4.48, we get

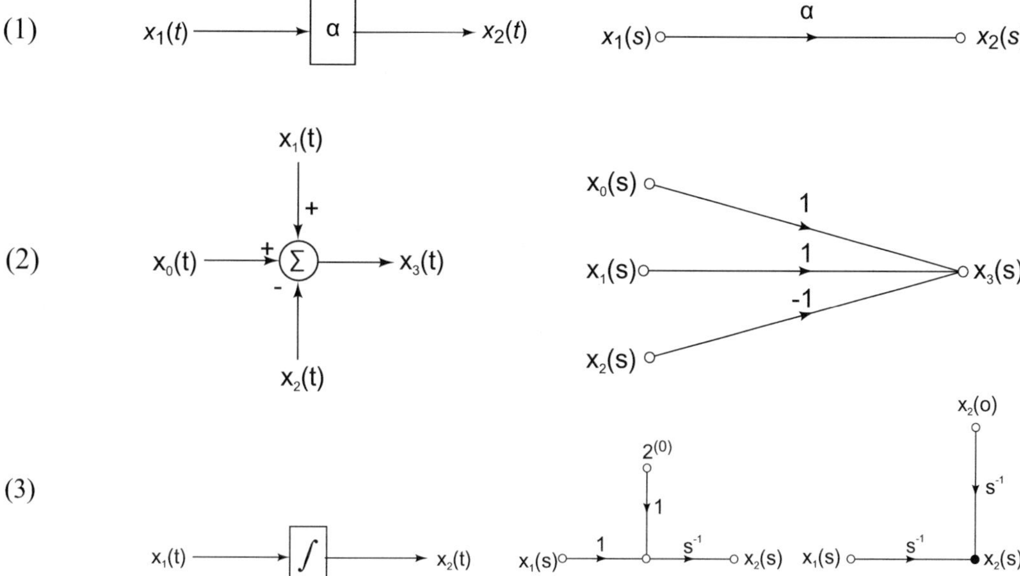

Fig. 4.1 Time and frequency domain representations of basic operations in an analog computer

$$\Phi_{11}(s) = \text{gain from node } x_1(t_0) \text{ to } X_1(s) \\ = \frac{(s+3)}{(s+1)(s+2)} \tag{4.53}$$

$$\Phi_{12}(s) = \text{gain from } x_2(t_0) \text{ to } X_1(s) \\ = \frac{1}{(s+1)(s+2)} \tag{4.54}$$

$$\Phi_{21}(s) = \text{gain from node } x_1(t_0) \text{ to } X_2(s) \\ = \frac{-2}{(s+1)(s+2)} \tag{4.55}$$

and

$$\Phi_{22}(s) = \text{gain from node } x_2(t_0) \text{ to } X_2(s) \\ = \frac{s}{(s+1)(s+2)} \tag{4.56}$$

The relation between $X_1(s)$ and $U(s)$ is given by the gain, from node $U(s)$ to $X_1(s)$ and is equal to

$$X_1(s)/U(s) = \frac{1}{(s+1)(s+2)} \tag{4.57}$$

The relation between $X_2(s)$ and $U(s)$ is given by the gain from node $U(s)$ to $X_2(s)$ and is equal to

$$X_2(s)/U(s) = \frac{s}{(s+1)(s+2)} \tag{4.58}$$

Thus

$$\begin{bmatrix} X_1(s) \\ X_2(s) \end{bmatrix} = \begin{bmatrix} \frac{s+3}{(s+1)(s+2)} & \frac{1}{(s+1)(s+2)} \\ \frac{-2}{(s+1)(s+2)} & \frac{s}{(s+1)(s+2)} \end{bmatrix} \begin{bmatrix} x_1(t_0) \\ x_2(t_0) \end{bmatrix} + \begin{bmatrix} \frac{1}{(s+1)(s+2)} \\ \frac{s}{(s+1)(s+2)} \end{bmatrix} U(s) \tag{4.59}$$

The state transition equation is obtained by taking the inverse Laplace transform of Eq. 4.59

$$\begin{bmatrix} x_1(t) \\ x_2(t) \end{bmatrix} = \begin{bmatrix} 2e^{-t} - e^{-2t} & e^{-t} - e^{-2t} \\ -2e^{-t} + 2e^{-2t} & -e^{-t} + 2e^{-2t} \end{bmatrix} \begin{bmatrix} x_1(t_0) \\ x_2(t_0) \end{bmatrix} + \mathcal{L}^{-1}\left\{ \begin{bmatrix} \frac{1}{(s+1)(s+2)} \\ \frac{s}{(s+1)(s+2)} \end{bmatrix} \right\} U(s) \tag{4.60}$$

The fundamental matrix in this result checks with the one found out earlier in Eq. 4.39 using the inverse matrix process. The extra benefit that

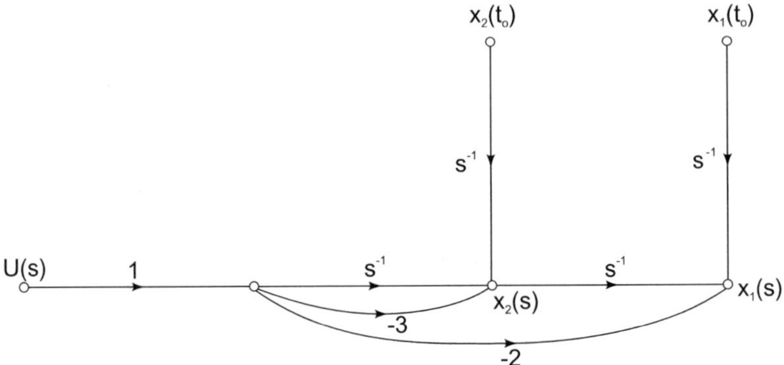

Fig. 4.2 State transition flow graph of the system of Example 7

has been derived from the state transition flow graph is that the effect of the input signal has been included. It is a distinct advantage that we have avoided the necessity of evaluating the integral

$$\int_{t_0}^{t} \phi(t-\tau)\mathbf{Bu}(\tau)\,\mathrm{d}\tau \tag{4.61}$$

For example, consider the case where the system is subjected to a unit step applied at $t = t_0$, i.e. $u\,(t) = \mu(t - t_0)$ and the initial conditions are zero. From Eq. 4.60, we easily obtain

$$\left.\begin{aligned} x_1(t) &= \mathcal{L}^{-1}\frac{(1/s)\mathrm{e}^{-st_0}}{(s+1)(s+2)} = \tfrac{1}{2} - \mathrm{e}^{-(t-t_0)} + \tfrac{1}{2}\mathrm{e}^{-2(t-t_0)} \\ x_2(t) &= \mathcal{L}^{-1}\frac{\mathrm{e}^{-st_0}}{(s+1)(s+2)} = \mathrm{e}^{-(t-t_0)} - \mathrm{e}^{-2(t-t_0)} \end{aligned}\right\} \tag{4.62}$$

These check with the earlier obtained results [see Eqs. 4.42 and 4.43].

The dominant advantage of the signal flow graph method of obtaining the state transition equation is that it does not become more difficult or tedious as the order of the system increases. This assertion will be clear from the next example, where a third-order system is considered.

Example 8 Consider a third-order system represented by the differential equation

$$\frac{\mathrm{d}^3w}{\mathrm{d}t^3} + 3\frac{\mathrm{d}^2w}{\mathrm{d}t^2} + 2\frac{\mathrm{d}w}{\mathrm{d}t} = u(t) \tag{4.63}$$

The state variables are chosen as

$$x_1 = w, x_2 = \dot{w} \text{ and } x_3 = w^{(2)} \tag{4.64}$$

Then, the set of first-order differential equations describing the system is

$$\left.\begin{aligned} \dot{x}_1 &= x_2 \\ \dot{x}_2 &= x_3 \\ \dot{x}_3 &= -2x_2 - 3x_3 + u(t) \end{aligned}\right\} \tag{4.65}$$

In matrix form, this set of equations becomes

$$\dot{\mathbf{x}} = \begin{bmatrix} 0 & 1 & 0 \\ 0 & 0 & 1 \\ 0 & -2 & -3 \end{bmatrix}\mathbf{x} + \begin{bmatrix} 0 \\ 0 \\ 1 \end{bmatrix}u(t) \tag{4.66}$$

The state transition flow graph of this system is shown in Fig. 4.3.

The characteristic function of the system is

$$\begin{aligned} \Delta(s) &= 1 - \left(-3s^{-1} - 2s^{-1}s^{-1}\right) = 1 + \frac{3}{s} + \frac{2}{s^2} \\ &= \frac{(s+1)(s+2)}{s^2} \end{aligned} \tag{4.67}$$

Using Mason's gain formula, we obtain

$$\begin{aligned} X_1(s) = {} & \frac{s^{-1}\Delta_1(s)}{\Delta(s)}x_1(t_0) + \frac{s^{-2}\Delta_2(s)}{\Delta(s)}x_2(t_0) \\ & + \frac{s^{-3}\Delta_3(s)}{\Delta(s)}x_3(t_0) + \frac{s^{-3}\Delta_4(s)}{\Delta(s)}U(s) \end{aligned} \tag{4.68}$$

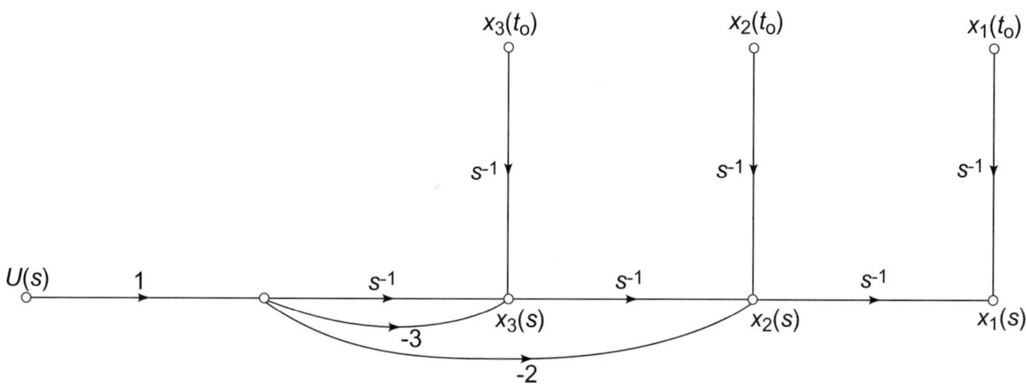

Fig. 4.3 State transition flow graph of the system of Example 8

Now

$$\left.\begin{aligned} \Delta_1(s) &= 1-(-3s^{-1}-2s^{-2}) = \Delta(s) \\ \Delta_2(s) &= 1-(-3s^{-1}) = 1+3/s \\ \Delta_3(s) &= 1 \\ \Delta_4(s) &= 1 \end{aligned}\right\} \tag{4.69}$$

Therefore

$$X_1(s) = \frac{1}{s}x_1(t_0) + \frac{(s+3)}{(s+1)(s+2)}x_2(t_0) + \frac{1}{s(s+1)(s+2)}x_3(t_0) + \frac{1}{s(s+1)(s+2)}U(s) \tag{4.70}$$

Continuing the process, we obtain the state transition equation as given by Eq. 4.72.

$$\mathbf{x}(t) = \mathcal{L}^{-1}\begin{bmatrix} \frac{1}{s} & \frac{s+3}{sq(s)} & \frac{1}{sq(s)} \\ 0 & \frac{s+3}{q(s)} & \frac{1}{q(s)} \\ 0 & \frac{-2s}{q(s)} & \frac{1}{q(s)} \end{bmatrix}\mathbf{x}(t_0) + \mathcal{L}^{-1}\begin{bmatrix} \frac{U(s)}{sq(s)} \\ \frac{U(s)}{q(s)} \\ \frac{sU(s)}{q(s)} \end{bmatrix} \tag{4.71}$$

where $q(s) = s^2 + 3\,s + 2$. For an input step of magnitude $u(t_0)$, we get

$$\begin{aligned} \mathbf{x}(t) = &\begin{bmatrix} \mu(t) & \frac{3}{2}\mu(\tau) - 2\varepsilon^{-\tau} + \frac{1}{2}\varepsilon^{-2\tau} & \frac{1}{2}\mu(\tau) - \varepsilon^{-\tau} + \frac{1}{2}\varepsilon^{-2\tau} \\ 0 & 2\varepsilon^{-\tau} - \varepsilon^{-2\tau} & \varepsilon^{-\tau} - \varepsilon^{-2\tau} \\ 0 & 2\varepsilon^{-\tau} - 4\varepsilon^{-2\tau} & -\varepsilon^{-\tau} + 2\varepsilon^{-2\tau} \end{bmatrix}\mathbf{x}(t_0) \\ &+ \begin{bmatrix} -\frac{3}{4}\mu(\tau) + \frac{1}{2}\tau - \frac{1}{2}\varepsilon^{-2\tau} + 2\varepsilon^{-\tau} \\ \frac{1}{2}\mu(\tau) - \varepsilon^{-\tau} + \frac{1}{2}\varepsilon^{-2\tau} \\ \varepsilon^{-\tau} - \varepsilon^{-2\tau} \end{bmatrix} u(t_0), \end{aligned} \tag{4.72}$$

where $\tau = t - t_0$ and $\mu(t)$ stands for the unit step function.

Concluding Discussion

As stated in the introduction in Part I, the aim of this presentation was to introduce the reader to the fundamentals of state variable characterization of linear systems. We started with the concept of state and discussed the choice of state variables for a given system. Next, we attempted to solve the state equation and we discovered the importance of the state transition or the fundamental matrix. Special methods were shown to be convenient for evaluating the fundamental matrix in simple cases. In the general case of a high-order system, one must take resort to the state transition flow graph, the properties and applications of which have been briefly explained.

The references cited below should be useful to readers interested in further exploration of state variables in characterization, analysis and synthesis of systems.

In the appendix, we have given a short, but comprehensive review of matrix algebra for ready reference on the notations, operations, properties and types of matrices.

Appendix on Review of Matrix Algebra

Notations

A matrix **A** of dimension $m \times n$ is a rectangular array of mn elements arranged in m rows and n columns as follows:

$$\mathbf{A} = \begin{bmatrix} a_{11} & a_{12} & \cdots & a_{1n} \\ a_{21} & a_{22} & \cdots & a_{2n} \\ \cdots & \cdots & \cdots & \cdots \\ a_{m1} & a_{m2} & \cdots & a_{mn} \end{bmatrix} = [a_{ij}] \quad (4.73)$$

The elements a_{ij} are called scalars; they may be real or complex numbers, or functions of real or complex variables.

A square matrix has the same number of rows and columns and can be associated with a determinant, usually written as det **A** or |**A**|.

A minor of an $n \times n$ determinant, denoted by M_{ij}, is the $(n-1) \times (n-1)$ determinant formed by crossing out the *i*th row and the *j*th column. The corresponding co-factor Δ_{ij} is

$$\Delta_{ij} = (-1)^{i+j} M_{ij} \quad (4.74)$$

Operations

Addition

Two matrices of the same size are added by summing the corresponding elements, i.e. if

$$\mathbf{A} + \mathbf{B} = \mathbf{C} \quad (4.75)$$

then

$$c_{ij} = a_{ij} + b_{ij} \quad (4.76)$$

Multiplication of a matrix by a scalar

If b is a scalar quantity and

$$b\mathbf{A} = \mathbf{A}b = \mathbf{C} = [C_{ij}] \quad (4.77)$$

then the new matrix **C** has the elements

$$\mathrm{C_{ij}} = ba_{\mathrm{ij}} \quad (4.78)$$

Multiplication of matrices

The product **AB** may be formed if the number of columns in **A** is equal to the number of rows in **B**. Such matrices are said to be conformable in the order stated. Thus, if **A** is $m \times p$ and **B** is $p \times n$, then **AB** = **C** is an $m \times n$ matrix defined by

$$C_{ij} = \sum_{k=1}^{p} a_{ij} b_{kj} \quad (4.79)$$

Matrix multiplication is associative [i.e. (**AB**) **C** = **A**(**BC**)], distributive with respect to addition [i.e. **A** (**B** + **C**) = **AB** + **AC**] but not, in general, commutative [i.e. $\mathbf{AB} \neq \mathbf{BA}$ in general]. In the product **AB**, **A** is said to pre-multiply **B**; an equivalent statement is that **B** post-multiplies **A**.

Transpose of a matrix

The transpose of **A**, denoted as $\mathbf{A}^T$, is a matrix formed by interchanging the rows and columns of **A**, i.e.

$$[a_{ij}]^T = [a_{ji}] = \mathbf{A}^T \tag{4.80}$$

Adjoint of a matrix

The adjoint of **A**, denoted as Adj **A**, is a matrix formed by replacing each element in **A** by its co-factor and then taking the transpose of the result, i.e.

$$\mathrm{Adj}\,\mathbf{A} = [\Delta_{ij}]^T \tag{4.81}$$

Conjugate matrix

The conjugate of **A**, denoted as $\bar{\mathbf{A}}$, is formed by replacing each element of **A** by its complex conjugate, i.e.

$$\bar{\mathbf{A}} = [\bar{a}_{ij}] \tag{4.82}$$

Inversion

The inverse of **A**, denoted by $\mathbf{A}^{-1}$ is defined as that matrix which, when pre-multiplied or post-multiplied by **A**, gives the unit or identity matrix (see Eq. 4.74) **I**, i.e.

$$\mathbf{A}\mathbf{A}^{-1} = \mathbf{A}^{-1}\mathbf{A} = \mathbf{I} \tag{4.83}$$

It can be shown that

$$\mathbf{A}^{-1} = \frac{\text{adj A}}{\det \text{A}} = \frac{[\Delta_{ij}]^T}{\Delta} \tag{4.84}$$

Properties

Equality

The matrices **A** and **B** are said to be equal if and only if $a_{ij} = b_{ij}$ for all i and j.

Equivalents

The matrices **A** and **B** are said to be equivalent if and only if non-singular matrices **P** and **Q** exist such that

$$\mathbf{B} = \mathbf{PAQ} \tag{4.85}$$

Rank

The rank of **A** is defined as the dimension of the largest square sub-matrix in **A** whose determinant does not vanish.

Inverse and transpose of a product

It can be shown that

$$\begin{aligned}(\mathbf{A}\,\mathbf{B})^{-1} &= \mathbf{B}^{-1}\mathbf{A}^{-1}\\ (\mathbf{AB})^T &= \mathbf{B}^T\mathbf{A}^T\end{aligned} \tag{4.86}$$

Cayley–Hamilton theorem

The matrix ($\mathbf{A} - \lambda\,\mathbf{I}$) is called the characteristic matrix of **A**. If **A** is a square matrix, then the equation

$$\det(\mathbf{A}-\lambda\mathbf{I}) = \varphi(\lambda) = 0 \tag{4.87}$$

is called its characteristic equation. Cayley–Hamilton theorem states that every square matrix **A** satisfies its characteristic equation, i.e. $\varphi\,(\mathbf{A}) = 0$

Types

Real and complex

A is real or complex according to whether a_{ij}'s are real or complex.

Diagonal scalar and unit or identity

A is diagonal if $a_{ij} = 0,\ i \neq j$. **A** is a scalar if **A** is diagonal and diagonal elements are all equal, i.e. $\mathbf{A} = k\mathbf{I}$. **A** is unit or identity matrix **I** if **A** is diagonal and all the diagonal elements are unity.

Hermitian and Skew-Hermitian

A is Hermitian if $\mathbf{A}^T = \mathbf{A}$, i.e. $a_{ij} = \bar{a}_{ij}$. Obviously, the diagonal elements of a Hermitian **A** are real numbers.

A is skew-Hermitian if $\mathbf{A}^T = -\bar{\mathbf{A}}$. Obviously, the diagonal elements in this case are either zero or pure imaginaries.

Orthogonal and unitary

A is orthogonal $\mathbf{A}^T = \mathbf{A}^{-1}$. **A** is unitary if **A** is both Hermitian and orthogonal, i.e. if $\bar{\mathrm{A}}^T = \mathbf{A}^{-1}$.

Positive definite

A real symmetric **A** is positive definite if

$$\mathbf{x}^T\mathbf{A}\mathbf{x} = \sum_{i=j=1}^{n} a_{ij}x_ix_j > 0 \tag{4.88}$$

for all non-trivial **x**. A necessary and sufficient condition is that the characteristic roots of **A** are positive.

Singular

A is singular if det **A** = 0, i.e. no inverse exists.

Symmetric

A is symmetric if it is square and **A** = $\mathbf{A}^T$, i.e. $a_{ij} = a_{ji}$.

Vector, column or row

A is a column vector if it has one column and a row vector if it has one row.

Zero

A is zero if $a_{ij} = 0$.

Problems

These problems are slightly more difficult than those in the previous chapters. But have no fear—if you have followed the contents of this chapter, then these should not be an issue. You will sail through comfortably.

P.1. Determine the fundamental state transition matrix for P.1 circuit of the previous chapter.

P.2. Same for P.2 circuit of the previous chapter.

P.3. Same for P.3 circuit of previous chapter.

P.4. Draw the state transition flow graph for the circuit of P.4 of the previous chapter.

P.5. Same for P.3 circuit of the previous chapter.

Bibliography

1. L.A. Zadeh, C.A. Desoer, *Linear System Theory: A State Space Approach* (McGraw-Hill, 1963)
2. S. Seeley, *Dynamic System Analysis* (Reinhold Publishing Co, 1964)

Carry Out Partial Fraction Expansion of Functions with Repeated Poles

5

This chapter aims to simplify partial fraction expansion with repeated poles—presented here are some techniques which should make this topic considerably easier.

Keywords

Partial fraction expansion · Repeated poles
New method

As you are well aware, the function

$$F(s) = \frac{N(s)}{(s - s_0)^n D(s)} \quad (5.1)$$

can be expanded in partial fractions as follows:

$$F(s) = \frac{K_0}{(s - s_0)^n} + \frac{K_1}{(s - s_0)^{n-1}} + \frac{K_2}{(s - s_0)^{n-2}} + \cdots + \frac{K_{n-1}}{(s - s_0)} + \frac{N_1(s)}{D(s)} \quad (5.2)$$

For finding the constants $K_0, K_1, \ldots K_{n-1}$, most of the textbooks (see, e.g. [1]) recommend a procedure based on differentiation of the function

$$F_1(s) = (s - s_0)^n F(s) N(s)/D(s) \quad (5.3)$$

The general formula is, in fact

$$K_r = \frac{1}{r!} \frac{d^r F_1(s)}{ds^r}\bigg|_{s=s_0} \quad r = 0 \text{ to } n - 1 \quad (5.4)$$

For even a moderate value of n, this can become quite tedious. Several alternative procedures have therefore been suggested in the literature—mostly in journals on circuits, systems and controls—and a few of them have been mentioned in some recent textbooks. While making a critical survey of all these procedures,

Source: S. C. Dutta Roy, "Carry Out Partial Fraction Expansion of Functions with Repeated Poles without Tears," *Students' Journal of the IETE*, vol. 26, pp. 129–131, October 1985.

S. C. Dutta Roy, *Circuits, Systems and Signal Processing*,
https://doi.org/10.1007/978-981-10-6919-2_5

the author has come to the conclusion that a procedure given in [2] and credited to Professor Leonard O. Goldstone of the Polytechnic Institute of Brooklyn, is the best. The method will be described here with a slight modification; it is hoped that you will readily appreciate its merits as compared to the differentiation or any other procedure you may have come across so far, and will adopt it in your future work.

The Method

Look at Eq. 5.2 and note that if $s - s_0$ is replaced by $\frac{1}{p}$,[1] then it becomes

$$G(p) = K_0 p^n + K_1 p^{n-1} + \cdots + K_{n-1} p + G_1(p), \tag{5.5}$$

where for convenience, we have called $F(s) = F(s_0 + 1/p)$ as $G(p)$ and $N_1(s_0 + 1/p)/D(s_0 + 1/p)$ as $G_1(p)$. Thus, if one transforms $F(s)$ to $G(p)$ and carries out a long division to extract, in the quotient, all powers of p from n to 1 then the constants $K_0, K_1, \ldots K_{n-1}$ will automatically be found out! The remaining function $G_1(p)$ should then be transformed back to the s-variable, and expanded for other simple or multiple poles.

An Example

Now consider a specific example for illustration.
Let

$$F(s) = \frac{s+2}{(s+1)^2(s+3)} \tag{5.6}$$

Putting $s + 1 = 1/p$ and simplifying gives

$$G(p) = \frac{p^3 + p^2}{2p+1} \tag{5.7}$$

On long division, we get

$$\begin{array}{r|l} 2p+1 & p^3 + p^2 \quad \left(\frac{p^2}{2} + \frac{p}{4} \right. \\ & p^3 + \frac{p^2}{2} \\ \hline & \frac{p^2}{2} \\ & \frac{p^2}{2} + \frac{p}{4} \\ \hline & -\frac{p}{4} \end{array}$$

Thus

$$G(p) = \tfrac{1}{2}p^2 + \tfrac{1}{4}p - \frac{p/4}{2p+1} \tag{5.8}$$

Now restoring s, we get

$$G(p) = F(s) = \frac{\frac{1}{2}}{(s+1)^2} + \frac{\frac{1}{4}}{s+1} - \frac{\frac{1}{4}}{s+3} \tag{5.9}$$

which completes the expansion.

Another Example

Consider, next, a more complicated example.
Let

$$F(s) = \frac{2s^2 + s + 1}{(s+1)(s+2)^2(s+3)^3} \tag{5.10}$$

Putting $s + 3 = 1/p$ and simplifying, we get

$$2s^2 + s + 1 = (2/p^2) - (11/p) + 16 \tag{5.11}$$

and

$$(s+1)(s+2)^2 = (1/p^3) - (4/p^2) + (5/p) - 2 \tag{5.12}$$

We have found the transformed form of $(s + 1)(s + 2)^2$ separately because we shall need it later, while expanding in terms of the multiple pole at $s = -2$. Combining Eqs. 5.10, 5.11 and 5.12, we get, after elementary simplifications

[1] This is the modification;

$$F(s) = G(p) = \frac{16p^6 - 11p^5 + 2p^4}{-2p^3 + 5p^2 - 4p + 1} \qquad (5.13)$$

The long division proceeds as follows:

$$\begin{array}{l}
-2p^3 + 5p^2 - 4p + 1 \overline{)\,16p^6 + 11p^5 + 2p^4} \\
\qquad \left(-8p^3 - \frac{29}{2}p^2 - \frac{85}{4}p\right. \\
\underline{16p^6 - 40p^5 + 32p^4 - 8p^3} \\
29p^5 - 30p^4 + 8p^3 \\
\underline{29p^5 - \frac{145}{2}p^4 + 58p^3 - \frac{29}{2}p^2} \\
\frac{85}{2}p^4 - 50p^3 + \frac{29}{2}p^2 \\
\underline{\frac{85}{2}p^4 - \frac{425}{4}p^3 + 85p^2 - \frac{85}{4}p} \\
\frac{225}{4}p^3 - \frac{141}{2}p^2 + \frac{85}{4}p
\end{array}$$

Thus

$$G(p) = -8p^3 - \frac{29}{2}p^2 - \frac{85}{4}p + G_1(p), \qquad (5.14)$$

where

$$G_1(p) = \frac{\frac{225}{4}p^3 - \frac{141}{2}p^2 + \frac{85}{4}p}{-2p^3 + 5p^2 - 4p + 1} \qquad (5.15)$$

Expansion in terms of the repeated pole at $s = -3$ is complete. The remainder function is now to be expanded in terms of the repeated pole at $s = -2$. In order to accomplish this, $G_1(p)$ has to be transformed back to a function of s, which we shall call $F_2(s)$; the job is not difficult because dividing both numerator and denominator of Eq. 5.15 by p^3, and taking help of Eq. 5.12, we simply get

$$G_1(p) = F_2(s) = \frac{\frac{225}{4} - \frac{141}{2}(s+3) + \frac{85}{4}(s+3)^2}{(s+1)(s+2)^2} \qquad (5.16)$$

Now, put $s + 2 = 1/q$ in Eq. 5.16 and call the resulting function as $G_2(q)$. After a bit of simplification, the following result is obtained:

$$G_2(q) = G_1(p) = \frac{7q^3 - 28q^2 + \frac{85}{4}q}{-q + 1} \qquad (5.17)$$

Carrying out the long division gives

$$\begin{array}{l}
-q+1\overline{)\,7q^3 - 28q^2 + \frac{85}{4}q}\left(-7q^2 + 21q\right. \\
\underline{7q^3 - 7q^2} \\
-21q^2 + \frac{85}{4}q \\
\underline{-21q^2 + 21q} \\
\frac{1}{4}q
\end{array}$$

Thus

$$G_2(q) = -7q^2 + 21q + \frac{\frac{1}{4}q}{-q + 1} \qquad (5.18)$$

From Eqs. 5.14 and 5.18, we get

$$\begin{aligned} F(s) = {} & -8p^3 - \frac{29}{2}p^2 - \frac{85}{4}p \\ & - 7q^2 + 21q + \frac{\frac{1}{4}q}{-q + 1} \end{aligned} \qquad (5.19)$$

Finally substituting $p = 1/(s+3)$ and $q = 1/(s+2)$ in Eq. 5.19 gives the desired partial fraction expansion:

$$\begin{aligned} F(s) = {} & -\frac{8}{(s+3)^3} - \frac{29/2}{(s+3)^2} - \frac{85/4}{s+3} \\ & - \frac{7}{(s+2)^2} + \frac{21}{s+2} + \frac{\frac{1}{4}}{s+1} \end{aligned} \qquad (5.20)$$

Problems

P.1. Expand the function

$$F(s) = \frac{s+3}{(s+1)^3(s+2)^2}$$

P.2. Do the same for

$$F(s) = \frac{s+2}{(s+1)^4}$$

P.3. Do the same for

$$F(s) = \frac{s+3}{(s+1)^3(s+2)^4}$$

P.4. Do the same for

$$F(s) = \frac{(s+1)^3(s+2)}{s^4}$$

and find $f(t)$. Note: $f(t)$ will contain a δ-function. Beware!

P.5. Do the same for

$$F(s) = \frac{s+3}{s^2(s+1)^2}$$

References

1. M.E. van Valkenburg, *Network Analysis* (Prentice-Hall of India, New Delhi, pp. 186–187) (1974)
2. F.F. Kuo, *Network Analysis and Synthesis* (Wiley, New York, pp. 153–154)

6 A Very Simple Method of Finding the Residues at Repeated Poles of a Rational Function in z^{-1}

If you have followed the last chapter carefully, this one would be a cakewalk! The two discussions are similar except for the variables. A very simple method is given for finding the residues at repeated poles of a rational function in z^{-1}. Compared to the multiple differentiation formula given in most textbooks, and several other alternatives, this method appears to be the simplest and the most elegant. It requires only a long division preceded by a small amount of processing of the given function.

Keywords
Partial fraction expansion · Repeated poles
New method

Introduction

Let $Y(z)$, a proper rational function in z^{-1}, have a pole at $z = p$, where p may be real or complex, of multiplicity q. Then, $Y(z)$ can be expanded in partial fractions as follows:

$$\begin{aligned} Y(z) &= P_1(z)/[(1-pz^{-1})^q Q_1(z)] \\ &= [A_1/(1-pz^{-1})] + [A_2/(1-pz^{-1})^2] + \ldots \\ &\quad + [A_q/(1-pz^{-1})^q] + Y_1(z), \end{aligned} \tag{6.1}$$

where A_i's are the residues, $i = 1$ to q, and $Y_1(z)$ contains terms due to other poles. Textbooks (see, e.g. [1]) usually give the following formula for finding A_i's:

$$\begin{aligned} A_i = {} & \{1/[(q-i)!(-p)^{q-i}]\} \cdot |d^{q-i}/d(z^{-1})^{q-i} \\ & [(1-pz^{-1})^L Y(z)]|_{pz}^{-1} = 1. \end{aligned} \tag{6.2}$$

This expression, involving multiple differentiations, is indeed formidable, and students invariably make mistakes in calculation. In a recent paper [1, 2], three alternative methods were outlined. These are

(1) Multiply both sides of Eq. 6.1 by $[(1-pz^{-1})^q Q_1(z)]$, simplify the right-hand side, equate the coefficients of powers of z^{-1} on both sides to get a set of linear equations in the unknown constants, and solve them.

Source: S. C. Dutta Roy, "A Very Simple Method of Finding the Residues at Repeated Poles of a Rational Function in z^{-1}," *IETE Journal of Education*, vol. 56, pp 68–70, July–December 2015.

S. C. Dutta Roy, *Circuits, Systems and Signal Processing*,
https://doi.org/10.1007/978-981-10-6919-2_6

(2) Put arbitrary specific values of z^{-1} on both sides, like 0, $\pm\frac{1}{4}$, ± 1, ± 2 etc., and solve the resulting set of linear algebraic equations.
(3) Obtain A_q as

$$A_q = (1 - pz^{-1})^q Y(z)|_{pz}^{-1} = 1 \quad (6.3)$$

and find the rational function,

$$Y_2(z) = Y(z) - [A_q/(1-pz^{-1})^q] \quad (6.4)$$

Clearly, $Y_2(z)$ will have a multiple pole of order $q - 1$ at $z = p$. Now, A_{q-1} can be obtained as

$$A_{q-1} = (1-pz^{-1})^{q-1} Y_2(z)|_{pz}^{-1} = 1 \quad (6.5)$$

The process can now be repeated till all the A_i's are obtained and the remainder function $Y_q(z)$ can then be handled.

Here, we present yet another method, which does not appear to be known to teachers and students, and to the best of knowledge of the author, it has not appeared in any literature. The method is based on a small amount of preprocessing of Eq. 6.1 followed by a long division; besides elegance, it can claim to be the simplest of all known methods.

The method presented here is an adaptation of a similar one for Laplace transforms, given in the previous chapter and Kuo's book [3] which was published long back, in 1966, and a tutorial paper on the method [1] appeared in 1985, the method did not figure in any textbook so far, and is not known to teachers and students. The author has used this method routinely in his courses on network theory and signals and systems and has found it to be well received by students.

The Method

In the method under discussion, first make a change of variable from z^{-1} to

$$x = (1/p) - z^{-1} \quad (6.6)$$

and let, in general,

$$F(z)|_{(1/p)-z^{-1}} = x = F'(x) \quad (6.7)$$

Then, Eq. 6.1 becomes

$$Y'(x) = P_1'(x)/[x^q Q_1'(x)] = [(A_1/p)/x] + [(A_2/p^2)/x^2] + \ldots + [(A_q/p^q)/x^q] + Y_1'(x). \quad (6.8)$$

Multiply both sides by x^q, so that Eq. 6.8 becomes

$$P_1'(x)/Q_1'(x) = A_1' x^{q-1} + A_2' x^{q-2} + \ldots + A_q' + x^q Y_1'(x), \quad (6.9)$$

where

$$A_i' = A_i/p^i. \quad (6.10)$$

Now make a long division of $P_1'(x)$ by $Q_1'(x)$, starting with the lowest powers. Then, the quotients will give all the required residues and the remainder gives the numerator of the function $Y_1'(x)$ which can then be analysed for the residues at the other poles.

We shall now illustrate the method by an example.

Example

Let

$$\begin{aligned} Y(z) &= [1-(1/8)z^{-1}]/[1-(1/2)z^{-1}]^3[1-(1/4)z^{-1}]\} \\ &= \{A_1/[1-(1/2)z^{-1}]^3\} + \{A_2/[1-(1/2)z^{-1}]^2\} \\ &\quad + \{A_3/[1-(1/2)z^{-1}]\} + \{A_4/[1-(1/4)z^{-1}]\}. \end{aligned} \quad (6.11)$$

Equation 6.11 can be rewritten as

$$\begin{aligned} Y(z) &= 4(8-z^{-1})/[(2-z^{-1})^3(4-z^{-1})] \\ &= [8A_1/(2-z^{-1})^3] + [4A_2/(2-z^{-1})^2] \\ &\quad + [2A_3/(2-z^{-1})] + [4A_4/(4-z^{-1})]. \end{aligned} \quad (6.12)$$

Put $x = 2 - z^{-1}$ and multiply both sides of Eq. 6.12 by x^3. The result is

$$4(6+x)/(2+x) = 8A_1 + 4A_2x + 2A_3x^2 + [4A_4x^3/(2+x)]. \quad (6.13)$$

Now, make a long division of 24 + 4x by 2 + x as shown below.

$$\begin{array}{r|l} 2+x & 24+4x \quad (12-4x+2x^2 \\ & \underline{24+12x} \\ & -8x \\ & \underline{-8x-4x^2} \\ & 4x^2 \\ & \underline{4x^2+2x^3} \\ & -2x^3 \end{array}$$

Comparing the result with the right-hand side of Eq. 6.13, we get

$$A_1 = 3/2, A_2 = -1, A_3 = 1, \text{ and } A_4 = -1/2. \quad (6.14)$$

Substituting these values in Eq. 6.11 and using the inversion formula

$$Z^{-1}[1/(1-pz^{-1})^q] = [1/(q-1)!](n+1)(n+2)\ldots(n+q-1)p^n u(n), \quad (6.15)$$

we finally get, after simplification,

$$y(n) = Z^{-1}Y(z) = \{[(3n^2+5n+6)/4](1/2)^n - (1/2)(1/4)^n\}u(n) \quad (6.16)$$

$u(n)$ being the unit step function.

Conclusion

A very simple method, claimed to be the simplest and the most elegant, has been presented for finding the residues at multiple poles of a rational function in z^{-1}.

Problems

All problems concern partial fraction expansion and finding the residues.

P.1. $F(z) = \frac{5+4z^{-1}-0.9z^{-2}}{(1-0.6z^{-1})^2(1+0.5z^{-1})}$

P.2. $F(z) = \frac{4-1.2z^{-1}+0.48z^{-2}}{(1-0.4z^{-1})^3}$

P.3. $F(z) = \frac{4z^{-4}-3.2z^3-5.2z^2+4.5z^{-5}}{(z-0.5)^2(z+2.1)^2(z-3)(z+4.5)}$

Clue: First convert this $F(z)$ into a rational function in z^{-1}

P.4. $F(z) = \frac{2z^2+4.5z-5}{(z+0.6)^3(z-2.1)}$

Clue: Same as in P.3

P.5. $F(z) = \frac{4z^{-4}-3.2z^{-3}-5.2z^{-2}-4.5z^{-1}-5}{(z^{-1}-0.5)^2(z^{-1}+2.1)^2(z^{-1}-3)(z^{-1}+4.5)}$

Clue: First convert the denominator factors into $(1 - \alpha z^{-1})$ forms.

References

1. S.C. Dutta Roy, Comments on fair and square computation of inverse z-transforms of rational functions. IEEE Trans. Educ. **58**(1), 56–57 (Feb 2015)
2. S.C. Dutta Roy, Carry out partial fraction expansion of rational functions with multiple poles—without tears. Stud. J. IETE. **26**, 129–31 (Oct 1985)
3. F.F. Kuo, in *Network Analysis and Synthesis* (Wiley, New York, 1966), pp. 153–154

Part II
Passive Circuits

This is the largest of the four parts and contains 16 chapters. As in Part I, the topics are interrelated here also. We deal with passive circuits and two distinct aspects of it, viz. analysis and synthesis. Most curricula today emphasize only the first aspect, viz. analysis, which deals with the problem of finding the response of a given circuit to a given excitation. The synthesis aspect deals with designing a circuit to perform in a specified way; by far, synthesis is more exciting than analysis. Analysis, however difficult it may be, is always possible. However, a synthesis problem may or may not have a solution. In real life, it is synthesis that is required more than analysis. Synthesis is an art and not a science. The beauty of synthesis is that if one solution exists, then there exists an indefinite number of solutions. For analysis, the solution is unique. Synthesis, therefore, facilitates choice. From among a variety of solutions, you can select the one that is the best for your situation.

The first 11 chapters of this part are concerned with analysis. Circuit analysis can be performed with ease in the case of linear circuits with the help of transforms—Fourier or Laplace—which transports the problem from the time domain to the frequency domain. In the latter, there is no differentiation or integration; instead, there are only algebraic manipulations, *viz.* addition, multiplication and division. That is the reason why you always prefer to work in the frequency domain. However, frequency domain analysis has its own demerits, as you have observed in Chap. 2. A differential (or difference) equation cannot be solved for all times by transforming it to an algebraic equation. There comes the difficulty of initial conditions which do not match. So time domain solutions are comprehensive; they do not give rise to such anomalies. The other reason why time domain cannot be divorced once for all is that most undergraduate curricula treat transform techniques later, may be in the second year. On the other hand, a basic Electrical Engineering course is taught in the very first year. Hence, one has to work in the time domain. This is how it was and is still is at IIT Delhi. Your curriculum would be no exception. I faced a difficulty while teaching the first-year course on basic Electrical Engineering. Circuit analysis forms a large part of the course, and I found that most textbooks either bring in the Laplace transform there itself or avoid it by introducing an artificial excitation e^{st}. Then follows the concept of poles, zeroes and inversion of a function in s. When the student goes to the second year, it is difficult for him or her to accept Laplace transform in place of e^{st} excitation. To remove this difficulty, I prepared a set of notes by remaining solely in the time domain and discarded the prescribed textbook. Students found these notes very useful and my other colleagues asked for them when they were assigned to teach this course. This is the genesis of Chap. 7.

Chapter 8 deals most comprehensively with the RLC circuit analysis in the time domain. The three cases of damping, viz. under-damping, critical damping and over-damping, are thoroughly analysed, and simple methods are devised for finding the response easily. Chapter 11 deals in the same circuit with Fourier transforms. Many types of resonances that may occur by taking the output across various elements or a combination of them are treated in Chap. 12.

Chapters 9 and 10 deal with two problems which counteract the popular belief that seeing is believing. Chapter 9 gives a circuit paradox which I hope you will find interesting, while in Chap. 10, I talk of the problem of initial values in an inductor.

The parallel-T RC network is an important circuit and finds application in many situations. Primarily, it is a null network, but it can be used to make a selective amplifier, a measuring instrument or a frequency discriminator. Chapter 13 presents several different methods for analysing this network and gives some simple methods to avoid the messy mesh or node equations. Chapter 14 goes deeper into the performance of the network with regard to selectivity and spells out design conditions to achieve maximum selectivity.

A perfect transformer does not exist in practice, but the concept facilitates design and applications of this important component. There arise some peculiar phenomena, like current discontinuity from $t=0-$ to $t=0+$, and degeneracy. These are of great theoretical interest and help you to understand the limitations of the device. Read this with attention to realize that what you take for granted with this simple device is not in general practical.

Resistive ladders are very commonly treated in the first course of circuits. Infinite ladders are particularly important because they bring in some irrational numbers and functions in their performance characteristics. Such networks can be analysed by step-by-step analysis and clever reasoning rather than mesh and node equation, which would be infinite in number. Difference equation formulation, however, gives an easier method, and z-transforms can be easily applied to them. This is what forms the contents of Chap. 17.

Chapters 18 and 19 deal with synthesis. In Chap. 18, we start with the driving-point synthesis of a relatively simple function, viz. one of the third order. Chapter 19 shows how the practical problem of interference rejection in an ultra-wideband system reduces to the synthesis of an LC driving-point function. We work out several alternative solutions to facilitate a choice of the best one.

A filter design problem has to be solved in several steps. First, from the specification of cut-off frequency(ies) and pass- and stop-band tolerances, you choose a magnitude function. Then by the process of analytic continuation, find the corresponding transfer function. Finally, realize the transfer function by using inductors and capacitors. The second step involves finding the poles and zeroes of the transfer function. Chapter 20 gives some shortcuts for moving from the magnitude function to the transfer function, by simple coefficient matching. This saves a lot of calculation. Also, normally you consider the Butterworth type because of its simplicity.

Design of band-pass (BP) and band-stop (BS) filters usually proceeds from a corresponding normalized low-pass (LP) transfer function. After realizing the LP function, you apply the frequency transformation technique to obtain the actual BP/BS components. There exists some confusion regarding this particular transformation. Chapter 21 attempts to remove these confusions.

Chapter 22 deals with passive differentiators. It gives, starting with the well-known RC differentiators, other circuits which give improved linearity of the frequency response. This contains a number of innovations and the new circuits are truly new, not available in textbooks.

7 Circuit Analysis Without Transforms

Is it simpler? In most cases, it is. Remember the difficulty you faced by working solely in the time domain, in the previous chapter (Chap. 2), in solving a differential equation with impulsive excitation? Except for these odd cases, time domain analysis is usually simpler. In this chapter, we discuss how linear circuits can be completely analysed without using Laplace or Fourier transforms. Is this analysis simpler than that using transform techniques? You should judge for yourself to realize.

Keywords
Circuit analysis · Differential equation
Time domain · Force-free response
First-order circuit · Second-order circuit
Damping · Root locus · Impedance
Natural response · Natural frequencies
DC and sinusoidal excitation · Pulse response
Impulse response

Analysis of electrical circuits forms part of a core course for all engineering students which is usually offered in the very first semester of the curriculum. At this stage, you are not exposed to Laplace and Fourier transforms, and hence you cannot appreciate how they simplify circuit analysis. This chapter is an attempt to show that circuits can indeed be completely analysed without the help of transform techniques.

We first introduce the concepts of natural response, forced response, transient response and steady-state response, and deal with typical examples of force-free response. The concepts of impedance, admittance, poles and zeros are then introduced through the artifice of e^{st} excitation. It is shown that the natural frequencies of a circuit depend upon the kind of forcing function, and that they are related to the poles and zeros of an impedance function.

Next, we deal with forced response to exponential excitation, and in particular to sinusoidal excitation; introduce the concept of phasors; and demonstrate how steady-state sinusoidal response can be found out only through phasors and impedances. Several examples of complete response are worked out. The chapter concludes with an introduction to the impulse and an example of impulse response of a circuit.

Throughout the chapter, the emphasis is on understanding through examples, rather than on intricate theories. Once you learn the technique through simple, heuristic and common sense arguments, detailed justification through transform techniques, an exposure to which will come

Source: S. C. Dutta Roy, "Circuit Analysis without Transforms," *IETE Journal of Education*, vol. 39, pp. 111–127, April–June 1998.

S. C. Dutta Roy, *Circuits, Systems and Signal Processing*,
https://doi.org/10.1007/978-981-10-6919-2_7

to you later in the curriculum, will make you a master of the same.

Let us start with an example.

An Example

Consider the circuit in Fig. 7.1, where we wish to find $i(t)$ due to an excitation by a current source $i_s(t)$. Note that C may have an initial voltage $v_c(0) = V$, say, and L may have an initial current $i(0) = I$, say, it being assumed that $i_s(t)$ has been switched on at $t = 0$. By KCL, we get

$$i_s(t) = C\frac{\mathrm{d}v_c(t)}{\mathrm{d}t} + i(t) \tag{7.1}$$

Also

$$v_c(t) = L\frac{\mathrm{d}i(t)}{\mathrm{d}t} + Ri(t) \tag{7.2}$$

Substituting Eq. 7.2 in Eq. 7.1 gives

$$i_s(t) = LC\frac{\mathrm{d}^2 i}{\mathrm{d}t^2} + RC\frac{\mathrm{d}i}{\mathrm{d}t} + i, \tag{7.3}$$

where we have dropped the argument t from $i(t)$ for brevity. Equation 7.3 can be written more succinctly as

$$\frac{\mathrm{d}^2 i}{\mathrm{d}t^2} + \frac{R\mathrm{d}i}{L\mathrm{d}t} + \frac{1}{LC}i = \frac{i_s(t)}{LC} \tag{7.4}$$

This is a second-order differential equation with constant coefficients and as you are aware, the solution will consist of two parts—the complementary function and the particular integral. The complementary function $i_c(t)$ is the solution to the homogeneous equation

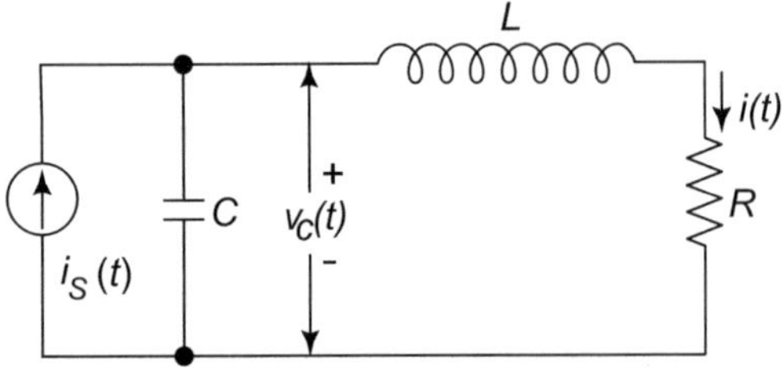

Fig. 7.1 An *RLC* circuit excited by a current source

$$\frac{\mathrm{d}^2 i}{\mathrm{d}t^2} + \frac{R\,\mathrm{d}i}{L\,\mathrm{d}t} + \frac{1}{LC}i = 0 \tag{7.5}$$

In general, for a second-order equation like Eq. 7.5, $i_c(t)$ will be of the form

$$i_c(t) = A_1 i_l(t) + A_2 i_2(t), \tag{7.6}$$

where A_1 and A_2 are constants. The particular integral $i_p(t)$ of Eq. 7.4 is any function which satisfies the equation as it is. For example, if $i_s(t) = \mathrm{e}^{st}$, then $i_p(t) = K\mathrm{e}^{st}$ where, by substitution in Eq. 7.4, we get

$$K\left(s^2 + \frac{R}{L}s + \frac{1}{LC}\right) = \frac{1}{LC} \tag{7.7}$$

or

$$K = \frac{1/(LC)}{s^2 + \frac{R}{L}s + \frac{1}{LC}} \tag{7.8}$$

In general, therefore, the complete solution to Eq. 7.4 is given by

$$i(t) = A_1 i_1(t) + A_2 i_2(t) + i_p(t) \tag{7.9}$$

It is emphasized that $i_p(t)$ will *not* contain any unknown constants. The unknown constants here are A_1 and A_2, which are to be evaluated from the two initial conditions. The condition $i(0) = I$ gives

$$I = A_1 i_1(0) + A_2 i_2(0) + i_p(0) \tag{7.10}$$

The other condition $v_c(0) = V$ gives, from Eq. 7.2,

$$V = Ri(0) + L\frac{\mathrm{d}i}{\mathrm{d}t}\bigg|_{t=0} \tag{7.10a}$$

or, combining with Eq. 7.9.

$$V = RI + L[A_1 i_1'(0) + A_2 i_2'(0) + i_p'(0)], \tag{7.11}$$

where the symbol $f'(0)$ has been used for $\mathrm{d}f/\mathrm{d}t|_{t=0}$. In Eqs. 7.10 and 7.11, everything else is known except A_1 and A_2, and hence they can be found out.

Some Nomenclatures

In the previous example, the term $i_c(t)$ of Eq. 7.6 arises due to initial energy in the capacitor as well as the inductor. However, Eq. 7.6 only gives the form of the response due to the initial conditions; the values of the constants depend upon the particular solution $i_p(t)$ and its derivative at $t = 0$.

This part of the response is called the *natural response*. Had the circuit been force free, i.e. if $i_s(t) = 0$, then obviously A_1 and A_2 would be determined from the initial conditions only. Hence, although the form of the natural response remains the same in force free as well as forced cases, the actual values will be different.

The part $i_p(t)$ of the total response depends solely on the forcing function and the parameters and the architecture of the circuit; it is independent of the initial conditions. $i_p(t)$ is called the *forced response* and is usually of the same form as the forcing function. An example has already been given for $i_s(t) = e^{st}$. As another example, if $i_s(t) = I_s$, a *dc*, then $i_p(t)$ is also a *dc*, I_p, whose value is obtained from Eq. 7.4 as $I_p = I_s$, because

$$\frac{dI_p}{dt} = 0 \text{ and } \frac{d^2I_p}{dt^2} = 0.$$

The superposition of natural response (in form only) and the forced response gives the total or complete response.

The natural response of the circuit, in the force-free case, usually decays with time and becomes negligible as $t \to \infty$ the term 'usually' indicates the practical situation where dissipation is invariably present. In the dissipationless case, it is possible that the natural response does not decay with time. The forced response can be maintained indefinitely by an appropriate forcing function, but for some forcing functions, (like $e^{-\alpha t}$, $\alpha > 0$), it may also decay with time. That part of the total response which decays with time is called the *transient response*. The value of the response as $t \to \infty$ (or in practice, after the transient part has become negligible) is called the *steady state response*. It must be emphasized that the transient response and natural response are not necessarily the same, neither are the steady-state response and forced response.

Force-Free Response: General Considerations

First, let us study some force-free cases. Obviously, the equation to be solved will be a homogeneous equation of order determined by the number of energy storage elements. In the example given earlier in the chapter (page 62), the order was two because of one L and one C. In general, our equation will be of the form

$$\frac{d^n x}{dt^n} + a_{n-1}\frac{d^{n-1}x}{dt^{n-1}} + \cdots + a_1\frac{dx}{dt} + a_0 = 0, \quad (7.12)$$

where x is a voltage or a current and $a_{n-1}, a_{n-2}, \ldots a_1$ and a_0 are constants determined by the parameters of the circuit.

Although there are various methods of solving Eq. 7.12, we find it convenient to assume a solution of the form $x = Ae^{st}$, substitute it in Eq. 7.12, and obtain the following algebraic equation in s:

$$s^n + a_{n-1}s^{n-1} + \cdots + a_1 s + a_0 = 0 \quad (7.13)$$

This is called the *characteristic equation* of the circuit and its roots are called the *characteristic roots* or *natural frequencies* of the circuit. Since Eq. 7.13 has n roots—s_1, s_2, … s_n, naturally, our desired solution will be of the form

$$x = A_1 e^{s_1 t} + A_2 e^{s_2 t} + \ldots + A_n e^{s_n t}, \quad (7.14)$$

where the constants A_1, A_2, ⋯, A_n, are to be determined from the n initial conditions—one on each energy storage element.

We now consider some specific examples.

Force-Free Response of a Simple RC Circuit

Consider the situation depicted in Fig. 7.2, where the switch S is in position a for a long time so that C is fully charged to the voltage V. At $t = 0$,

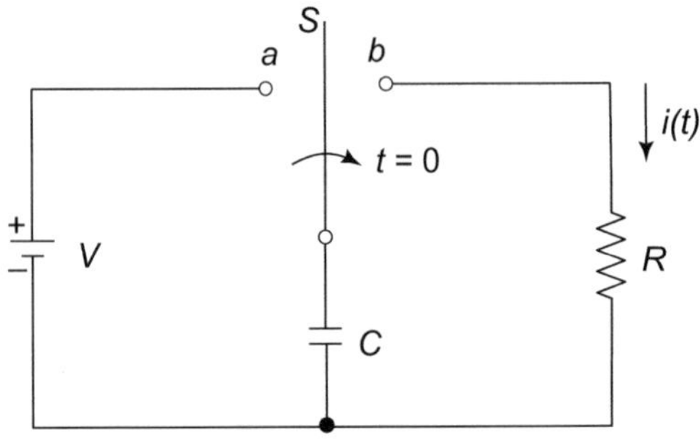

Fig. 7.2 A simple RC circuit

the switch is shifted to the position b. We then have a force-free RC circuit (R and C are in series or parallel?) with an initial condition on C, as shown in Fig. 7.3. KVL around the loop in Fig. 7.3 gives

$$Ri + \frac{1}{C}\int_0^t i\mathrm{d}t - V = 0 \qquad (7.15)$$

The sign is negative on V because V and i oppose each other, i.e. i tends to charge C to a voltage whose polarity is opposite to that of V. Equation 7.15 is an integral equation and can be converted to a homogeneous differential equation by differentiation. The result is:

$$R\frac{\mathrm{d}i}{\mathrm{d}t} + \frac{1}{C}i = 0 \qquad (7.16)$$

Assuming $i = A\mathrm{e}^{st}$, we get the characteristic equation

$$s + \frac{1}{RC} = 0 \qquad (7.17)$$

Thus, there is only one characteristic root or natural frequency at $s = -\frac{1}{RC}$; this is consistent with the fact that the circuit has only one energy storage element, viz. C. Hence, our solution is

$$i(t) = A\mathrm{e}^{-t/(RC)} \qquad (7.18)$$

To evaluate A, we take help of the fact that $i(0)\ R = V$. Here, the argument of i is to be interpreted as 0^+, i.e. immediately after the switch has been shifted from a to b [note that $i(0^-) = 0$]. Hence

$$\frac{V}{R} = A \qquad (7.19)$$

and finally

$$i(t) = \frac{V}{R}\mathrm{e}^{-t/(RC)} \qquad (7.20)$$

A plot of Eq. 7.20 is shown in Fig. 7.4. The product RC has the dimension of time, is denoted by T, and is called the time constant of the circuit. It is the time after which the current decays to $\frac{1}{\mathrm{e}} = 0.368$ times the initial value. At $t = 5T$, the current value is $\frac{V}{R}\mathrm{e}^{-5} \cong 0.0067\frac{V}{R}$ which is only 0.67% of the initial value and hence can be neglected. It is therefore said that the current has a life time of $5T$.

Physically, the current i represents the rate of decay of charge in the capacitor. As $t \to \infty$, the charge tends to zero and the current also tends to zero.

Fig. 7.3 Circuit in Fig. 7.2 at $t \geq 0$, with $i(0)\ R = V$

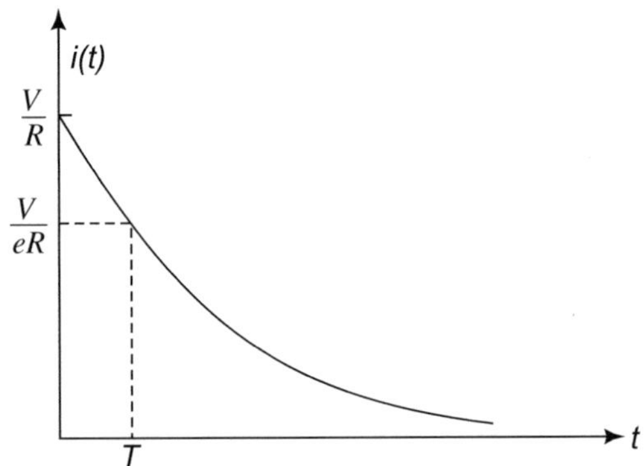

Fig. 7.4 Plot of Eq. 7.20

Force-Free Response of a Simple RL Circuit

Now consider the circuit in Fig. 7.5, where the switch S is in the position a for a long time, so that a steady current V/r is established in the inductor, and then at $t = 0$, S is shifted to position b. We wish to find the current $i(t)$, $t \geq 0+$. The differential equation is, by KVL,

$$L\frac{\mathrm{d}i}{\mathrm{d}t} + Ri = 0 \tag{7.21}$$

and by following a similar procedure as in the previous example, we get

$$i(t) = \frac{V}{r}\mathrm{e}^{-Rt/L} \tag{7.22}$$

Here, the time constant is $T = L/R$.

First-Order Circuits with More Than One Energy Storage Element

Both of the circuits in the previous two examples are first-order circuits, because they are governed by a first-order differential equation. It is not necessary that there should be only one energy storage element for a first-order circuit. There can be more than one capacitor, for example, in an RC circuit, but they must be connected in such a fashion that the capacitors can be combined into one. As an example, the circuits in Fig. 7.6 are first-order circuits, but not the ones in Fig. 7.7.

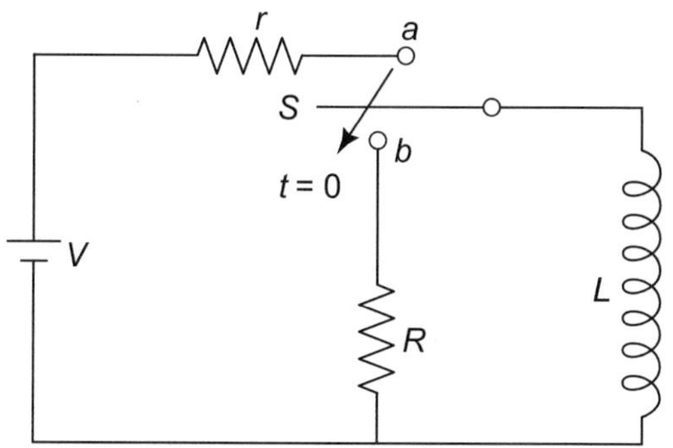

Fig. 7.5 A simple RL circuit

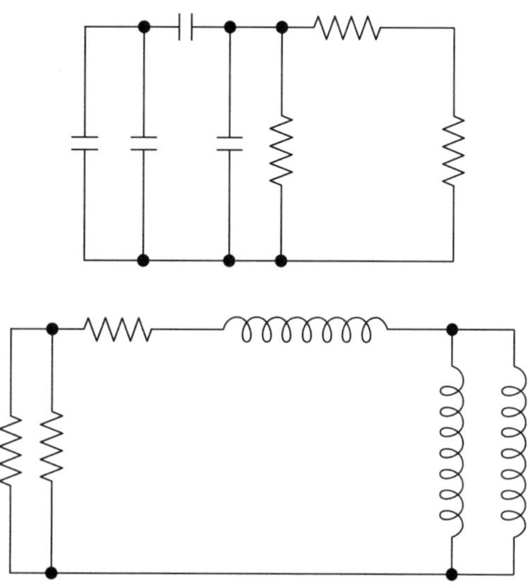

Fig. 7.6 More than one C or L but still first-order

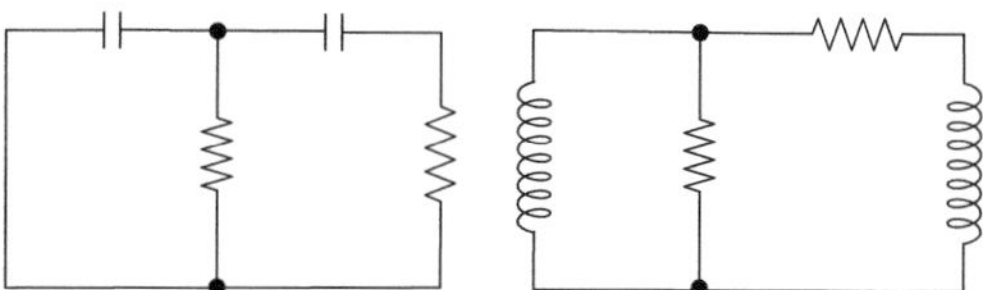

Fig. 7.7 Second order circuits

Force-Free Response of a Second-Order Circuit

Consider a charged capacitor C, with an initial voltage V_0 to be connected across a series RL combination at $t = 0$ as shown in Fig. 7.8. Assume the inductor to be initially relaxed. Application of KVL gives

$$L\frac{\mathrm{d}i}{\mathrm{d}t} + Ri + \frac{1}{C}\int_0^t i\mathrm{d}t - V_0 = 0 \tag{7.23}$$

Differentiating both sides gives

$$\frac{\mathrm{d}^2 i}{\mathrm{d}t^2} + \frac{R\mathrm{d}i}{L\mathrm{d}t} + \frac{i}{LC} = 0 \tag{7.24}$$

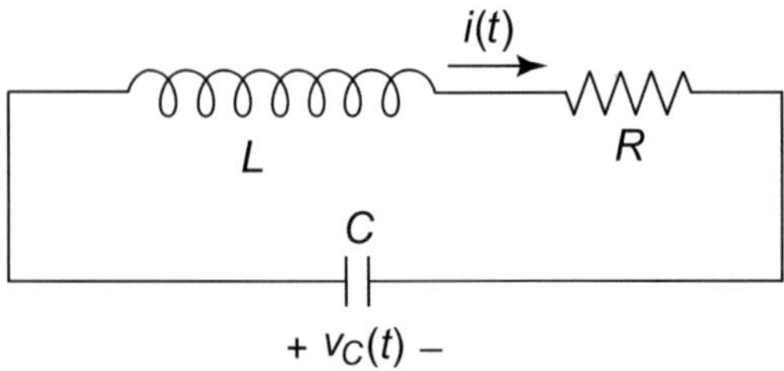

Fig. 7.8 A second-order circuit

which is the same as Eq. 7.5. The characteristic equation is

$$s^2 + \frac{R}{L}s + \frac{1}{LC} = 0 \tag{7.25}$$

which has the roots

$$s_{1,2} = -\frac{R}{2L} \pm \sqrt{\left(\frac{R}{2L}\right)^2 - \frac{1}{LC}} \tag{7.26}$$

The most general solution is, therefore,

$$i(t) = A_1 e^{s_1 t} + A_2 e^{s_2 t} \tag{7.27}$$

Clearly, the nature of the solution will depend upon whether $\left(\frac{R}{2L}\right)^2 - \frac{1}{LC}$ is >, = or <0. Accordingly, we shall have three cases. But first let us evaluate A_1 and A_2. The condition $i(0) = 0$ gives, from Eq. 7.27,

$$A_1 + A_2 = 0 \tag{7.28}$$

Also, from Eq. 7.23, putting $t = 0$, we get

$$L\frac{di}{dt}\bigg|_{t=0} = V_0 \tag{7.29}$$

which, combined with Eq. 7.27, gives

$$s_1 A_1 + s_2 A_2 = V_0/L \tag{7.30}$$

Solving Eqs. 7.28 and 7.30, we get

$$A_1 = \frac{V_0}{L(s_1 - s_2)} = -A_2 \tag{7.31}$$

Hence, the solution becomes

$$i(t) = \frac{V_0}{L(s_1 - s_2)}(e^{s_1 t} - e^{s_2 t}) \tag{7.32}$$

Case I:

$$\left(\frac{R}{2L}\right)^2 - \frac{1}{LC} > 0$$

In this case, the roots s_1 and s_2 are real, negative and distinct and the solution is

$$\left.\begin{aligned} i(t) &= V_0 \frac{e^{-\alpha t}}{2\beta L}(e^{\beta t} - e^{-\beta t}) \\ &= \frac{V_0 e^{-\alpha t} \sinh \beta t}{\beta L} \end{aligned}\right\}, \tag{7.33}$$

where

$$\alpha = +\frac{R}{2L} \text{ and } \beta = \sqrt{\frac{R^2}{4L^2} = \frac{1}{LC}} \tag{7.34}$$

Case II:

$$\left(\frac{R}{2L}\right)^2 - \frac{1}{LC} < 0$$

In this case, the roots will be complex conjugates of each other:

$$s_{1,2} = -\alpha \pm j\omega, \tag{7.35}$$

where α is defined in Eq. 7.34 and

$$\omega = \sqrt{\frac{1}{LC} - \frac{R^2}{4L^2}} = \sqrt{\omega_n^2 - \alpha^2} \tag{7.36}$$

ω_n being given by

$$\omega_n^2 = 1/(LC) \tag{7.37}$$

Hence, the complete solution is

$$i(t) = e^{-\alpha t}[A_1 e^{j\omega t} + A_2 e^{-j\omega t}] \tag{7.38}$$

which can be simplified to the form

$$i(t) = Ae^{-\alpha t} \sin(\omega t + \theta) \tag{7.39}$$

This is a damped sinusoid with the damping coefficient α (per second), natural frequency of oscillation ω and initial phase θ. A and θ are to be determined, as in the earlier case, from $i(0) = 0$ and $\left.\frac{\mathrm{d}i}{\mathrm{d}t}\right|_{t=0} = \frac{V_0}{L}$. Applied to Eq. 7.39, it gives

$$\theta = 0 \tag{7.40}$$

and

$$A = \frac{V_0}{\omega L} \tag{7.41}$$

Thus, finally,

$$i(t) = \frac{V_0}{\omega L} \mathrm{e}^{-\alpha t} \sin \omega t \tag{7.42}$$

Note that Eq. 7.39 is the most general form; it has reduced to Eq. 7.42 because we took the initial current in the inductor as zero. A plot of Eq. 7.42 is shown in Fig. 7.9.

Case III:

$$\left(\frac{R}{2L}\right)^2 = \frac{1}{LC}$$

In this case, the roots are real and equal. Case II is called the underdamped case, because there are oscillations, while case I is called the overdamped case because the response decays monotonically with time after reaching the first maximum. In the background of this nomenclature, the case under consideration should be called the critically damped case. From the theory of differential equations, the solution to this case will be since $s_1 = s_2 = -\alpha$,

$$i(t) = \mathrm{e}^{-\alpha t}(A_1 + A_2 t) \tag{7.43}$$

While this is the general form, $i(0) = 0$ indicates that $A_1 = 0$. Hence,

$$i(t) = A_2 t\ \mathrm{e}^{-\alpha t} \tag{7.44}$$

The other initial condition, viz. $\left.\frac{\mathrm{d}i}{\mathrm{d}t}\right|_{t=0} = V_0/L$ gives

$$A_2 = \frac{V_0}{L} \tag{7.44a}$$

Hence, finally,

$$i(t) = \frac{V_0}{L} t\mathrm{e}^{-\alpha t} \tag{7.44b}$$

a plot of which is shown in Fig. 7.10. Note that $i(t) = 0$ at $t = 0$ as well as at $t = \infty$, and that there is a maximum at $t = 1/\alpha$, the maximum value being

$$I_M = \frac{V_0}{\alpha L e} \tag{7.45}$$

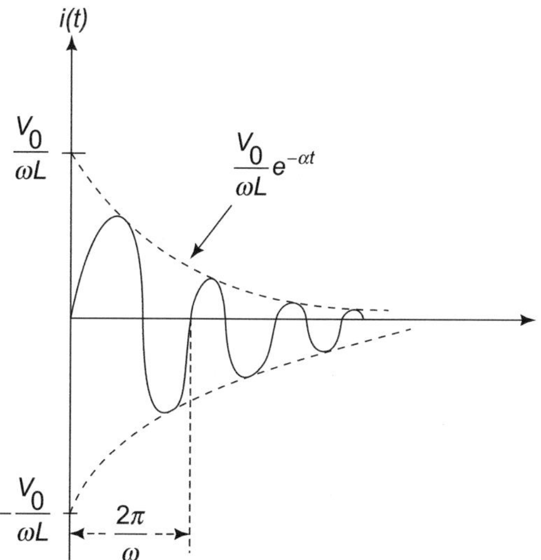

Fig. 7.9 Plot of Eq. 7.42

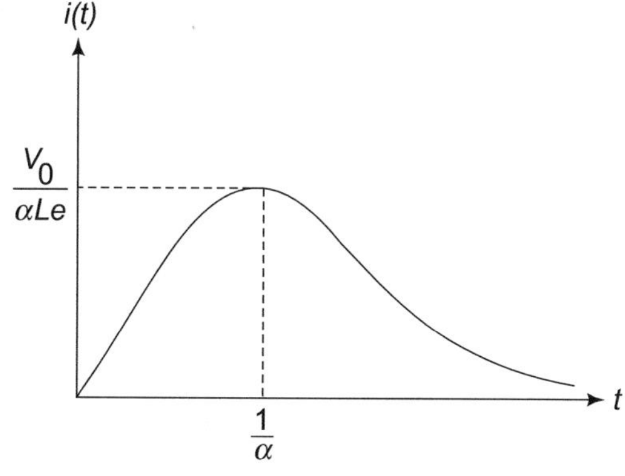

Fig. 7.10 Plot of Eq. 7.44b

It is worth mentioning here that the plot of Eq. 7.33 (for the overdamped case) will be similar to that of Fig. 7.10 except that the maximum will be smaller and the decay will be slower.

Root Locus of the Second-Order Circuit

The roots of the characteristic equation of the circuit considered in the preceding section move in the complex plane $s = \sigma + j\omega$ as R is varied. Referring to Eq. 7.26, we see that

(i) when $R = 0$ (note R cannot be negative), the roots are purely imaginary

$$s_{1,2} = \pm j/\sqrt{LC} = \pm j\omega_n \tag{7.46}$$

(ii) When $0 < R < 2\sqrt{L/C}$, the roots are complex:

$$s_{1,2} = -\alpha \pm j\omega, \tag{7.47}$$

where

$$\alpha^2 + \omega^2 = \omega_n^2, \tag{7.47a}$$

i.e. roots move in a circle, starting from $+j\omega_n$ and $-j\omega_n$, towards the negative real axis; the centre of the circle is at the origin of the complex plane and its radius is ω_n, as shown in Fig. 7.11;

(iii) when $R = 2\sqrt{\frac{L}{C}}$, both the roots coalesce at $s = -\alpha$, i.e. on the negative real axis (point P in Fig. 7.11); and

(iv) when R increases beyond $2\sqrt{\frac{L}{C}}$, we get the overdamped case, with both roots on the negative real axis, one (s_1) going towards the origin and the other (s_2) moving towards $-\infty$.

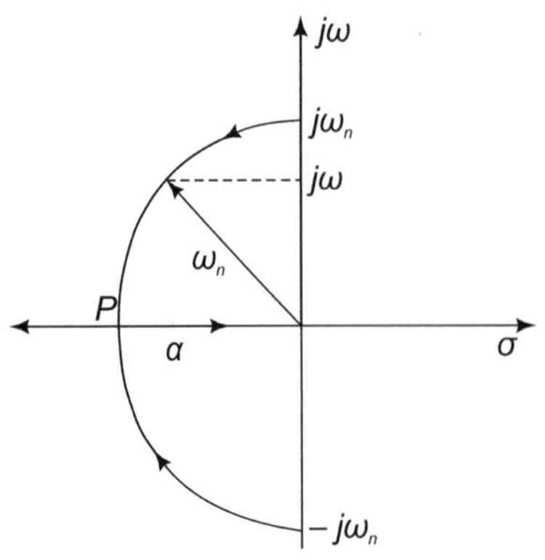

Fig. 7.11 Root locus of the series RLC circuit

Natural Frequencies of Circuits with a Forcing Function

Consider the example in Fig. 7.1 again, we have seen that the natural response satisfies Eq. 7.5, which is same as Eq. 7.25. The natural response is therefore of the form Eq. 7.27 with the natural frequencies given by Eq. 7.26. Now suppose instead of a current source $i_s(t)$, we connect a voltage source $v_s(t)$ across C, as shown in Fig. 7.12. The differential equation now becomes

$$L\frac{\mathrm{d}i}{\mathrm{d}t} + Ri = v_s(t) \tag{7.48}$$

so that the complementary function $i_c(t)$ satisfies the following homogeneous equation

$$L\frac{\mathrm{d}i}{\mathrm{d}t} + Ri = 0 \tag{7.49}$$

The characteristic equation is, therefore,

$$s + (R/L) = 0 \tag{7.50}$$

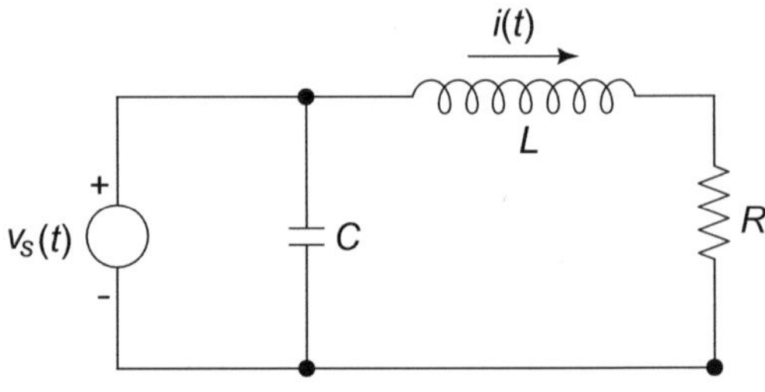

Fig. 7.12 Circuit of Fig. 7.1 driven by a voltage source

and the natural frequency is

$$s_0 = -R/L \tag{7.51}$$

This is quite different from Eq. 7.26 and leads us to the conclusion that the natural frequencies of a circuit under a forcing function depend upon the nature of the latter. We next consider a general technique based on the concept of impedance for finding the natural frequencies. But before that, note that $s_{1,\,2}$ of Eq. 7.26 can be called *open-circuit natural frequencies* because $i_s(t) = 0$ in Fig. 7.1 makes C look at an open circuit to the left, while s_0 of Eq. 7.51 qualifies to be called a *short-circuit natural frequency* because $v_s(t) = 0$ makes C look at a short circuit.

Concept of Impedance

Suppose a current e^{st} passes through an inductor L, then the voltage developed across it will be

$$v = L\frac{di}{dt} = sLe^{st} \tag{7.52}$$

The ratio of v to i is

$$Z_L = sL \tag{7.53}$$

and is called the *impedance* of the inductor. Similarly, if the voltage across a capacitor C is e^{st}, then the current through it is given by

$$i = C\frac{dv}{dt} = sCe^{st} \tag{7.54}$$

so that the ratio of v to i in this case is

$$Z_c = \frac{1}{sC} \tag{7.55}$$

Z_c is called the impedance of the capacitor. For a resistance R, the ratio of v to i is simply R, independent of the form of v or i. It follows that Z_c and Z_L have the same dimension as that of R, and that as far as an exponential excitation is concerned, they perform the same role as that of a resistance. Impedances can be combined exactly as resistances. For example, consider a series combination of R, L and C carrying a current e^{st}, then the voltage drop across the combination is

$$v = \left(R + sL + \frac{1}{sC}\right)e^{st} \tag{7.56}$$

Thus, the ratio of v to i is

$$Z(s) = R + sL + \frac{1}{sC} \tag{7.57}$$

which is simply the addition of the individual impedances. For the circuit in Fig. 7.1, the impedance seen by $i_s(t)$ is

$$Z(s) = \frac{\frac{1}{sC}(R + sL)}{R + sL + \frac{1}{sC}} \tag{7.58}$$

Relation Between Impedance and Natural Frequencies

Equation 7.58 can be simplified to as follows:

$$Z(s) = \frac{1}{C}\frac{(s + R/L)}{s^2 + s\frac{R}{L} + \frac{1}{LC}} \tag{7.59}$$

and can be rewritten as

$$Z(s) = \frac{1}{C}\frac{(s - s_0)}{(s - s_1)(s - s_2)}, \tag{7.60}$$

where s_0 is the same as the natural frequency with voltage excitation, given in Eq. 7.51, and s_1 and s_2 are the natural frequencies with current excitation, given by Eq. 7.26. These observations hold in general, too. That is, the open-circuit natural frequencies are the value of s at which $Z(s) \to \infty$; these values are called the *poles* of $Z(s)$. Similarly, the short-circuit natural frequencies are those values of s at which $Z(s) = 0$; these values are given the name *zeros*. Clearly, both

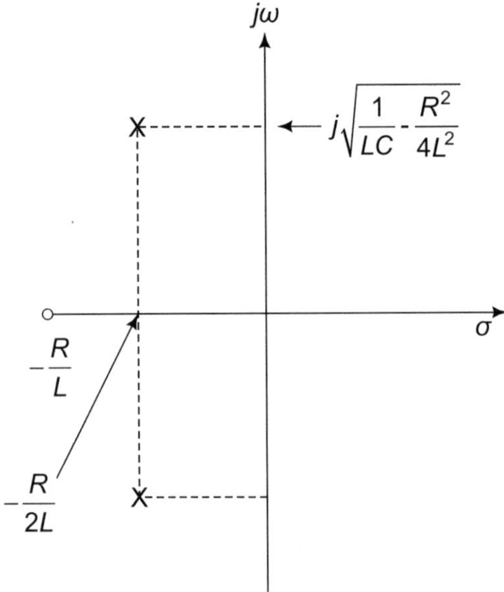

Fig. 7.13 Pole-zero plot for the circuit in Fig. 7.1 with $R^2 < 4L/C$

poles and zeros can, in general, be complex quantities. In the $s = \sigma + j\omega$ plane, a zero is indicated by a small circle, while a pole is indicated by a cross and the picture so obtained is called the pole-zero plot. For the circuit in Fig. 7.1, the pole-zero plot for a typical underdamped case is shown in Fig. 7.13.

An admittance is defined as the reciprocal of an impedance and is denoted by Y. The poles and zeros of Y are obviously the zeros and poles, respectively, of Z.

Forced Response to an Exponential Excitation

We shall illustrate the calculation of forced response to an exponential excitation by an example. Consider the circuit in Fig. 7.14. We are interested in finding the voltage v. The impedance faced by the current source is

$$\begin{aligned} Z(s) &= 2 + \frac{(s+3)\frac{1}{0.5s}}{s+3+\frac{1}{0.5s}} \\ &= 2 + \frac{2(s+3)}{(s+2)(s+1)} \end{aligned} \tag{7.61}$$

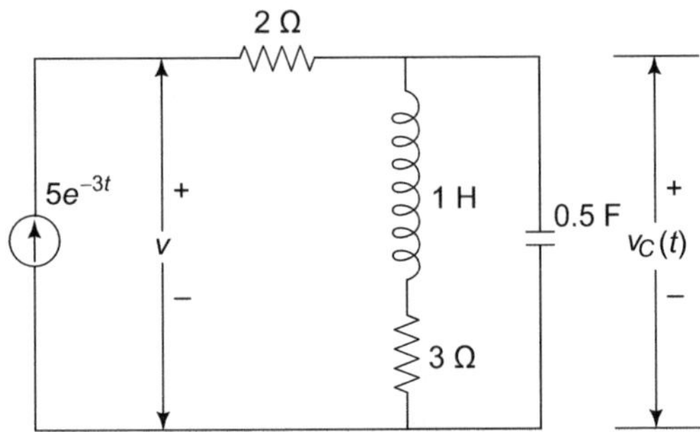

Fig. 7.14 An example of exponential excitation

Since $Z(s) = v/i$ when v or i is of the form e^{st} or Ae^{st}, it follows that

$$\begin{aligned} v(t) &= i(t)Z(s)|_s = -3 \\ &= 5e^{-3t} \times 2 = 10e^{-3t} \end{aligned} \tag{7.62}$$

If we are interested in finding the voltage across the capacitor, then we can use potential division, i.e.

$$v_c(t) = v(t) \cdot \left.\frac{\frac{2(s+3)}{(s+2)(s+1)}}{2 + \frac{2(s+3)}{(s+2)(s+1)}}\right|_{s=-3} = 0 \tag{7.63}$$

Forced Response Due to DC

A DC can be considered as an exponential excitation with $s = 0$. Then, the impedances of R, L and C become R, 0 and ∞, respectively. Hence for DC excitation, an inductor acts as a short circuit and a capacitor behaves as an open circuit.

Forced Response to a Sinusoidal Excitation

The forced response to an excitation of the form e^{st} is of the same form, except for multiplication by a constant. The forced response to dc is also a dc. If $s = j\omega$, then the forced response will be of the form $Ae^{i\omega x}$ where A can be a complex quantity. Let $A = |A|e^{j\theta}$; then, symbolically, we can write

$$e^{j\omega t} \rightarrow |A|e^{j(\omega t+\theta)} \quad (7.64)$$

Since $e^{j\omega t}$ is the superposition of cos ωt and j sin ωt, and since we are considering only linear circuits, it follows that

$$\begin{aligned}\mathrm{Im}(e^{j\omega t}) &= \sin\omega t \rightarrow \mathrm{Im}[|A|e^{j(\omega t+\theta)}] \\ &= |A|\sin(\omega t+\theta)\end{aligned} \quad (7.65)$$

and

$$\begin{aligned}\mathrm{Re}(e^{j\omega t}) &= \cos\omega t \rightarrow \mathrm{Re}[|A|e^{j(\omega t+\theta)}] \\ &= |A|\cos(\omega t+\theta)\end{aligned} \quad (7.66)$$

This suggests a methodology for finding the forced response of a general circuit to sinusoidal excitation. Suppose, we have a voltage excitation of the form

$$v(t) = V_m\cos(\omega t+\theta) = \sqrt{2}V\cos(\omega t+\theta), \quad (7.67)$$

where V is the rms value of the voltage. We can write

$$v(t) = \mathrm{Re}[V_m e^{j\theta} e^{j\omega t}] = \mathrm{Re}[Ae^{j\omega t}], \quad (7.68)$$

where A is the complex quantity $V_m\, e^{i\theta}$. We can then find the response of the circuit to the exponential $Ae^{j\omega t}$ by using impedance concepts. In this case, the impedance of R, L and C will be $Z_R = R$

$$Z_L = j\omega L = jX_L \quad (7.69)$$

$$Z_C = \frac{1}{j\omega C} = jX_C$$

$X_L = \omega L$ and $X_c = -1/(\omega C)$ are called the *reactances* of L and C, respectively.

Suppose the response is $Be^{j\omega t}$ where B is another complex quantity $|B|e^{j\phi}$. Then

$$\begin{aligned}Ae^{j\omega t} &\rightarrow Be^{j\omega t} \\ \mathrm{Re}[Ae^{j\omega t}] &\rightarrow \mathrm{Re}[Be^{j\omega t}]\end{aligned}$$

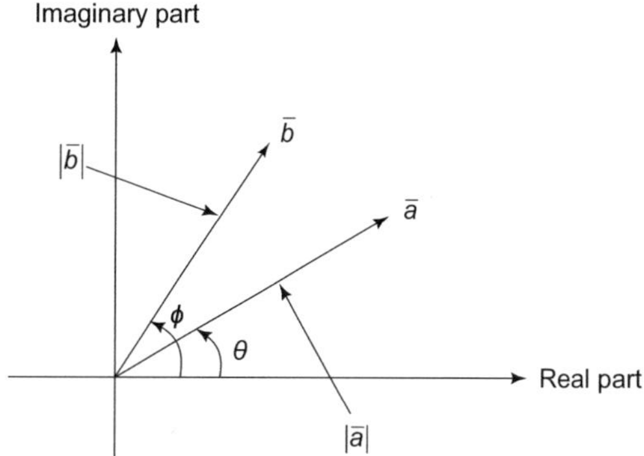

Fig. 7.15 Phasor representation

or

$$V_m\cos(\omega t+\theta) \rightarrow |B|\cos(\omega t+\phi) \quad (7.70)$$

Thus, one can divorce the time dependence and work only in terms of $A = |A|e^{j\theta}$ and impedances to find $B = |B|e^{j\phi}$. It is conventional to work in terms of rms rather than peak values, so that the given excitation will be represented by $\bar{a} = A/\sqrt{2}$ instead of A and the response obtained will then be in the form $\bar{b} = B/\sqrt{2}$ rather than B. Each of these quantities is called a phasor and can be represented as a vector in the complex plane as shown in Fig. 7.15.

From the convention just discussed, obviously a current 10 cos ωt will be represented by the phasor $(10/\sqrt{2})e^{j0}$, i.e. cos ωt has been taken to define the reference direction for angle. (It must be mentioned that there is nothing sacred about this convention. In fact, we shall, as illustrated later, sometime find it convenient to use sinωt as the reference phasor for angles.)

As examples, the current 10 cos (ωt + π/6) will be represented by the phasor $(10/\sqrt{2})e^{j\pi/6}$ while the current 10 sin (ωt + π/6) will be represented by the phasor $\frac{10}{\sqrt{2}}e^{j}\left(\frac{\pi}{6}-\frac{\pi}{2}\right) = \frac{10}{\sqrt{2}}e^{-j\pi/3}$ because

$$\sin(\omega t+\pi/6) = \cos(\omega t+\pi/6-\pi/2) \quad (7.71)$$

Phasors can be added or subtracted like vectors. Suppose, for example, we wish to find $v = v_1 - v_2$ where $v_1 = 100\sqrt{2}\,\cos(100\pi t - \pi/6)$

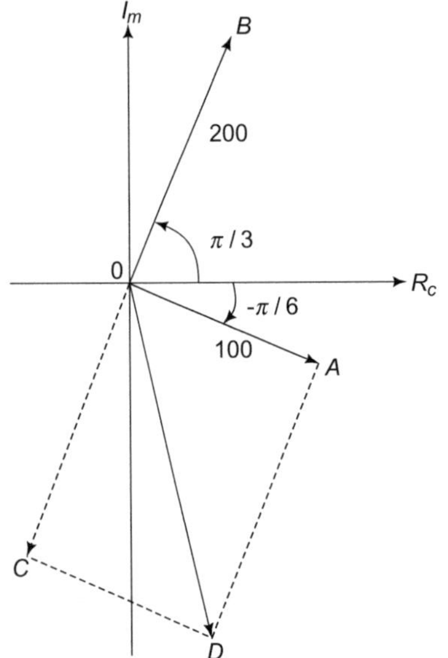

Fig. 7.16 Phasor addition

and $v_2 = 200\sqrt{2}\cos(100\pi t + \pi/3)$. The phasors corresponding to these two voltage are $\bar{V}_1 = 100e^{-j\pi/6}$ and $\bar{V}_2 = 200e^{+j\pi/3}$, which are shown in Fig. 7.16 by the vectors *OA* and *OB*, respectively. The vector *OC* represents the negative of *OB* and *OD* and the vector sum of *OA* and *OC* represents the phasor corresponding to $v_1 - v_2$. One can find the phasor corresponding to *OD* by measurements or by geometrical formulas. It is, however, most convenient to compute this by converting both *OA* and *OB* to rectangular coordinates. Thus

$$\left.\begin{aligned}\bar{V}_1 &= 100\cos\pi/6 - j100\sin\pi/6\\ &= 50\sqrt{3} - j50\end{aligned}\right\} \tag{7.72}$$

$$\left.\begin{aligned}\bar{V}_2 &= 200\cos\pi/3 - j200\sin\pi/3\\ &= 100 + j100\sqrt{3}\end{aligned}\right\} \tag{7.73}$$

Hence

$$\bar{V}_1 - \bar{V}_2 = (50\sqrt{3} - 100) - j(50 + 100\sqrt{3}) \tag{7.74}$$

whose magnitude *M* and phase θ can be easily found out. The quantity $v_1 - v_2$ will then be $M\sqrt{2}\cos(100\pi t + \theta)$.

Basic Elements and Their *V–I* Relationships for Sinusoidal Excitation

Note that if a current $i = \sqrt{2}I\cos\omega t$ flows through a resistance, the voltage across it is $\sqrt{2}IR\cos\omega t$. Thus, the current and the voltage phasors are Ie^{j0} and IRe^{j0}. Thus, the voltage and current are in phase.

If the same current flows through an inductance *L*, the voltage across it is $-\sqrt{2}\,LI\omega\sin\omega t = \sqrt{2}\omega\,LI\cos(\omega t + \pi/2)$. The phasor corresponding to this is $\omega L\,I\,e^{j\pi/2} = j\omega LI$.

The impedance of the inductor $j\omega L$ is therefore simply the ratio of voltage to current phasor. Also note that the voltage leads the current by 90° or the current lags the voltage by 90°.

Similarly for a capacitor *C*, with sinusoidal excitation, the impedance is $\frac{1}{j\omega C}$ and the voltage across it lags the current by 90°.

In general, if a current phasor $\bar{I}$ flows that an impedance $Z(j\omega)$, then the voltage phasor across the latter is

$$\bar{V} = \bar{I}Z \tag{7.75}$$

which is often referred to as Ohm's law for sinusoidal excitation. Phasors, like actual voltages/currents, obey KCL and KVL, provided, of course, the circuit is linear, for which all currents and voltages will be of the same frequency.

An Example of the Use of Phasors and Impedances

Let a current source $i(t) = 10\sqrt{2}\cos(1000t + \pi/4)$ be connected across a parallel combination of *R*, *L* and *C* as shown in Fig. 7.17. It is required to determine the voltage $v(t)$ across the combination and the currents through the individual branches.

The admittance (=1/impedance) of the combination is

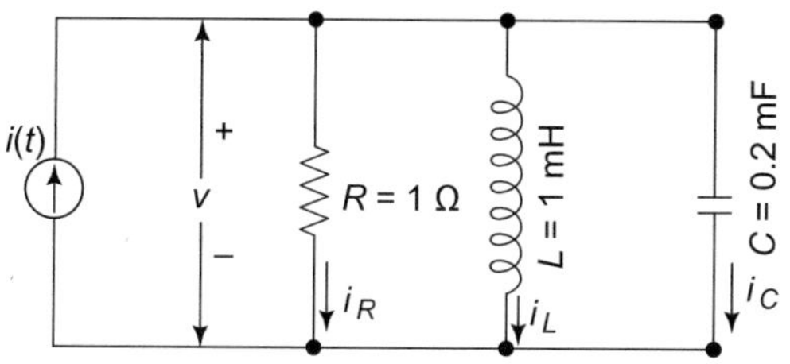

Fig. 7.17 A parallel *RLC* circuit with sinusoidal excitation

$$\left.\begin{aligned} Y &= \tfrac{1}{R} + \tfrac{1}{j\omega L} + j\omega C \\ &= 1 - j\tfrac{1}{1000\times 10^{-3}} + j \times 1000 \times 2 \times 10^{-3} \\ &= 1 + j \end{aligned}\right\} \quad (7.76)$$

Hence, the impedance is

$$Z = \frac{1}{1+j} = \frac{1-j}{2} \quad (7.77)$$

The current phasor is

$$\bar{I} = 10e^{j\pi/4} = \frac{10}{\sqrt{2}}(1+j) \quad (7.78)$$

Hence, the voltage phasor will be

$$\bar{V} = \bar{I}Z = \frac{10}{2\sqrt{2}}(1-j)(1+j) = \frac{10}{\sqrt{2}} \quad (7.79)$$

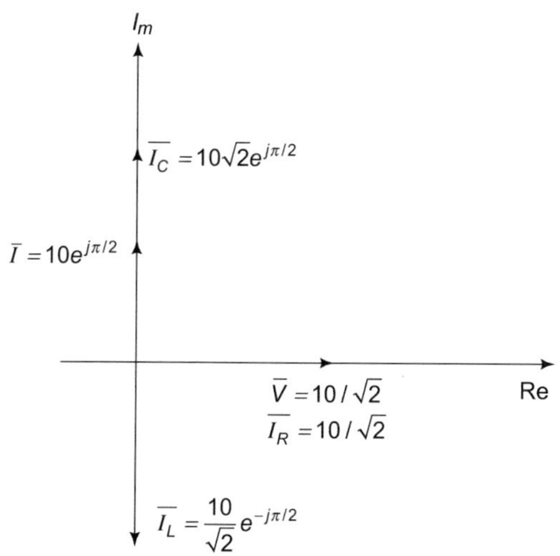

Fig. 7.18 Phasor diagram for the circuit in Fig. 7.17

Thus

$$\left.\begin{aligned} \bar{I}_R &= \tfrac{10}{\sqrt{2}}A \\ \bar{I}_L &= \tfrac{10/\sqrt{2}}{j\omega L} = \tfrac{10}{\sqrt{2}}e^{-j\pi/2} \\ \bar{I}_c &= (10/\sqrt{2})j\omega C = j10\sqrt{2} = 10\sqrt{2}e^{j\pi/2} \end{aligned}\right\} \quad (7.80)$$

These various phasors are shown in Fig. 7.18. In the time domain, the expressions would be

$$\left.\begin{aligned} v(t) &= 10\cos 1000t \\ i_R(t) &= 10\cos 1000t \\ i_L(t) &= 10\sin 1000t \\ i_c(t) &= -20\sin 1000t \end{aligned}\right\} \quad (7.81)$$

Back to Complete Response

We now return to the complete response of a given circuit to a given excitation. Consider first a simple *RL* circuit to which a sinusoidal voltage source $V_m \cos \omega t$ is connected in series at $t = 0$, as shown in Fig. 7.19. The differential equation is

$$L\frac{di}{dt} + Ri = V_m \cos \omega t \quad (7.82)$$

whose complete solution is

$$i(t) = i_c(t) + i_p(t), \quad (7.83)$$

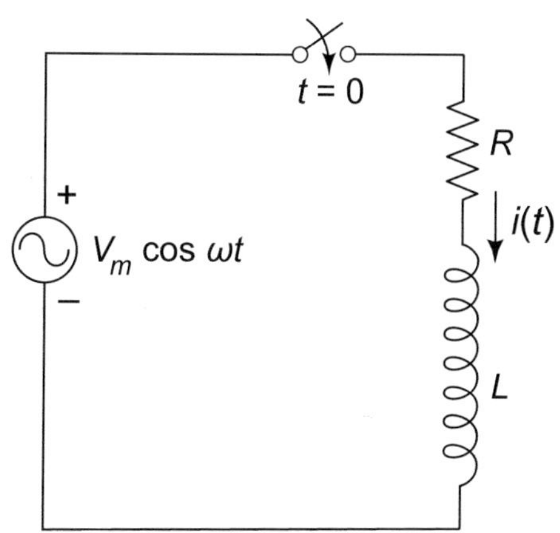

Fig. 7.19 A series *RL* circuit with sinusoidal excitation

where

$$i_c(t) = Ae^{-Rt/L} \quad (7.84)$$

and

$$i_p(t) = I_m \cos(\omega t + \theta) \quad (7.85)$$

The particular solution parameters I_m and θ can be easily obtained from phasor analysis. The voltage phasor is

$$\bar{V} = \frac{V_m}{\sqrt{2}} e^{jo} \quad (7.86)$$

The impedance is

$$Z = R + j\omega L \quad (7.87)$$

Hence, the current phasor is

$$\bar{I} = \frac{\frac{V_m}{\sqrt{2}}}{R + j\omega L} = \frac{V_m}{\sqrt{2(R^2 + \omega^2 L^2)}} e^{-j\tan^{-1}(\omega L/R)} \quad (7.88)$$

Hence

$$I_m = \frac{V_m}{\sqrt{(R^2 + \omega^2 L^2)}} \quad (7.89)$$

and

$$\theta = -\tan^{-1}\frac{\omega L}{R} \quad (7.90)$$

Thus

$$i(t) = Ae^{-Rt/L} + \frac{V_m}{\sqrt{R^2 + \omega^2 L^2}} \cos\left(\omega t - \tan^{-1}\frac{\omega L}{R}\right) \quad (7.91)$$

To obtain A, one appeals to the initial condition. Let $i(0) = 0$; then from Eq. 7.91,

$$A = -\frac{V_m}{\sqrt{R^2 + \omega^2 L^2}} \cos\left(\tan^{-1}\frac{\omega L}{R}\right) \quad (7.92)$$

As is evident from Fig. 7.20.

$$\cos\left(\tan^{-1}\frac{\omega L}{R}\right) = \frac{R}{\sqrt{R^2 + \omega^2 L^2}} \quad (7.93)$$

Hence, finally

$$i(t) = \frac{V_m}{\sqrt{R^2 + \omega^2 L^2}} \left[\cos\left(\omega t - \tan^{-1}\frac{\omega L}{R}\right) - \frac{R}{\sqrt{R^2 + \omega^2 L^2}} e^{-Rt/L}\right] \quad (7.94)$$

The first term in square brackets is the forced response, while the second term is the natural response. Note that the natural response form depends on R and L only, but the actual value is affected by the forcing function. Also note that the natural response here is also the transient part while the forced response is also the steady-state response.

In general, if the forcing function is not sinusoidal, one finds the particular solution by assuming it to be of the same form as the forcing function, and substituting it in the complete differential equation to determine the unknown constant.

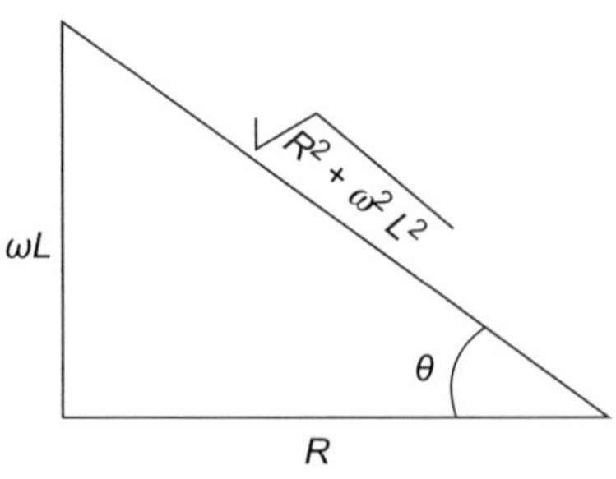

Fig. 7.20 Computing $\cos\left(\tan^{-1}\frac{\omega L}{R}\right)$

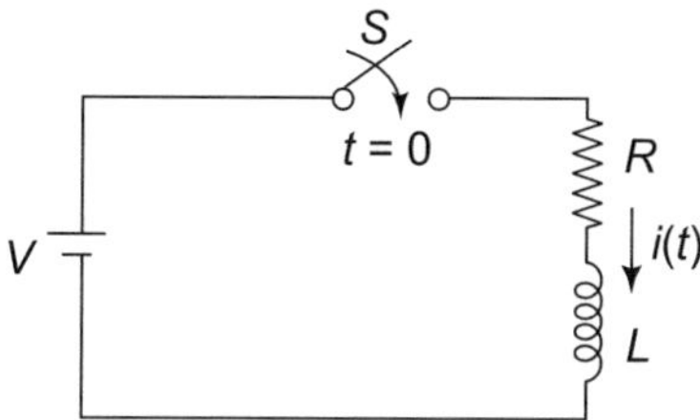

Fig. 7.21 An *RL* circuit excited by a DC voltage source

We now consider several other examples of complete response.

Step Response of an RL Circuit

Consider the circuit in Fig. 7.21, where S was open for a long time and switched on at $t = 0$. Hence, there is no initial current in L, i.e. $i(0^+) = i(0^-) = 0$. (≡continuity of current in an inductor). From the differential equation

$$L\frac{\mathrm{d}i}{\mathrm{d}t} + Ri = V,\ t > 0^+ \tag{7.95}$$

it follows that

$$\begin{aligned} i(t) &= i_c(t) + i_p(t) \\ &= A\mathrm{e}^{-Rt/L} + \frac{V}{R} \end{aligned} \tag{7.96}$$

because the natural frequency is the zero of $Z(s) = R + sL$ and the forced response is a constant whose value is determined by Eq. 7.95 itself. Note that i = constant means that $\mathrm{d}i/\mathrm{d}t = 0$ so that the particular solution is V/R. To find A, take help of the fact that $i(0^+) = 0$; thus

$$A = -V/R \tag{7.97}$$

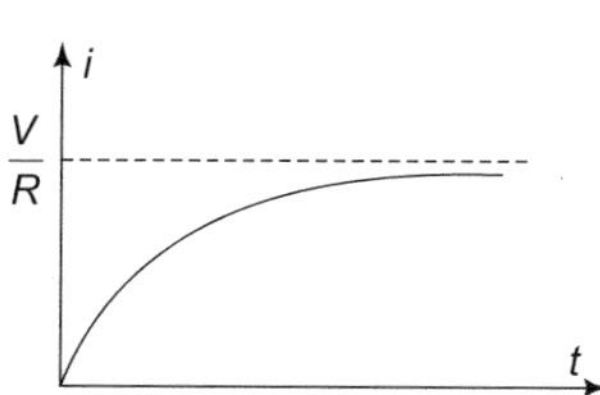

Fig. 7.22 Step response of an *RL* circuit

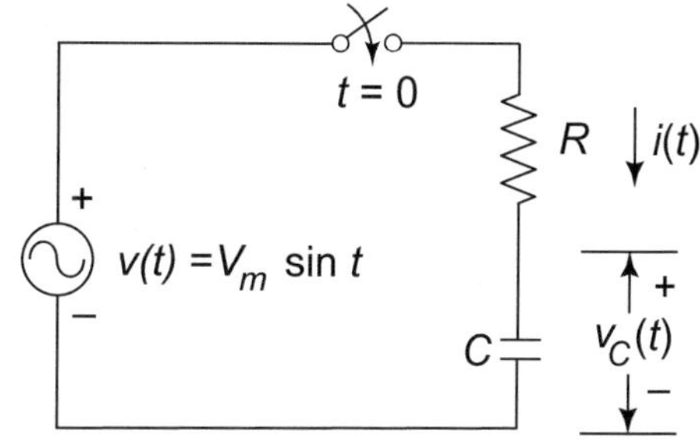

Fig. 7.23 A series *RC* circuit with sinusoidal excitation

and finally

$$i(t) = \frac{V}{R}(1 - \mathrm{e}^{-Rt/L}) \tag{7.98}$$

A plot of i is shown in Fig. 7.22.

Sinusoidal Response of a Series RC Circuit

Consider the situation shown in Fig. 7.23, where $v_c(0) = 0$. The impedance of the series RC circuit for e^{st} excitation is

$$Z(s) = R + \frac{1}{sC} \tag{7.99}$$

and since the excitation is a voltage source, the natural frequency will be the zero of $Z(s)$, occurring at $s = -1/(CR)$. Also, the forced current response can be found from the voltage phasor

$$\bar{V} = (V_m/\sqrt{2})\mathrm{e}^{j0} \tag{7.100}$$

(where we have taken sin ωt as the reference phasor, instead of cos ωt; this does not change the result) and the impedance for $\mathrm{e}^{j\omega t}$ excitation;

$$\begin{aligned} Z(j\omega) &= R + \frac{1}{j\omega C} \\ &= \sqrt{R^2 + \frac{1}{\omega^2 C^2}}\angle - \tan^{-1}\frac{1}{\omega CR} \end{aligned} \tag{7.101}$$

Note that instead of writing $\mathrm{e}^{-\tan^{-1} 1/(\omega CR)}$, we have indicated $\angle\tan^{-1} 1/(\omega CR)$; these are interchangeable notations. Hence, the current phasor will be

$$\bar{I} = \frac{\bar{V}}{Z(j\omega)} = \frac{V_m}{\sqrt{2\left(R^2 + \frac{1}{\omega^2 C^2}\right)}} \angle \tan^{-1}\frac{1}{\omega CR} \tag{7.102}$$

Thus, the forced response is

$$i_p(t) = \frac{V_m}{\sqrt{R^2 + \frac{1}{\omega^2 C^2}}} \sin\left(\omega t + \tan^{-1}\frac{1}{\omega CR}\right) \tag{7.103}$$

and the complete response is

$$i(t) = Ae^{-t/(RC)} + \frac{V_m}{\sqrt{R^2 + \frac{1}{\omega^2 C^2}}} \sin\left(\omega t + \tan^{-1}\frac{1}{\omega CR}\right) \tag{7.104}$$

To evaluate A, note that $v_c(0^-) = v_c(0^+)$ has been assumed to be zero so that $i(0^+) = v(0)/R = 0$. Hence

$$A = -\frac{V_m}{\sqrt{R^2 + \frac{1}{\omega^2 C^2}}} \sin\left(\tan^{-1}\frac{1}{\omega CR}\right) \tag{7.105}$$

$$= -\frac{V_m}{\sqrt{R^2 + \frac{1}{\omega^2 C^2}}} \frac{1}{\sqrt{1+\omega^2 C^2 R^2}} \tag{7.106}$$

from Fig. 7.24. Hence, finally

$$i(t) = \frac{V_m}{\sqrt{R^2 + \frac{1}{\omega^2 C^2}}} \left[\sin\left(\omega t + \tan^{-1}\frac{1}{\omega CR}\right) - \frac{e^{-t/(RC)}}{\sqrt{1+\omega^2 C^2 R^2}}\right] \tag{7.107}$$

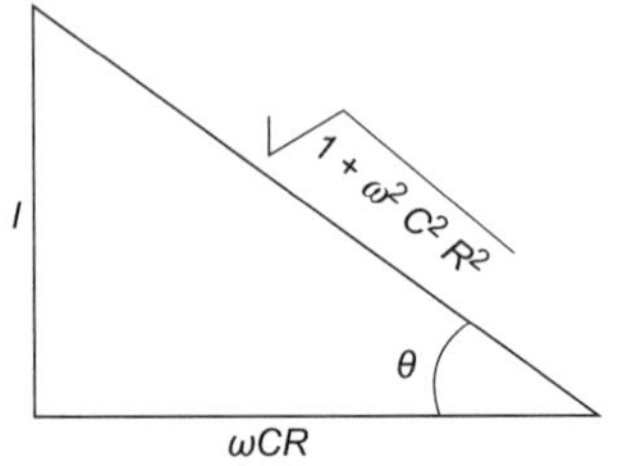

Fig. 7.24 Computing $\sin\left(\tan^{-1}\frac{1}{\omega CR}\right)$

Response of an RC Circuit to an Exponential Excitation

In the circuit shown in Fig. 7.25, let $v_c(0) = V$ and the current source be exponentially decaying, i.e.

$$i_s(t) = Ie^{-\alpha t}, \tag{7.108}$$

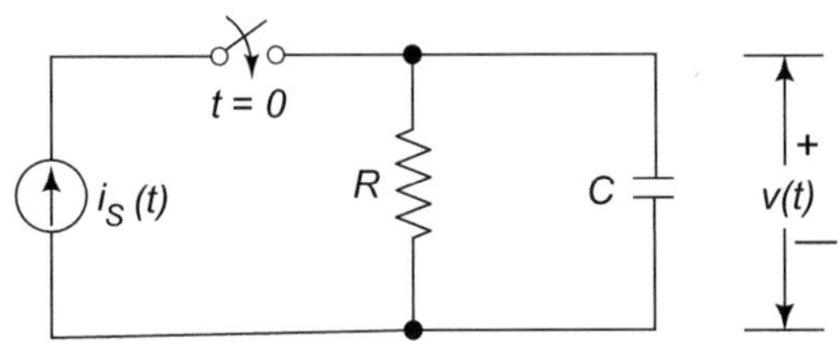

Fig. 7.25 Current excited parallel *RC* circuit

where α is real, positive and not equal to $1/(CR)$. The impedance to e^{st} excitation is

$$Z(s) = \frac{R \times \frac{1}{sC}}{R + \frac{1}{sC}} = \frac{1}{C}\frac{1}{s + \frac{1}{CR}} \tag{7.109}$$

and therefore the natural frequency is given by

$$s = -\frac{1}{CR} \tag{7.110}$$

the pole of $Z(s)$. The forced response will be of the form $v_p(t) = Ke^{-\alpha t}$ and referring to the differential equation

$$C\frac{dv}{dt} + \frac{v}{R} = i_s(t) \tag{7.111}$$

we get

$$-CK\alpha + \frac{K}{R} = I \tag{7.112}$$

or

$$K = \frac{I}{\frac{1}{R} - C\alpha} = \frac{I}{C}\frac{1}{\frac{1}{CR} - \alpha} \tag{7.113}$$

Note that an easier method for calculating K would have been to refer to the fact that $Z(s) = v/i_s(t)$ for e^{st} excitation so that for $Ie^{-\alpha t}$ excitation, we would have

$$v_p(t) = Ie^{-\alpha t}Z(-\alpha) = \frac{I}{C}\frac{1}{\frac{1}{CR} - \alpha}e^{-\alpha t} \tag{7.114}$$

This gives the same value of K as in Eq. 7.113. Hence, we have the complete response of the circuit as

$$v(t) = Ae^{-t(CR)} + \frac{I}{C}\frac{1}{\frac{1}{CR} - \alpha}e^{-\alpha t} \tag{7.115}$$

To find A, put $v(0) = V$; this gives

$$V = A + \frac{I}{C\left(\frac{1}{CR} - \alpha\right)} \tag{7.116}$$

or,

$$A = V - \frac{I}{C\left(\frac{1}{CR} - \alpha\right)} \tag{7.117}$$

This can be substituted in Eq. 7.115 to get the complete solution. The result is

$$v(t) = \frac{I}{C\left(\frac{1}{CR} - \alpha\right)}\left(e^{-\alpha t} - e^{-\frac{1}{CR}}\right) + Ve^{-\frac{1}{CR}} \tag{7.118}$$

We have written $v(t)$ in this form to illustrate the case when $\alpha = \frac{1}{CR}$. In general, by series expansion, we can write

$$\begin{aligned} v(t) &= e^{-\frac{1}{CR}}\left[V + \frac{I}{C\left(\frac{1}{CR} - \alpha\right)}\left\{e^{t\left(\frac{1}{CR} - \alpha\right)} - 1\right\}\right] \\ &= e^{-\frac{1}{CR}}\left[V + \frac{I}{C\left(\frac{1}{CR} - \alpha\right)}\left\{t\left(\frac{1}{CR} - \alpha\right) + \frac{t^2\left(\frac{1}{CR} - \alpha\right)^2}{2!} + \cdots\right\}\right] \end{aligned} \tag{7.119}$$

When $\alpha \to \frac{1}{CR}$, Eq. 7.119 gives

$$v(t) = e^{-\alpha t}\left(V + \frac{I}{C}t\right) \tag{7.120}$$

This result can also be derived from the fact that if the characteristic root of a first-order equation ($s = -\alpha$) is also contained in the forcing exponential function, then the particular solution is of the form $Kte^{-\alpha t}$. Substituting this in Eq. 7.111 gives

$$CK[t(-\alpha)e^{-\alpha t} + e^{-\alpha t}] + \frac{1}{R}Kte^{-\alpha t} = Ie^{-\alpha t} \tag{7.121}$$

in which the first and the third terms on the left-hand side cancel, giving

$$K = \frac{I}{C} \tag{7.122}$$

Hence, $v(t)$ is of the form

$$v(t) = \left(A + \frac{I}{C}t\right)e^{-\alpha t} \tag{7.123}$$

Putting $t = 0$ in this gives

$$A = V \tag{7.124}$$

Step Response of an RLC Circuit

In the second-order circuit of Fig. 7.26, as we have seen, the natural response will be of the form $A_1e^{s_1t} + A_2e^{s_2t}$, where

$$\left.\begin{array}{l} s_{1,2} = -\frac{R}{2L} \pm \sqrt{\frac{R^2}{4L^2} - \frac{1}{LC}} \\ \underline{\Delta} - \alpha \pm j\omega \end{array}\right\} \tag{7.125}$$

while the forced response will be a constant. From the circuit as well as from this differential equation, it is clear that the forced response is zero. Then, the total response is

$$i(t) = A_1e^{s_1t} + A_2e^{s_2t}, \tag{7.126}$$

where A_1 and A_2 are to be found from the initial conditions on L and C. Depending on the value of R, we can have three cases. For real and distinct $s_{1,2}$ (overdamped case), Eq. 7.126 is the general form. For $s_1 = s_2$ (critically damped case), the solution becomes

$$i(t) = (A_1 + A_2t)e^{s_1t} \tag{7.127}$$

For complex $s_{1,2} = -\alpha \pm j\omega$ (underdamped case), the solution can be put in the form

$$i(t) = Ae^{-\alpha t}\sin(\omega t + \theta) \tag{7.128}$$

In each case, there are two constants to be determined from the initial conditions. Let L and C be initially relaxed. Then $i(0) = 0$ and $v_c(0) = 0$. The first condition gives, for the overdamped case,

$$A_1 + A_2 = 0 \tag{7.128a}$$

while the second gives

$$v_L(0) = V - i(0)R - v_c(0) = V, \tag{7.128b}$$

i.e.

$$L\left.\frac{di}{dt}\right|_{t=0} = V \tag{7.128c}$$

or

$$s_1A_2 + S_2A_2 = \frac{V}{L} \tag{7.129}$$

Hence, A_1 and A_2 can be determined. The other two cases can be similarly dealt with.

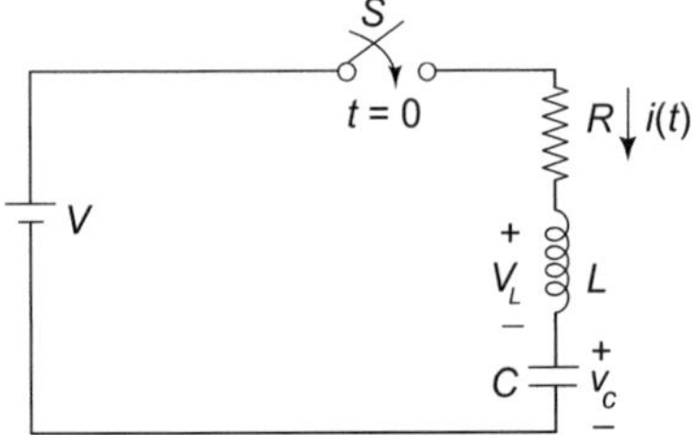

Fig. 7.26 An *RLC* circuit with step excitation

Sinusoidal Response of an RLC Circuit

Consider the circuit in Fig. 7.27, where the switch has been in the closed position for a long time so that $v(0) = 0$ and $i(0) = 0$ (why don't we write v

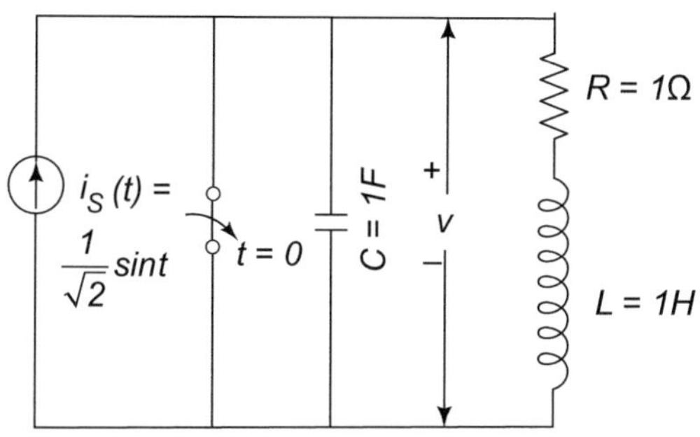

Fig. 7.27 *RLC* circuit excited by a sinusoidal current

(0) = i(0) = 0?) Since the excitation is a current generator, the poles of $Z(s)$ or the zeros of $Y(s)$ will determine the natural frequencies. We have

$$Y(s) = sC + \frac{1}{R+sL} = s + \frac{1}{s+1} = \frac{s^2+s+1}{s+1} \tag{7.130}$$

The natural frequencies are, therefore, the roots of

$$s^2+s+1=0, \tag{7.130a}$$

i.e.

$$s_{1,2} = -\frac{1}{2} \pm j\frac{\sqrt{3}}{2} \tag{7.131}$$

The natural response is thus of the form $Ae^{-t/2}$ $\cos\left(\frac{\sqrt{3}}{2}t+\theta\right)$. The forced response phasor, $\bar{V}$, is given by the product of the current phasor corresponding to $i(t)$ and the impedance at $s = j1$. Hence

$$\begin{aligned}\bar{V} &= \left[\frac{1}{2}\angle 0^\circ\right]\frac{1+j1}{j1}\\ &= \frac{1}{2}\angle 0^\circ(1-j1) = \frac{1}{\sqrt{2}}\angle -\pi/4,\end{aligned} \tag{7.132}$$

where we have again taken sin ωt as the reference phasor, without any loss of accuracy. Hence, the forced response will be

$$v_P = \sin(t-\pi/4) \tag{7.133}$$

The complete solution is, therefore,

$$v(t) = Ae^{-t/2}\cos\left(\frac{\sqrt{3}}{2}t+\theta\right) + \sin(t-\pi/4) \tag{7.134}$$

To determine A and θ, take help of $v(0) = 0$ and $i(0) = 0$. The first condition gives

$$0 = A\cos\theta - \sin\pi/4 \tag{7.135}$$

or

$$A\cos\theta = \frac{1}{\sqrt{2}} \tag{7.136}$$

The second condition says that

$$C\left.\frac{dv}{dt}\right|_{t=0} = i_s(0) - i(0) \tag{7.137}$$

Substituting for $v(t)$ from Eq. 7.134 and simplifying, we get

$$A = \frac{\sqrt{2}}{\sqrt{3}\sin\theta + \cos\theta} \tag{7.138}$$

Combining Eq. 7.136 and Eq. 7.138, we get

$$\frac{1}{\sqrt{2}\cos\theta} = \frac{\sqrt{2}}{\sqrt{3}\sin\theta+\cos\theta} \tag{7.139}$$

or

$$\sqrt{3}\sin\theta = \cos\theta \tag{7.140}$$

or

$$\theta = \frac{\pi}{6} \tag{7.141}$$

Hence, from Eq. 7.136,

$$A = \sqrt{\frac{2}{3}} \tag{7.142}$$

Finally, therefore, the solution for v is

$$v(t) = \sqrt{\frac{2}{3}}\mathrm{e}^{-t/2}\cos\left(\frac{\sqrt{3}}{2}t + \frac{\pi}{6}\right) + \sin(t - \pi/4) \tag{7.143}$$

The first term is the transient response while the second term represents the steady-state response.

Pulse Response of an RC Circuit

For the circuit in Fig. 7.28, where $v(t)$ is a pulse as shown in Fig. 7.29, we can write

$$v(t) = V[u(t) - u(t-T)], \tag{7.144}$$

where $u(t)$ is the unit step function. We know the response to $Vu(t)$ to be

$$i_1(t) = \frac{V}{R}\mathrm{e}^{-\frac{1}{RC}}u(t), \tag{7.145}$$

where the multiplication by $u(t)$ indicates that $i_1(t) = 0$ for $t < 0$. The response to the delayed step $Vu(t - T)$ will be, from the principle of time invariance (which states that if the excitation is delayed, so will the response be),

$$i_2(t) = \frac{V}{R}\mathrm{e}^{-\frac{t-T}{RC}}u(t - T), \tag{7.146}$$

where multiplication by $u(t - T)$ indicates that $i_2(t) = 0$ for $t < T$. Hence, by superposition, the complete response will be

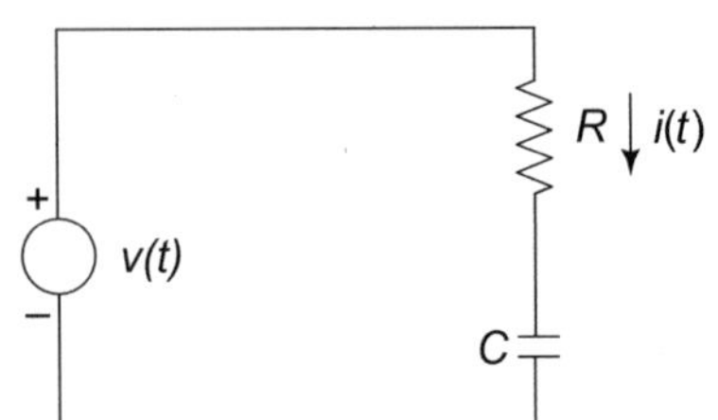

Fig. 7.28 An *RC* circuit excited by the pulse of Fig. 7.29

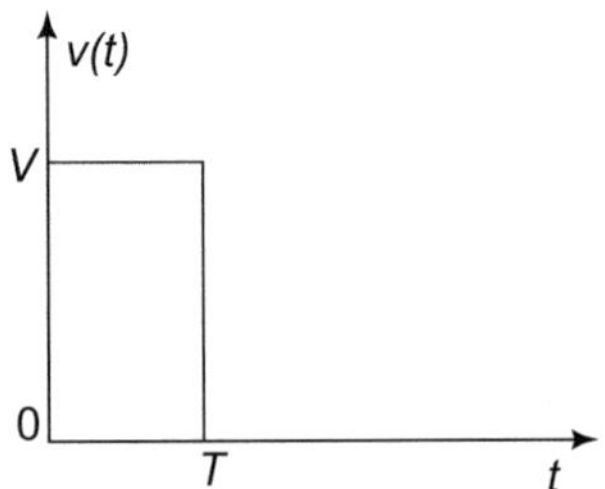

Fig. 7.29 Pulse excitation of the circuit in Fig. 7.28: A gate

$$\begin{aligned} i(t) &= i_1(t) - i_2(t) \\ &= \frac{V}{R}\left[\mathrm{e}^{-\frac{1}{RC}}u(t) - \mathrm{e}^{-\frac{t-T}{RC}}u(t - T)\right] \end{aligned} \tag{7.147}$$

A sketch of $i(t)$ is shown in Fig. 7.30.

What about the voltage across the capacitor? It can be found by integrating $i(t)$ from 0 to t. Physically, it is not difficult to argue that C charges according to the relation

$$v_c(t) = V[1 - \mathrm{e}^{-t/(CR)}] \tag{7.148}$$

for $0 < t \leq T$. At $t = T$, the voltage source becomes zero, and current direction reverses. Hence, C discharges through R from the value $V[1 - \mathrm{e}^{-T/(CR)}]$ to zero exponentially. Hence, $v_c(t)$ will vary with time as shown in Fig. 7.31. The expression for $v_c(t)$ is

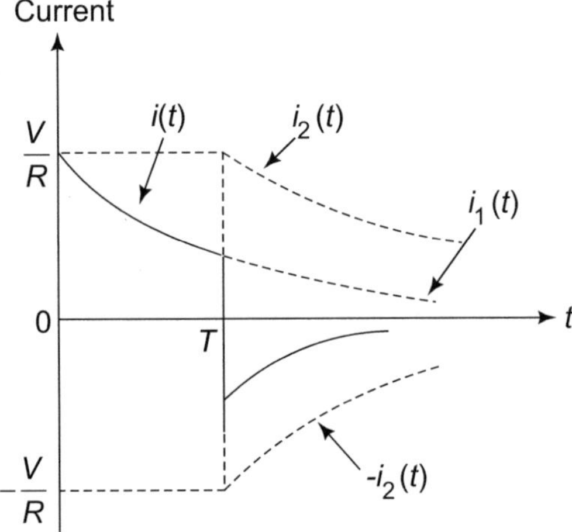

Fig. 7.30 Pulse response of the circuit in Fig. 7.28

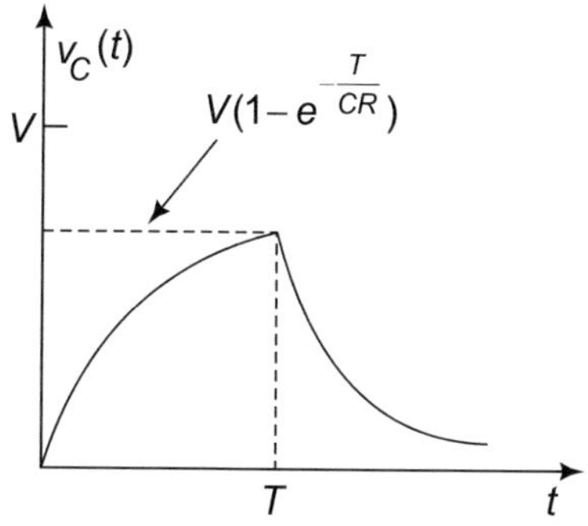

Fig. 7.31 Plot of $v_c(t)$ for the circuit in Fig. 7.28

$$v_c(t) = \begin{cases} V\left(1 - e^{-\frac{t}{CR}}\right), & 0 \le t \le T \\ V\left(1 - e^{-\frac{T}{CR}}\right) e^{-\frac{t-T}{CR}} & t \ge T \end{cases} \quad (7.149)$$

Impulse Response

Consider the pulse shown in Fig. 7.32 whose area is unity. Let T be decreased to $T/2$, and the height increased to $2/T$ so that the area is still unity. Let $T \to 0$ and the height $\to \infty$ in such manner that the area is still unity. This limiting condition of the pulse, whose duration is zero and the height is infinite such that the area under it is unity, is called *a unit impulse*. A unit impulse function is denoted by $\delta(t)$, and since infinite height is not a determinate quantity, we define $\delta(t)$ by the integral

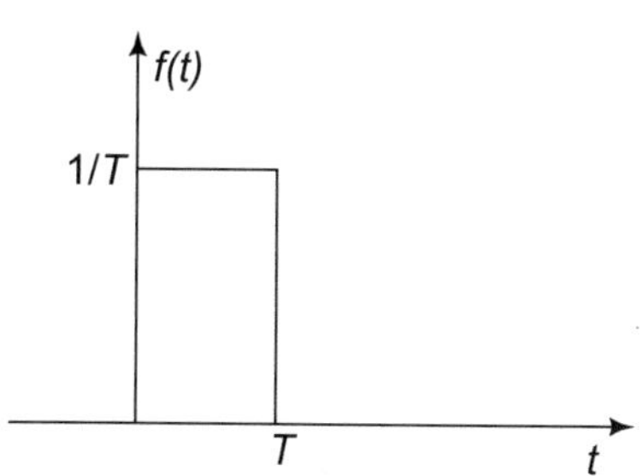

Fig. 7.32 A pulse which becomes a unit impulse when $T \to 0$

$$\int_{0-}^{0+} \delta(t)\mathrm{d}t = 1 \quad (7.150)$$

It follows that $\delta(t - a)$ is a unit impulse occurring at $t = a$ and that

$$\int_{a-}^{a+} \delta(t-a)\mathrm{d}t = 1 \quad (7.151)$$

If the range of integral does not include the value of t at which the impulse occurs, then the integral will be zero, e.g.

$$\int_{0+}^{\infty} \delta(t)\mathrm{d}t = 0 = \int_{-\infty}^{0-} \delta(t)\mathrm{d}t \quad (7.152)$$

Also

$$f(t)\delta(t-a) = f(a)\delta(t-a) \quad (7.153)$$

and

$$\int_{a-}^{a+} f(t)\delta(t-a)\mathrm{d}t = f(a) \quad (7.154)$$

Impulse is, of course, a hypothetical function, but is a useful concept in analysing circuits and systems. Let an impulse $Q(t)$ be applied to an RC circuit as shown in Fig. 7.33 with $v(0-) = 0$. Then, the infinite current $Q(t)$ flows through C and establishes a voltage

$$v(0^+) = \frac{1}{C}\int_{0-}^{0+} Q\delta(t)\,\mathrm{d}t = \frac{Q}{C} \quad (7.155)$$

At $t \ge 0+$, $Q(t)$ acts as an open circuit; hence $v(t)$ decays with time according to

$$v(t) = \frac{Q}{C}e^{-t/(CR)} \quad (7.156)$$

Note that here $v(0-) \ne v(0+)$, this is so because an infinite amount of current flows through C for the short duration $0-$ to $0+$.

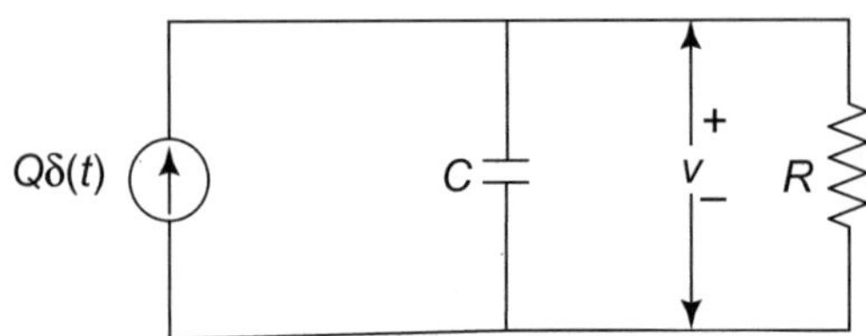

Fig. 7.33 An *RC* circuit excited by an impulsive current source

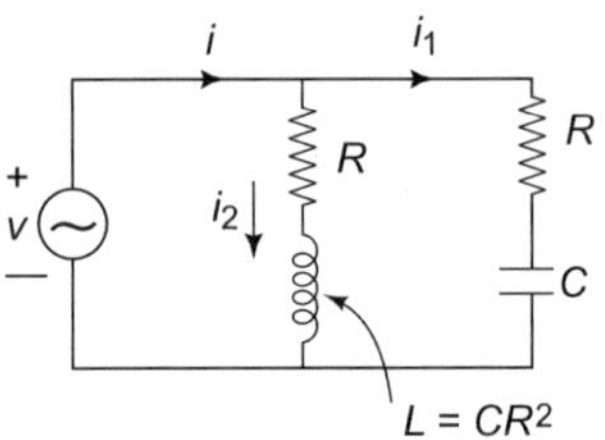

Fig. 7.34 Circuit for P.1

Similarly when a voltage excitation $\phi\delta(t)$ is applied across a series *RL* combination with no initial current in *L*, a current $i(0+) = \phi/L$ is established. Here also $i(0-) \neq i(0+)$ in an inductor. But for these exceptions, which are, of course, hypothetical ones, the inductor current and capacitor voltage cannot change instantaneously.

Problems

P.1. Write the differential equation for the circuit shown below for i, i_1 as well as i_2 (Fig. 7.34).

P.2. Find the transfer impedance $Z_T(s) = \frac{V(s)}{I(s)}$ for the circuit of P.1 and sketch its poles and zeros.

P.3. Write the differential equation governing the circuit in Fig. 7.33 in the text. Clue: Chap. 8.

P.4. Determine the transfer function $Z_T(s) = \frac{V(s)}{I_s(s)}$ for the circuit in Fig. 7.27 of the text and sketch is poles and zeros.

P.5. Working in the frequency domain, find the unit step response of the circuit in Fig. 7.26 in the text. Find the inverse transform and verify that the response in the time domain, as determine in the text, is identical.

Transient Response of RLC Networks Revisited

8

As compared to the conventional approach of trial solutions for solving the differential equation governing the transient response of RLC networks, we present here a different approach which is totally analytical. We also show that the three cases of damping, viz. overdamping, critical damping and underdamping, can be dealt with in a unified manner from the general solution. Won't you appreciate my innovations? Please do and encourage me.

Keywords
Transient response of RLC network
Overdamped case · Underdamped case
Critically damped case

Introduction

In dealing with transients in *RLC* networks, most textbooks on circuit theory [1, 2] derive the governing second-order differential equation and then treat the three different cases of damping, viz. overdamping, critical damping and underdamping separately. A trial solution of the form Ae^{st} is shown to be suitable for the overdamping and underdamping cases while for critical damping, the trial solution assumes the form $(A_1 + A_2t)e^{st}$ because of repeated roots of the characteristic equation. To a beginner, it is not clear why one has to use trial solutions, and why they have to be of these specific forms, but since they work, he (or she) accepts the solutions faithfully (This illustrates the principle of the end justifying the means!). This chapter presents a different approach to the solution of the differential equation, which is totally analytical. Also, it is shown that the three cases, particularly, the critical damping one, need not be treated separately and that a unified treatment is possible from the general solution of the differential equation.

Example Circuit and the Differential Equation

For a clear exposition and understanding of the analytical approach, we shall consider the simple series *RLC* circuit shown in Fig. 8.1, where the switch is closed at $t = 0$ with $i\ (0^-) = 0$ and $v_C\ (0^-) = V$. KVL gives, for $t > 0$.

$$Ri + L\frac{di}{dt} + \frac{1}{C}\int_0^t i dt - V = 0. \tag{8.1}$$

Source: S. C. Dutta Roy, "Transient Response of RLC Networks Revisited," *IETE Journal of Education*, vol. 44, pp. 207–211, October–December 2003.

S. C. Dutta Roy, *Circuits, Systems and Signal Processing*,
https://doi.org/10.1007/978-981-10-6919-2_8

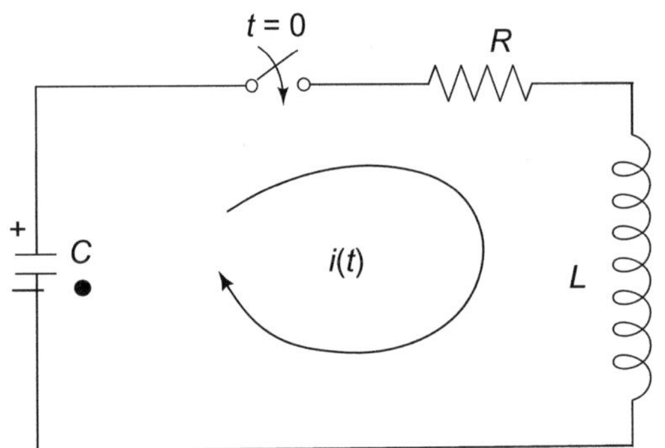

Fig. 8.1 The simple *RLC* circuit

Differentiating Eq. 8.1 and dividing both sides by L, we get

$$i'' + 2\alpha i' + \omega_0^2 i = 0, \tag{8.2}$$

where prime (′) denotes differentiation with respect to t,

$$2\alpha = R/L \quad \text{and} \quad \omega_0^2 = 1/(LC) \tag{8.3}$$

Analytical Solution of the Differential Equation

As already mentioned, the conventional approach to solving Eq. 8.2 is to try the solution $i = Ae^{st}$, with little or, at the best, heuristic justification. We take a different approach here, by introducing the new variable

$$y = i' - si, \tag{8.4}$$

where s is a constant to be chosen shortly. From Eq. 8.4, we have

$$i' = y + si \tag{8.5}$$

so that

$$i'' = y' + si' = y' + sy + s^2 i \tag{8.6}$$

Substituting Eqs. 8.5 and 8.6 in Eq. 8.2 and simplifying, we get

$$y' + (s + 2\alpha)y + (s^2 + 2\alpha s + \omega_0^2)i = 0 \tag{8.7}$$

Since s is our choice, let

$$s^2 + 2\alpha s + \omega_0^2 = 0 \tag{8.8}$$

Note that this is precisely the result we would have obtained if we tried the solution $i = Ae^{st}$ on Eq. 8.2, and as is well known, Eq. 8.8 is called the characteristic equation of the system, having the two roots

$$s_{1,2} = -\alpha \pm \beta, \tag{8.9}$$

where

$$\beta = \sqrt{\alpha^2 - \omega_0^2} \tag{8.10}$$

Combining Eqs. 8.7 and 8.8, we have

$$y' + (s + 2\alpha)y = 0, \tag{8.11}$$

which can be written in the form

$$\mathrm{d}y/y = -(s + 2\alpha)\mathrm{d}t \tag{8.12}$$

Integrating Eq. 9.12 and simplifying, we get

$$y = k_1 e^{-(s+2\alpha)t}, \tag{8.13}$$

where k_1 is a constant, as are all the k_1's in what follows. Combining Eqs. 8.13 with 8.4, we get the following first-order equation in i:

$$i' - si = k_1 e^{-(s+2\alpha)t} \tag{8.14}$$

Unlike Eq. 8.11, however, Eq. 8.14 is not a homogeneous equation. That should not be a cause for worry because we can use the well-known method of integrating factor, which in this case is e^{-st}. Multiplying both sides of Eq. 8.14 by this factor, the result can be put in the form

$$(ie^{-st})' = k_1 e^{-2(s+\alpha)t} \tag{8.15}$$

Integrating Eq. 8.15 and simplifying, we get

$$i = k_2 e^{st} + k_2 e^{-(s+2\alpha)t} \tag{8.16}$$

Now s has two possible values given by Eq. 8.9. If we choose $s = s_1 = -\alpha + \beta$ then $-(s + 2\alpha) = -\alpha - \beta = s_2$. Similarly, if we choose $s = s_2$, then $-(s + 2\alpha) = -\alpha + \beta = s_1$. Thus in either case, our solution is of the form

$$i = k_4 e^{s_1 t} + k_5 e^{s_2 t} \tag{8.17}$$

Evaluating the Constants

Now we go back to the example circuit. To evaluate the constants in Eq. 8.17, we invoke the two initial conditions, viz. $i(0^-) = 0$ and $v_C(0^-) = V$. Because of continuity of inductor currents and capacitor voltages, we have

$$i(0^+) = i(0^-) = 0 \text{ and } v_C(0^+) = v_C(0^-) = V \tag{8.18}$$

From Eq. 8.17 and the first condition in Eq. 8.18, we get

$$k_4 + k_5 = 0 \tag{8.19}$$

so that

$$i = k_4(e^{s_1 t} - e^{s_2 t}) \tag{8.20}$$

Also, referring to Eq. 8.1 and putting $t = 0^+$, we get

$$(di/dt)_{0^+} = V/L \tag{8.21}$$

Combining this with Eq. 8.20 and simplifying, we get

$$k_4 = V/[L(s_1 - s_2)] \tag{8.22}$$

From Eqs. 8.20, 8.22 and 8.9, the general solution can be written in the form

$$i = \frac{V}{2\beta L} e^{-\alpha t}(e^{\beta t} - e^{-\beta t}) \tag{8.23}$$

This single equation, as will be shown here, is adequate for considering all the cases of damping.

Overdamped Case

The *RLC* circuit is overdamped if $\alpha > \omega_0$, i.e. from Eq. 8.3, $R > 2\sqrt{(L/C)}$. Consequently, β is real and we can write Eq. 8.23 in the form

$$i = \frac{V}{\beta L} e^{-\alpha t} \sinh \beta t \tag{8.24}$$

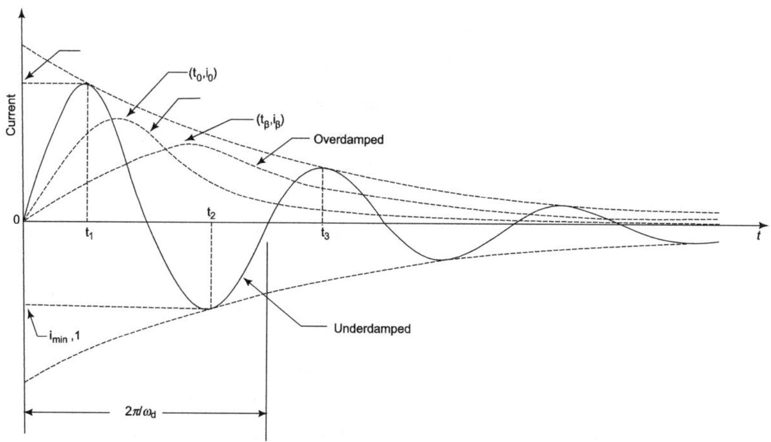

Fig. 8.2 Current response for various cases of damping

This gives $i = 0$ at $t = 0$, as it should; the same holds at $t = \infty$ also because $\beta < \alpha$. Hence, there must be a maximum at some value of t, say t. Differentiating Eq. 8.24 with respect to t and putting the result to zero gives, after simplification,

$$t_\beta = (1/\beta)\tanh^{-1}(\beta/\alpha) \quad (8.25)$$

The maximum value of i (= i) is obtained by combining Eqs. 8.24 and 8.25; using the identity

$$\sinh\theta = 1/\sqrt{1+\coth^2\theta} \quad (8.26)$$

to simplify the result, we get

$$i_\beta = V\sqrt{C/L}e^{-(\alpha/\beta)\tanh^{-1}(\beta/\alpha)} \quad (8.27)$$

The general nature of variation of the current is shown in Fig. 8.2. Using the series

$$\tanh^{-1}\theta = \theta + (\theta^3/3) + (\theta^5/5) + \ldots \text{to}\,\infty, |\theta| < 1, \quad (8.28)$$

It is easily shown that as β increases, t_β increases and i_β decreases. The time t_β is the smallest and the current i_β is the highest when $\beta = 0$, but we shall talk about this limiting situation later.

Underdamped Case

The *RLC* circuit is underdamped if $\alpha < \omega_0$, i.e. $R < 2\surd(L/C)$. In this case, β is imaginary. Let

$$\beta = j\sqrt{\omega_0^2 - \alpha^2} = j\omega_d \quad (8.29)$$

Putting this value in Eq. 8.23 and simplifying, we get

$$i = \frac{V}{\omega_d L}e^{-\alpha t}\sin\omega_d t \quad (8.30)$$

Obviously, the response is oscillatory but damped, as shown in Fig. 8.2. The envelope of the damped sine wave deceases exponentially, while its zero crossings occur at intervals of π/ω_d. The maxima and minima occur at times satisfying the equation $di/dt = 0$. Carrying out the required algebra, we get maxima at

$$t_{2n+1} = (1/\omega_d)[\tan^{-1}(\omega_d/\alpha) + 2n\pi], n = 0, 1, 2, \ldots \quad (8.31)$$

and minima at

$$t_{2n} = (l/\omega_d)[\tan^{-1}(\omega_d/\alpha) + (2n-l)\pi], \quad n = 1, 2, 3, \ldots \quad (8.32)$$

Combining Eqs. 8.30, 8.31 and the identity

$$\sin\theta = 1/\sqrt{1+\cot^2\theta}, \quad (8.33)$$

and simplifying, we get the value of the first maximum current as

$$i_{\min,1} = V\sqrt{C/L}e^{-(\alpha/\omega_d)\tan^{-1}(\omega_d/\alpha)} \quad (8.34)$$

Similarly, the value of the first minimum current can be found as

$$i_{\min,1} = -V\sqrt{C/L}e^{-(\alpha/\omega_d)[\tan^{-1}(\omega_d/\alpha)+\pi]} \quad (8.35)$$

All successive maxima (as well as minima) differ from each other in magnitude by the factor $e^{-2\alpha\pi/\omega_d}$.

Critically Damped Case

For this case, $\beta = 0$, i.e. $\alpha = \omega_0$ or $R = 2\surd(L/C)$, and as already mentioned, in the conventional approach, one notes that here $s_1 = s_2 = -\alpha$, and to obtain two independent solutions, one tries out a solution of the form $(A_1 + A_2 t)e^{st}$. One approach [3] for justifying this trial solution is to assume a solution of the form $f(t)\,e^{st}$; substitute it in Eq. 8.2; take help of the fact that $s + \alpha = 0$; and hence obtain $f''(t) = 0$, so that $f(t)$ has to be of the form $A_1 + A_2 t$. Here, we obtain the solution directly from the general solution Eq. 8.23 as the limiting case of $\beta \to 0$. Thus

$$i = \frac{V}{2L} e^{-\alpha t} \lim_{\beta \to 0} \frac{e^{\beta t} - e^{-\beta t}}{\beta}. \quad (8.36)$$

Using L'Hospital's rule, this becomes

$$i = \frac{V}{L} t e^{-\alpha t} \quad (8.37)$$

A plot of Eq. 8.37 is shown in Fig. 8.2. It is easily shown that the maximum occurs at the time $t = t_0$ where

$$t_0 = 1/\alpha, \quad (8.38)$$

the maximum value being

$$i_0 = (V/e)\sqrt{C/L}, \quad (8.39)$$

where use has been made of the fact that $R = 2\sqrt{(L/C)}$. Using the infinite series Eq. 8.28 and the following:

$$\sinh \theta = \theta + (\theta^3/3!) + (\theta^5/5!) + \ldots \text{ to } \infty, |\theta| < \infty, \quad (8.40)$$

it is easily shown that Eqs. 8.38 and 8.39 are the limiting values of Eqs. 8.25 and 8.27, respectively, for $\beta \to 0$.

The critically damped case could also have been treated ab initio. The procedure is the same up to Eq. 8.15, at which stage we take account of the fact that $s_1 = s_2 = -\alpha$. Equation 8.15 then becomes

$$(ie^{-st})' = k_1 \quad (8.41)$$

Integrating Eq. 8.41 gives

$$i = (k_1 t + k_2)e^{st} \quad (8.42)$$

k_1 and k_2 can now be evaluated from the initial conditions to arrive at precisely the same result as Eq. 8.37.

Concluding Comments

A totally analytical approach is given here for dealing with transients in second-order networks and systems, which does not require trial solutions. It deserves to be mentioned here that our method is an improvement over that using the operator concept, as given in some books [4]. In the latter, one uses the operator D for d/dt and D^2 for d^2/dt^2 so that Eq. 8.2 can be written as

$$(D^2 + 2\alpha D + \omega_0^2)i = 0 \quad (8.43)$$

One then treats $(D^2 + 2\alpha D + \omega_0^2)$ as an algebraic expression, finds its roots s_1 and s_2 (same as those given by Eq. 8.9) and rewrites Eq. 8.43 as

$$(D - s_1)(D - s_2)i = 0 \quad (8.44)$$

The conceptual difficulty of a student arises on two counts—first, in accepting D^2 for d^2/dt^2 and second, in treating $(D^2 + 2\alpha D + \omega_0^2)$ as an algebraic expression in D, i.e. in treating D as a variable instead of an operator. Of course, here also, as in the trial solution method, the end can be used to justify the means as follows:

$$(D - s_1)(D - s_2)i = (D - s_1)\left(\frac{di}{dt} - s_2 i\right)$$
$$= \frac{d^2 i}{dt^2} - s_2\frac{di}{dt} - s_1\frac{di}{dt} + s_1 s_2 i = i'' + 2\alpha i' + \omega_0^2 i, \quad (8.45)$$

which is the same as the left-hand side of Eq. 8.2. The method of solution is similar to ours in that one solves two first-order differential equations, viz. the homogeneous equation $(D - s_1)y = 0$ first, and then the nonhomogeneous equation $(D - s_2)i = y$. Here also, no special care is needed to deal with the critically damped case in which $s_1 = s_2 = -\alpha$.

Problems

P.1. Arrange an *RLC* circuit to give a band-stop response in as many ways as you can.

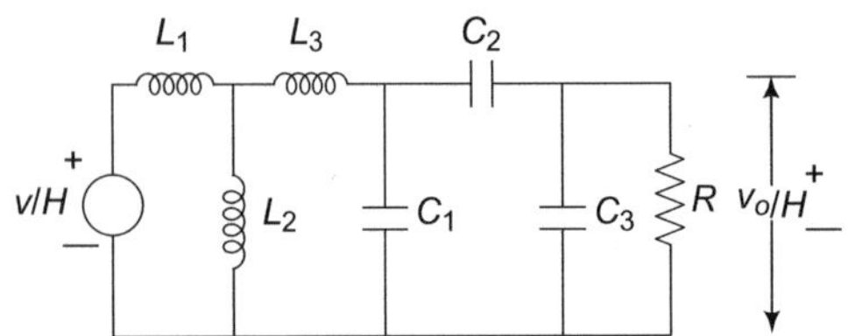

Fig. 8.3 For P.3

P.2. Find at least two circuits using a parallel *LC* and a series *LC* circuit to obtain a band-stop response. Find the poles and zeroes of the transfer function.

P.3. Obtain the differential equation governing the circuit Fig. 8.3.

P.4. Draw as many third-order circuits as possible and comment on the nature of the frequency response.

P.5. Take any one circuit of P.4 and determine its poles and zeros.

References

1. W.H Hayt, Jr, J. Kemmerly, *Engineering Circuit Analysis* (McGraw-Hill, 1978)
2. M.E Van Valkenburg, *Network Analysis* (Prentice Hall of India, 1974)
3. A.B. Carlson, *Circuits* (John Wiley, 1996)
4. D.K. Cheng, *Analysis of Linear Systems* (Addison Wesley, 1959)

9 Appearances Can Be Deceptive: A Circuit Paradox

How can a paradox be deceptive? What appears to be an obvious conclusion may not be correct after all! This is illustrated with the help of a differential amplifier circuit, whose gain is actually half of what it appears to be. This paradox is indeed deceptive. See for yourself and decide if you wish to agree or not.

Keywords
Paradox · DC analysis

The Illusion

Consider the differential amplifier circuit shown in Fig. 9.1, where the symbols v_0–v_4 are used for small signal ac voltages. As has been proved in standard textbooks [1–3], the gain of the internal differential amplifier comprising of matched transistors Q_3 and Q_4 is $A_d = v_0/(v_3 - v_4) \cong - g_m R_c$, where $g_m = g_{m3} = g_{m4}$. Since Q_1 as well as Q_2 basically act as emitter followers, it is logical to assume that $v_3 \cong v_1$ and $v_4 \cong v_2$ so that the overall gain of the circuit is the same as A_d. However convincing this logic may be, things turn out to be quite different in practice. In fact, the differential gain of the overall amplifier is half of A_d! Let us see how.

AC Analysis

First, recall that the gain of the simple follower (all gains are referred to the mid-band situation, of course) in Fig. 9.2, ignoring the effects of r_x and r_0 in the hybrid-π equivalent circuit, is [1–3].

$$\frac{v_0}{v_i} = \frac{(\beta+1)R_E}{r_\pi + (\beta+1)R_E} \tag{9.1}$$

$$\cong 1, \text{ if } \quad r_p \ll (\beta+1)R_E \tag{9.2}$$

The emitter follower Q_1 has a load of $r_{\pi 3}$ in the differential mode (in this mode the node E acts as virtual ac ground); hence its gain is

$$\frac{v_3}{v_1} = \frac{(\beta_1+1)r_{\pi 3}}{r_{\pi 1} + (\beta_1+1)r_{\pi 3}}, \tag{9.3}$$

where the subscripts on the parameters refer to the corresponding transistors. Now

Source: S. C. Dutta Roy, "Appearances can be Deceptive: A Circuit Paradox," *Students' Journal of the IETE*, vol. 37, pp. 79–81, July–September 1996.

S. C. Dutta Roy, *Circuits, Systems and Signal Processing*,
https://doi.org/10.1007/978-981-10-6919-2_9

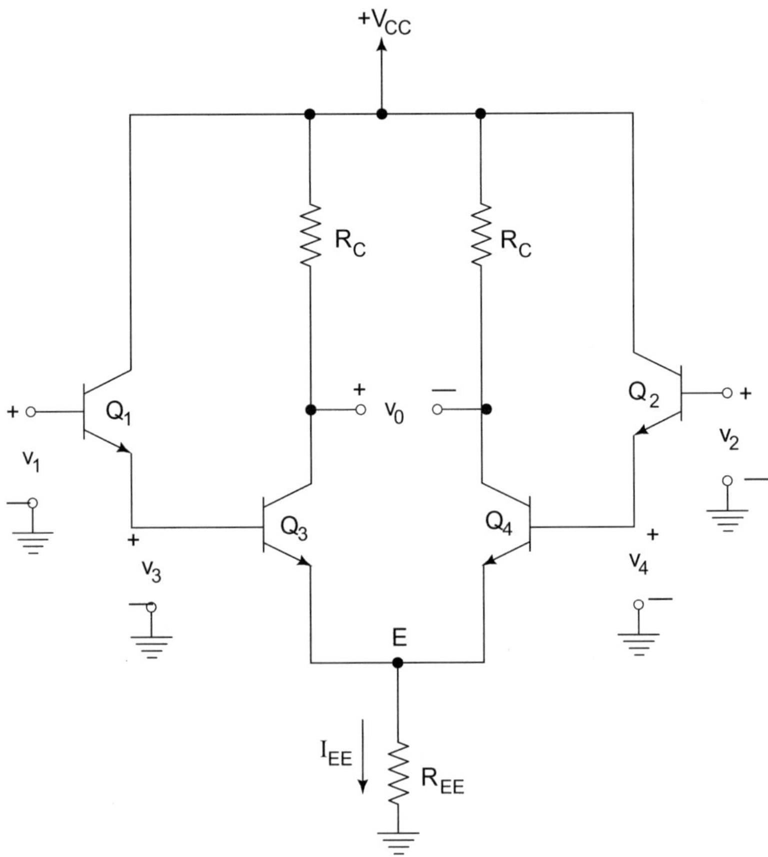

Fig. 9.1 The circuit under consideration

$$r_{\pi 1} = \frac{\beta_1}{g_{m1}} = \frac{\beta_1 V_T}{I_{C1}} = \frac{\beta_1 V_T}{\beta_1 I_{B1}} = \frac{V_T}{I_{B1}} \tag{9.4}$$

and, similarly,

$$r_{\pi 3} = \frac{V_T}{I_{B3}} = \frac{V_T}{I_{E1}} = \frac{V_T}{(\beta_1 + 1) I_{B1}} \tag{9.5}$$

$$= \frac{r_{\pi 1}}{\beta_1 + 1}, \tag{9.6}$$

where V_T stands for the thermal voltage ($kT/q \cong$ 25 mV at room temperature) and capital I with capital subscript stands for DC current. Substituting Eq. 9.6 in Eq. 9.3 gives

$$\frac{v_3}{v_1} = \frac{1}{2} \tag{9.7}$$

Hence, each of the emitter followers Q_1 and Q_2 has a gain of 1/2 instead of 1, and the actual differential gain of the overall circuit is

$$\frac{v_0}{v_1 - v_2} = \frac{v_0}{2(v_3 - v_4)} = -\frac{g_m R_c}{2}, \tag{9.8}$$

where $g_m = g_{m3} = g_{m4}$, as mentioned earlier.

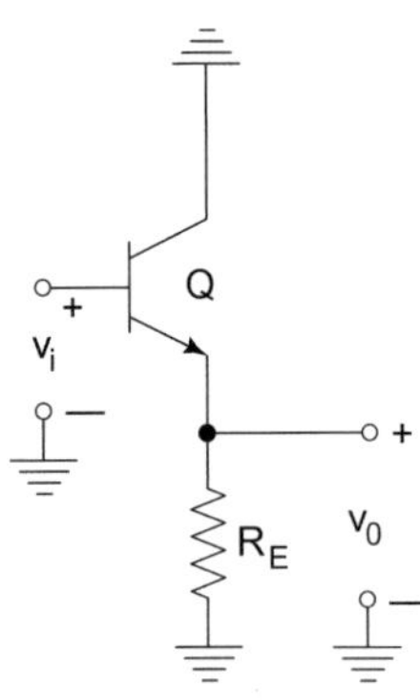

Fig. 9.2 AC equivalent of a simple emitter follower circuit

DC Analysis

The result obtained above can be corroborated by analysing the DC characteristics of the circuit. We wish to establish a relationship between

$$V_0 \triangleq V_{C3} - V_{C4} = (V_{CC} - I_{C3}R_C) - (V_{CC} - I_{C4}R_C) = (I_{C4} - I_{C3})R_C \quad (9.9)$$

and

$$V_i \triangleq V_{B1} - V_{B2} \quad (9.10)$$

Recall the basic current–voltage relationship of an active transistor

$$I_c \cong I_s \exp(V_{BE}/V_T) \quad (9.11)$$

Now,

$$\begin{aligned} V_i &= V_{BE1} + V_{BE3} + V_E - (V_{BE2} + V_{BE4} + V_E) \\ &= (V_{BE1} - V_{BE2}) + (V_{BE3} - V_{BE4}) \end{aligned} \quad (9.12)$$

Also, assuming $I_{S1} = I_{S2} = I_{S3} = I_{S4} = I_S$, we have

$$V_{BE3} = V_T \ln \frac{I_{C3}}{I_S} = V_T \ln = \frac{\beta_3 I_{B3}}{I_S}$$

$$= V_T \ln \frac{\beta_3 I_{E1}}{I_S} = V_T \ln = \frac{\beta_3(\beta_1 + 1)I_{C1}}{\beta_1 I_S} \quad (9.13)$$

Similarly,

$$V_{BE4} = V_T \ln \frac{\beta_4(\beta_2 + 1)I_{C2}}{\beta_2 I_S} \quad (9.14)$$

Thus,

$$V_{BE3} - V_{BE4} = V_T \ln\left[\frac{\beta_3(\beta_1 + 1)I_{C1}}{\beta_1} \frac{\beta_2}{\beta_4(\beta_2 + 1)/I_{C2}}\right] \quad (9.15)$$

If the transistors are matched, as is the case with IC fabrication, then Eq. 9.15 becomes

$$V_{BE3} - V_{BE4} = V_T \ln \frac{I_{C1}}{I_{C2}} = V_{BE1} - V_{BE2} \quad (9.16)$$

From Eqs. 9.12 and 9.16, we get

$$V_i = 2(V_{BE3} - V_{BE4}) = 2V_T \ln \frac{I_{C3}}{I_{C4}} \quad (9.17)$$

Solving for I_{C4}/I_{C3} from Eq. 9.17 gives

$$\frac{I_{C4}}{I_{C3}} = \exp\left(-\frac{V_i}{2V_T}\right) \quad (9.18)$$

so that

$$\frac{I_{C4} - I_{C3}}{I_{C4} + I_{C3}} = \frac{\exp\left(-\frac{V_i}{2V_T}\right) - 1}{\exp\left(-\frac{V_i}{2V_T}\right) + 1} \quad (9.19)$$

Also, note that $I_{C4} + I_{C3} = I_{EE}$ = constant. Hence from Eqs. 9.19 and 9.9, we get

$$V_0 = -I_{EE}R, \tanh \frac{V_i}{4V_T} \quad (9.20)$$

The differential gain, evaluated at $V_1 = 0$, is

$$\left.\frac{dV_0}{dV_i}\right|_{V_i=0} = -I_{EE}R_C\left(\operatorname{sech}^2 \frac{V_i}{4V_T}\right)\left.\frac{1}{4V_T}\right|_{V_i=0}$$

$$= -\frac{I_{EE}R_C}{4V_T} \tag{9.21}$$

For $V_i = 0$, $I_{C3} = I_{C4} = I_{EE}/2$; hence the gain is

$$-\frac{I_{C3}}{2V_T}R_C = -\frac{g_m R_C}{2}, \tag{9.22}$$

where $g_m = g_{m3} = g_{m4}$. This is exactly the same as the result derived under AC analysis.

Problems

P.1. What happens when $R_{EE} \to \infty$ in Fig. 9.1 of text. What about $R_{EE} = \infty$?

P.2. What happens when $R_E \to \infty$ in Fig. 9.2 of text. What about $R_E = \infty$?

P.3. Approximate exp. function in Eq. 9.19 of text by the first two terms and find I_{C4} in terms of I_{C3}.

P.4. What happens when β_1, β_2 & β_3 all $\to \infty$ in Eq. 9.15 of text? Comment on the result.

P.5 What happens when R_E is replaced by a current generator?

Acknowledgements Acknowledgement is due to my students in the EE204 N class, to whom I had posed this paradox as a challenge during the semester commencing January 1996. Special mention must be made of Ankur Srivastava, Atul Saroop and Ram Sadhwani whose enthusiastic participation in the resolution of the paradox made it an enjoyable experience for me.

References

1. S.G. Burns, P.R. Bond, *Principles of Electronic Circuits* (West Publishing Co, St Paul, USA, 1987)
2. A.S. Sedra, K.C. Smith, *Microelectronic Circuits* (Sanders College Publishing, Fort Worth, USA, 1992)
3. J. Millman, A. Grabel, *Microelectronics*, 2nd edn. (McGraw Hill, New York, 1987)

Appearances Can Be Deceptive: An Initial Value Problem

10

An initial value problem is posed and solved in a systematic way, illustrating the fact that what meets the eye may not be the truth!

Keyword
Initial value problem

The Problem

Consider the circuit shown in Fig. 10.1 in which the switch is closed from $t = -\infty$ and is opened at $t = 0$. The problem that is posed here is: what is $i_2(0^-)$ (This forms part of Problem 5.9 in [1])?

One way of looking at the problem is to realize that at $t = 0^-$, L_2 behaves as a short circuit and therefore, there are two short circuits in parallel. Since $i_1(0^-) = V/R$, is $i_2(0^-) = V/(2R)$? The answer, as we shall demonstrate here, is; no, not necessarily. In fact, we show that $i_2(0^-)$ can have an arbitrary value I_2.

Source: S. C. Dutta Roy, "Appearances can be Deceptive: An Initial Value Problem," *IETE Journal of Education*, vol. 45, pp. 31–32, January–March 2004.

Establishing $I_2(0^-)$: One Possibility

To investigate the problem, let us examine how $i_2(0^-)$ can be established in the circuit. One possible way is shown in Fig. 10.2, where the switch is closed at $t = -\infty$ with $i_1(-\infty^-) = 0$ and $i_2(-\infty^-) = I_2$, an arbitrary value. Since for $t > -\infty$, the circuit is equivalent to a series combination of V, R and L_1, i_1 increases exponentially from zero towards the asymptotic value V/R, reachable, at $t = 0$. All this time, can i_2 change from the value I_2? This is not possible, because any change in i_2 requires a corresponding change in the voltage across it (v_2), but since L_2 is short-circuited, v_2 is forced to remain zero. Thus, i_2 remains I_2 throughout, and I_2 can be arbitrary! The current through the short circuit keeps on changing from $-I_2$ at $t = -\infty^-$ to $(V/R) - I_2$ at $t = 0^-$.

Establishing I_2 (0^-): Another Possibility

Figure 10.3 shows another method of establishing $i_2(0^-)$. The switch is shorted at $t = -\infty$ with $i_1(-\infty^-) = i_2(-\infty^-) = V_1/R$. The current i_1 remains constant at V_1/R (why?) and so does i_2 at the same value, the current through the short

S. C. Dutta Roy, *Circuits, Systems and Signal Processing*,
https://doi.org/10.1007/978-981-10-6919-2_10

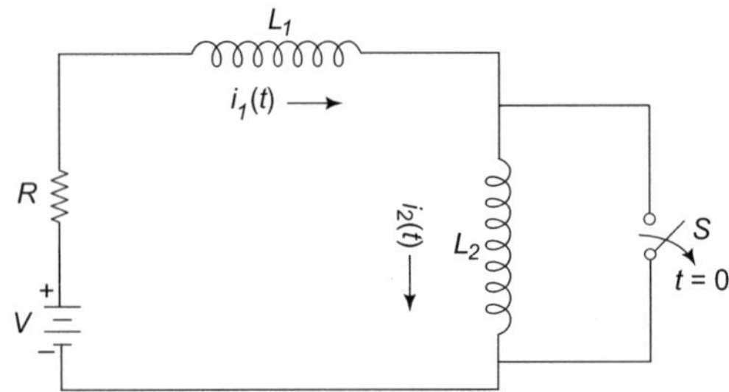

Fig. 10.1 The circuit under consideration

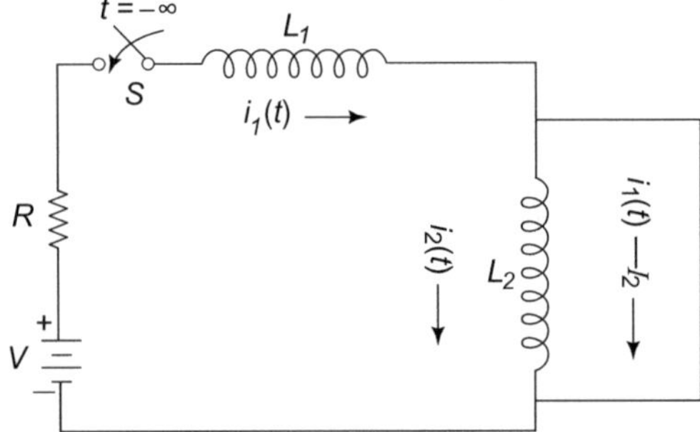

Fig. 10.2 One possible way of establishing $i_2\ (0^-)$

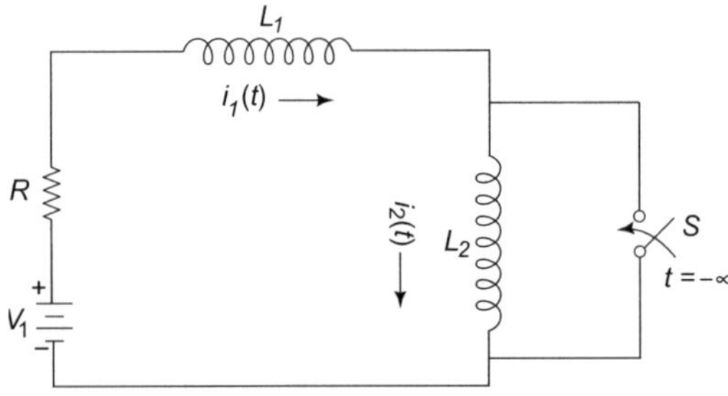

Fig. 10.3 Another possible way of establishing $i_2(0^-)$

circuit being zero all the time. Since V_1 is arbitrary, we see that $i_2\ (0^-)$ can also be arbitrary.

Solve the Circuit

We have established that $i_2(0^-)$ in the circuit of Fig. 10.1 can have an arbitrary value I_2. Hence, for solving for i_1 in this circuit, I_2 also needs to be specified (in [1], I_2 is not specified). The solution proceeds by invoking the principle of conservation of flux, viz.

$$L_1 i_1(0^-) + L_2 i_2(0^-) = L_1 i_1(0^+) + L_2 i_2(0^+). \tag{10.1}$$

Since $i_1(0^+) = i_2(0^+)$, $i_1(0^-) = V/R$ and $i_2(0^-) = I_2$, we get

$$i_1(0^+) = (VL_1/R + L_2 I_2)/(L_1 + L_2). \tag{10.2}$$

For $t \geq 0^+$, $i_1 = i_2$ so that differential equation governing the circuit becomes

$$Ri_1 + (L_1 + L_2)\,(\mathrm{d}i_1/\mathrm{d}t) = V. \tag{10.3}$$

This equation has a solution of the form

$$i_1(t) = A + B\mathrm{e}^{-Rt/(L_1+L_2)}. \tag{10.4}$$

Using the boundary conditions given by Eq. 10.2 and $i_1(\infty) = V/R$, one can find out the constants A and B. The final result is:

$$i_1(t) = \frac{V}{R} + \left(\frac{L_2 I_2 - VL_2/R}{L_1 + L_2}\right)\mathrm{e}^{-\frac{Rt}{L_1+L_2}} \tag{10.5}$$

Note that the circuit of Fig. 10.1 is an example of a situation where, upon switching, inductor currents are not continuous, i.e. $i_1(0^-) \neq i_1(0^+)$ and $i_2(0^-) \neq i_2(0^+)$.

Problems

P.1. What happens when the switch in Fig. 10.1 of text is closed for $t = 0$ to $t = T_0$ and then opened?

P.2. What happens when the switch in Fig. 10.1 is shifted to be across L_1?

P.3. Determine the response of the circuit of Fig. 10.1 when L_2 is replaced by a capacitor C.

P.4. Same when C of P.3 is shifted to be across L_1.

P.5. Same when C as well as the switch are shifted to be across L_1.

Reference

1. F F Kuo, *Network Analysis and Synthesis* (Wiley, 1966), p. 129

11 Resonance

In this chapter, we discuss the basic concepts of resonance in electrical circuits, and its characterization, and illustrate its application by an example. Several problems have been added at the end for the students to work out. Do work them out.

Keywords
Resonance · Figure of merit for coils and capacitors · Q · Series resonance · Parallel resonance · Impedance · Admittance Bandwidth

A one-port network containing resistors and inductors has the property that the current lags behind the voltage. For a resistor–capacitor one-port, on the other hand, current leads the voltage. If a one-port contains inductors as well as capacitors in addition to the inevitable dissipative element, *viz.* resistors, then the circuit may be inductive at some frequencies, capacitive at some other frequencies and, most interestingly, purely resistive at one or more frequencies. In the last situation, the current obviously is in phase with the voltage and the power factor in unity. It is also clear that this situation can arise only when the inductive and capacitive reactances (or susceptances) cancel each other. Such a situation is known as resonance, which plays a very important role in impedance matching, filtering, measurements, and many other applications.

Two types of resonance are distinguished, viz. series and parallel. A cancellation of reactances in series is referred to as series resonance, while if susceptances in parallel cancel, we call it parallel resonance. In either case, the condition of resonance is usually associated with extremum (maximum or minimum) of impedance or admittance magnitude, and voltage/current.

Q: A Figure of Merit for Coils and Capacitors

Dissipation, as already mentioned, is an inevitable phenomenon in nature, in general, and in electric circuits, in particular. In other words, you cannot make a pure inductor or capacitor in practice; there will always be some losses. The less the losses are, the better is the reactive element. A figure of merit, Q, is defined for reactive

Source: S. C. Dutta Roy, "Resonance," *Students' Journal of the IETE*, vol. 36, pp. 169–178, October–December 1995.

S. C. Dutta Roy, *Circuits, Systems and Signal Processing*,
https://doi.org/10.1007/978-981-10-6919-2_11

elements in terms of energy stored and dissipated in it, as follows:

$$Q = 2\pi \frac{\text{maximum energy stored per cycle}}{\text{energy disspated per cycle}}. \tag{11.1}$$

Consider a coil having an inductance L and a series resistance R, through which a current[1] $i(t) = I \sin \omega t$ flows. Then, the maximum energy stored in a cycle is $(1/2)LI_m^2$, while the average power dissipated is $(1/2)RI_m^2$. But power is energy per unit time, so that energy dissipated per cycle is (average power) × (time duration of one cycle) = $(1/2)RI_m^2 f$. Thus, the Q of an inductor is

$$Q = \frac{2\pi f L}{R} = \frac{\omega L}{R}. \tag{11.2}$$

Similarly, for a capacitor, usually represented by a pure capacitance C in parallel with a resistance R, you can show (Problem 1) that the Q is given by

$$Q = \omega CR. \tag{11.3}$$

As you can see from Eqs. 11.2 and 11.3, Q is a linear function of frequency. While this is very nearly the case in the case of an air capacitor, it is not so in the case of a coil or other types of capacitors. For a coil, the Q usually increases at low frequencies, attains a maximum at some frequency, and then decreases. This happens because of skin effect, due to which the resistance increases with frequency, and at sufficiently high frequencies, this increase is at a more rapid rate than the linear increase of frequency.

Series Resonance

Consider the circuit of Fig. 11.1 in which a series RLC circuit is excited by a sinusoidal voltage generator of variable frequency, represented by

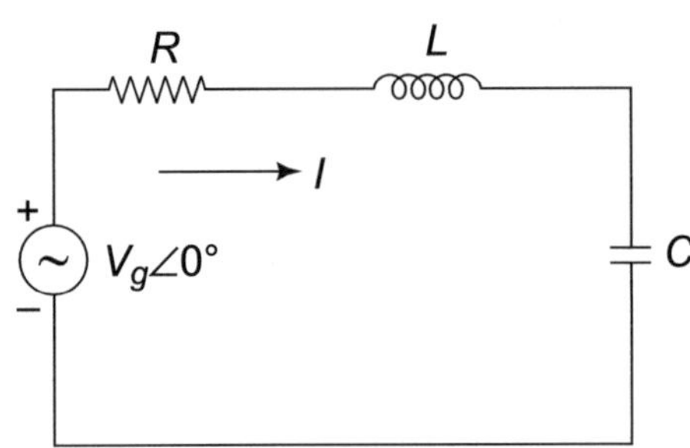

Fig. 11.1 A series *RLC* circuit

the phasor $V_g \angle 0°$. The current through the circuit, represented by the phasor $I = |I|\angle\theta$ is given by

$$I = \frac{V_g}{R + j\omega L + \frac{1}{j\omega C}}, \tag{11.4}$$

where the denominator represents the total impedance $Z(j\omega)$. Note that R includes the generator internal resistance, coil losses and any external resistance which may have been inserted. Taking the magnitude and phase of Eq. 11.4, we have

$$|I| = \frac{V_g}{\sqrt{R^2 + \left(\omega L - \frac{1}{\omega C}\right)^2}}. \tag{11.5}$$

$$\theta = -\tan^{-1} \frac{\omega L - \frac{1}{\omega C}}{R} \tag{11.6}$$

Note that at $\omega = 0$, i.e. *DC*, at which the capacitor acts as an open circuit, the current is zero; the same is true at $\omega = \infty$, at which the inductor acts as an open circuit. In between, the current will show a maximum at the frequency ω_0 at which the second term in the denominator of Eq. 11.5 vanishes, i.e.

$$\omega_o L = \frac{1}{\omega_o C} \text{ or } \omega_o = \frac{1}{\sqrt{LC}} \tag{11.7}$$

The maximum value of the current, denoted by I_o is

$$I_o = \frac{V_g}{R} \tag{11.8}$$

A sketch of $|I|$ versus ω is shown in Fig. 11.2a, while Fig. 11.2b shows a sketch of

[1]As a matter of notations, we shall use small i or v for instantaneous value, subscript m for maximum value, capital I or V for phasor representation, and $|I|$ or $|V|$ for the rms value.

the corresponding phase θ. It is easy to argue from Eq. 11.6 that the phase is $+\pi/2$ at *dc*, decreasing to 0° at $\omega = \omega_o$ and then to $-\pi/2$ as $\omega \to \infty$. This is a reflection of the fact that the impedance $Z(j\omega)$ is capacitive for $\omega < \omega_o$ purely resistive at $\omega = \omega_o$ and inductive for $\omega > \omega_o$. At the frequency ω_o, the current and voltage are in phase, and by definition, resonance occurs.

Taking the current as the reference, the phasor diagram for the series resonant circuit is shown in Fig. 11.3 for three frequencies, viz. (i) $\omega < \omega_o$, (ii) $\omega = \omega_o$ and (iii) $\omega > \omega_o$. In each of these diagrams,

$$V_R = IR,\ V_L = j\omega LI \text{ and } V_C = \frac{I}{j\omega C} \tag{11.9}$$

At $\omega < \omega_o, 1/\omega C > \omega L$ so that $|V_C| > |V_L|$. Consequently, the resultant of V_R and $V_C + V_L$ which gives V_g lagging behind *I*, i.e. *I* leads V_g. At $\omega > \omega_o$ the reverse is the case, while at $\omega = \omega_o$, $V_C = -V_L$ and V_g and V_R are the same.

It is interesting to observe that at resonance,

$$V_L = j\omega_o LI_o = j\frac{\omega_o L}{R}V_g = jQV_g. \tag{11.10}$$

where Q refers to that of the series *RL* combination at resonance. Similarly, at resonance

$$V_C = \frac{I_o}{j\omega_o C} = -j\omega_o LI_o = -jQV_g, \tag{11.11}$$

where use has been made of Eq. 11.7.[2] In most practical situations, Q is required to be greater than unity; thus the voltage across the capacitor or the inductor at resonance is greater than the input voltage, i.e. the resonant circuit can be used as a voltage magnifier.

[2]Equation 11.11 can also be written as $V_C = V_g/(j_o\, CR)$. Obviously, the Q of the series *RC* circuit at resonance is 1/($_o$ *CR*), in contrast to Eq. 11.3 for a parallel *RC* circuit (Problem 2).

Parallel Resonance

Figure 11.4 shows a parallel RLC circuit excited by a sinusoidal current generator of varying frequency. The voltage across the circuit is given by the product of I_g and total impedance, i.e.

$$V = \frac{I_g}{G + j\omega C + \frac{1}{j\omega L}}, \tag{11.12}$$

where $G = 1/R$. Note the similarity of this with Eq. 11.4; this is, in fact, expected because of the duality of the circuits of Figs. 11.1 and 11.4. Thus, all the results derived earlier will apply here also. For completeness, we summarize below the main results:

(i) If $I_g = I_g \angle 0°$, $V = |V| \angle \phi$, then

$$|V| = \frac{I_g}{\sqrt{G^2 + \left(\omega C - \frac{1}{\omega L}\right)^2}} \tag{11.13}$$

$$\phi = -\tan^{-1}\frac{\omega C - \frac{1}{\omega L}}{G} \tag{11.14}$$

(ii) $|V|$ is a maximum, equal to $I_g R$ at the resonance frequency ω_o given by Eq. 11.7

(iii) V leads I_g for $\omega < \omega_o$ lags I_g for $\omega > \omega_o$ and is in phase with I_g at $\omega = \omega_o$

(iv) At resonance, $I_L = -jQ\, I_g$ and $I_c = jQ\, I_g$, where $Q = \frac{R}{\omega_o L} = \omega_o CR$ (Problem 3).

The phasor diagram can be easily constructed, and is left to you as an exercise (Problem 4).

Impedance/Admittance Variation with Frequency: Universal Resonance Curves

From the similarity between series and parallel resonance equations, it follows that the two cases can be described by a single equation of the form

$$H(j\omega) = \frac{1}{1 + j\left(\omega\beta - \frac{1}{\omega\gamma}\right)}, \tag{11.15}$$

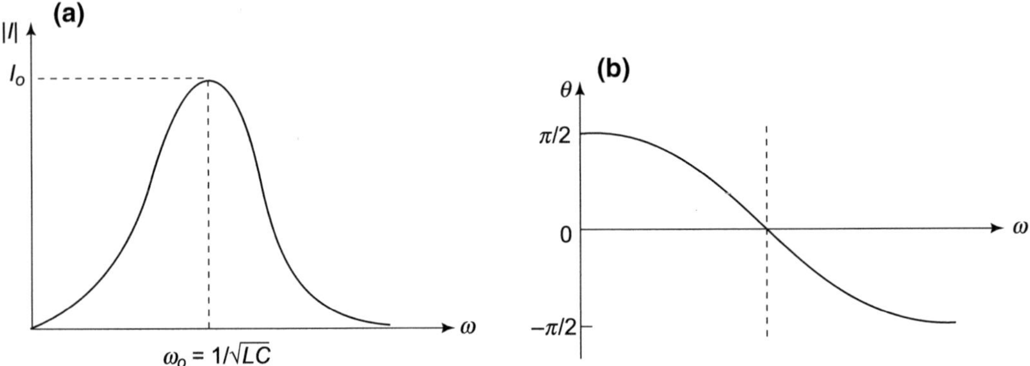

Fig. 11.2 Variation of **a** magnitude and **b** phase of current in a series RLC circuit

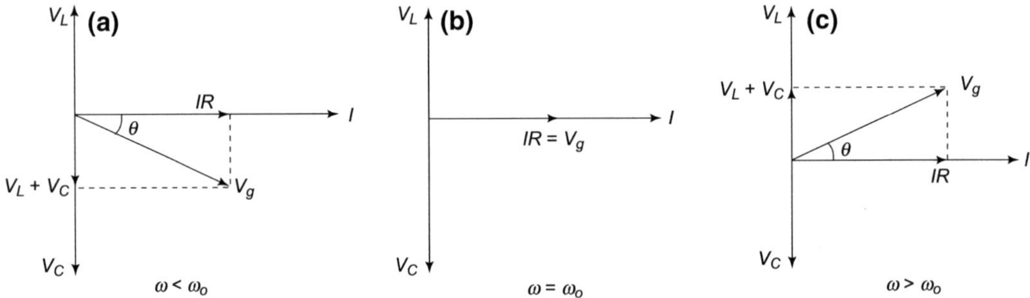

Fig. 11.3 Phasor diagrams for the series RLC circuit at frequencies **a** below, **b** at and **c** above the resonance frequency

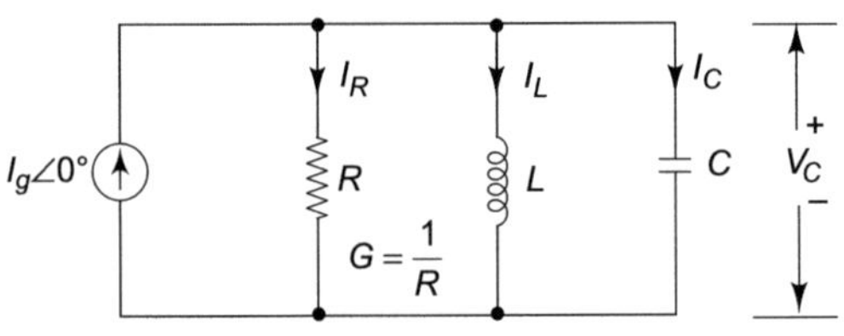

Fig. 11.4 A parallel RLC circuit

where the interpretation of the symbols for the two kinds of resonance are given in Table 11.1. Equation 11.15 can be written in a normalized form as follows:

$$A(j\omega) = \alpha H(j\omega) = \frac{1}{\alpha + \frac{j}{\alpha}\left(\omega\beta - \frac{1}{\omega\gamma}\right)}. \quad (11.16)$$

The interpretation of $A(j\omega)$ is also given in Table 11.1, where the subscript zero refers to the value at resonance.

Note that in either series or parallel resonance, the resonance frequency is $\omega_o = 1/\sqrt{\beta\gamma}$;

$$\omega\beta = \omega\sqrt{\beta\gamma}\sqrt{\frac{\beta}{\gamma}} = \sqrt{\frac{\beta}{\gamma}}\frac{\omega}{\omega_o} \quad (11.17)$$

Table 11.1 Interpretation of the symbols in Eq. 11.15 for series and parallel resonances

Type of resonance	$H(j\omega)$	α	β	γ	$A(j\omega)$
Series	$Y(j\omega)$	R	L	C	$\frac{Y(j\omega)}{Y_o} = \frac{I(j\omega)}{I_o}$
Parallel	$Z(j\omega)$	G	C	L	$\frac{Z(j\omega)}{Z_o} = \frac{V(j\omega)}{V_o}$

and similarly,

$$\omega\gamma = \sqrt{\frac{\gamma}{\beta}} \frac{\omega}{\omega_o} \tag{11.18}$$

introducing these in Eq. 11.16, we get

$$A(j\omega) = \frac{1}{1 + \frac{j}{\alpha}\sqrt{\frac{\beta}{\gamma}}\left(\frac{\omega}{\omega_o} - \frac{\omega_o}{\omega}\right)}. \tag{11.19}$$

Now, for series resonance,

$$\frac{1}{\alpha}\sqrt{\frac{\beta}{\gamma}} = \frac{1}{R}\sqrt{\frac{L}{C}} = \frac{L}{R\sqrt{LC}} = \frac{\omega_o L}{R} = Q. \tag{11.20}$$

Similarly for parallel resonance,

$$\frac{1}{\alpha}\sqrt{\frac{\beta}{\gamma}} = \frac{R}{\omega_o L} = Q. \tag{11.21}$$

Thus, Eq. 11.19 becomes

$$A(j\omega) = \frac{1}{1 + jQ\left(\frac{\omega}{\omega_o} - \frac{\omega_o}{\omega}\right)}. \tag{11.22}$$

This expression is independent of the type of resonance or the actual element values used in the circuit, and is, therefore, applicable to all resonant circuits of the form of Fig. 11.1 or Fig. 11.4. A family of curves can be drawn for the variation of $|A|$ and $\angle A$ with the normalized frequency $x = \frac{\omega}{\omega_o}$, taking Q as a parameter. Because of universal applicability, these are called universal resonance curves.

A more useful form of the universal resonance curve can be obtained if the behaviour at or near resonance is only of concern. Define the fractional deviation of source frequency from the resonance frequency as

$$\delta = \frac{\omega - \omega_o}{\omega_o}, \tag{11.23}$$

i.e.

$$\omega = \omega_o(1 + \delta) \tag{11.24}$$

Then Eq. 11.22 can be written as

$$A(j\omega) = \frac{1}{1 + jQ\delta(2 + \delta)/(1 + \delta)}. \tag{11.25}$$

Universal resonance curves are obtained by plotting $|A|$ and $\angle A$ versus δ with Q as a parameter. Note that $\delta = 0$ corresponds to $\omega = \omega_0$ or $x = \omega/\omega_0 = 1$ (Fig. 11.5).

Equation 11.25 is an exact expression; for $\delta \ll 1$, it can be approximated by

$$A(j\omega) \cong \frac{1}{1 + j2Q\delta}. \tag{11.26}$$

Bandwidth of Resonance

The sharpness of resonance, as we shall see, is determined by Q. A measure of the sharpness is the bandwidth, defined as the band of frequencies around ω_o at which the magnitude of $A(j\omega)$ is no less than $1/\sqrt{2}$

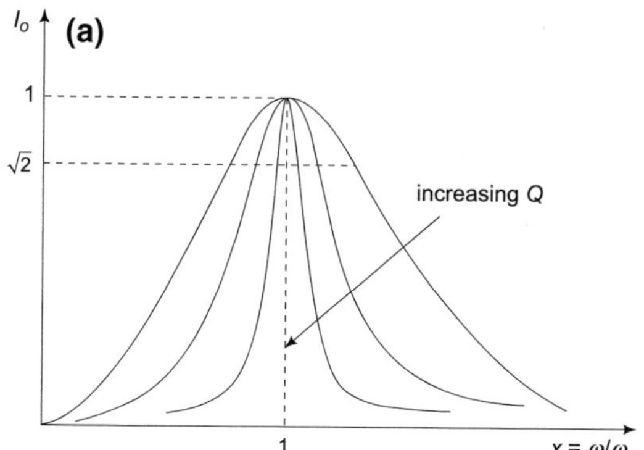

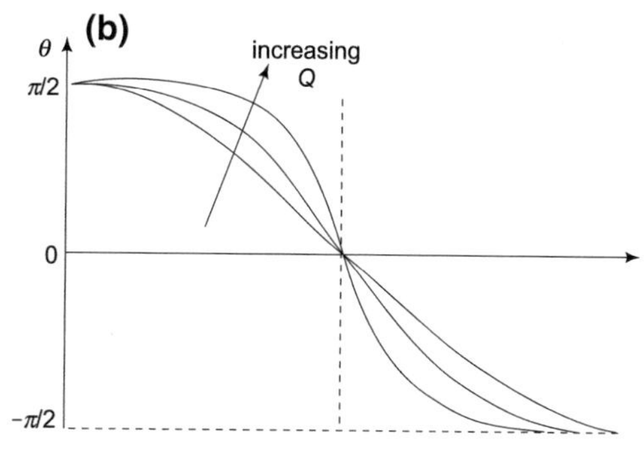

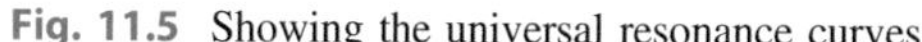
Fig. 11.5 Showing the universal resonance curves

Figure 11.5 shows the bandwidth, B, as

$$B = \omega_2 = \omega_1, \tag{11.27}$$

where ω_2 is called the 'upper half power frequency' and ω_1 is called the 'lower half power frequency'. The nomenclature 'half power' is derived from the following consideration. Consider, for example series resonance in which $A(j\omega) = I(j\omega)/I_o$. The power dissipated in R at $\omega_{1,2}$ will be $\left(I_o/\sqrt{2}\right)^2 R = I_o^2 R/2$, which is half of that dissipated at ω_o. A similar interpretation can be derived for parallel resonance.

To determine $\omega_{1,2}$ turn to Eq. 11.22 and notice that $|A(j\omega)| = \frac{1}{\sqrt{2}}$ implies

$$Q\left(\frac{\omega}{\omega_o} - \frac{\omega_o}{\omega}\right) = \pm 1 \tag{11.28}$$

Solving 11.28, we get four solutions for ω, two of which are negative while the other two are positive. Obviously, the latter are the acceptable ones. These are given by (Problem 7)

$$\omega_{1,2} = \omega_o\left[\sqrt{1 + \left(\frac{1}{2Q}\right)^2} \pm \frac{1}{2Q}\right]. \tag{11.29}$$

Thus

$$B = \omega_2 - \omega_1 = \frac{\omega_o}{Q}. \tag{11.30}$$

Thus, higher Q leads to a lower B and hence a sharper resonance.

In a general selective response curve of the type of Fig. 11.5, one often defines the 'sharpness of resonance' or 'selectivity' as the ratio (frequency of maximum response)/(bandwidth). Applied to the two cases of resonance discussed here, this quantity, as is evident from Eq. 11.3, is identical with Q.

Also note, from Eq. 11.29, that

$$\omega_1\omega_2 = \omega_0^2 \tag{11.31}$$

which shows that the resonance curve is geometrically symmetrical about ω_o, i.e. the response at $\omega = \omega_o/a$ is the same as that at $a\,\omega_o$. This is also obvious from Eq. 11.22.

Other Types of Resonant Circuits

The parallel resonant circuit of Fig. 11.4 cannot be made in practice because you cannot make a pure inductance (in contrast, almost pure capacitances can be readily obtained). A practical circuit is shown in Fig. 11.6, where R is the winding resistance of the coil, usually much less than the inductive reactance ωL at or around resonance. If this is true, then the total impedance of the circuit is

$$\begin{aligned} Z(j\omega) &= \frac{1}{j\omega C + \frac{1}{R + j\omega L}} \\ &= \frac{R + j\omega L}{1 + j\omega RC - \omega^2 LC} \\ &\cong \frac{j\omega L}{1 + j\omega RC - \omega^2 LC}. \end{aligned} \tag{11.32}$$

The last expression can be written as

$$Z(j\omega) = \frac{1}{\frac{RC}{L} + j\omega C + \frac{1}{j\omega L}} \tag{11.33}$$

which is of the form of Eq. 11.15, and hence all the results we have derived will apply. Notice that at resonance, the impedance attains a maximum value, given by

$$\begin{aligned} Z_o &\cong \frac{L}{RC} = \frac{\omega_o L}{R}\frac{1}{\omega_o C} \\ &= \frac{(\omega_o L)^2}{R} = R\left(\frac{\omega_o L}{R}\right)^2 = RQ^2, \end{aligned} \tag{11.34}$$

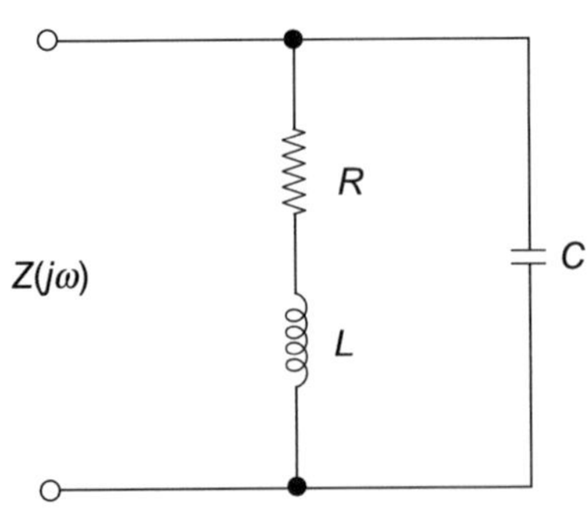

Fig. 11.6 A practical parallel resonant circuit

where Q is again $\omega_o L/R$ because R and L are in series. The condition $\omega L \gg R$ assumed for the approximations is usually taken to mean $Q \geq 10$, as a rule of thumb.

Another interesting resonant circuit is shown in Fig. 11.7, which has the property that if $R_1 = R_2 = \sqrt{L/C}$, then the impedance is R at all frequencies, i.e. this becomes an all-pass resonant circuit (Problem 8).

An Example

In concluding this discussion, we illustrate the analysis and design of a resonant circuit of practical importance.

Example 1 The voltage induced in a radio receiver aerial may be approximated by a voltage generator of internal resistance 2000 Ω and an emf containing equal amplitudes of the frequencies 1000 and 1050 kHz. It is desired to tune the receiver to the first frequency with the second frequency discriminated at least by a factor of 2. Design a resonant circuit of the form of Fig. 11.6 for the purpose.

Solution: The overall circuit is shown in Fig. 11.8. By superposition, the voltage V will be the sum of the voltages developed due to the two sources applied independently. Let us, therefore, consider the response due to a source $V_g \angle 0°$ of frequency ω; then

$$V = \frac{V_g Z(j\omega)}{R_g + Z(j\omega)} = \frac{V_g}{1 + R_g Y(j\omega)}, \qquad (11.35)$$

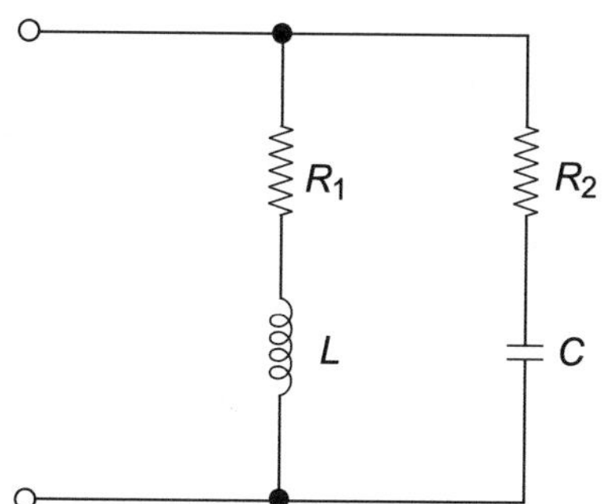

Fig. 11.7 A resonant circuit having resistances in both parallel branches

where

$$Z(j\omega) = \frac{1}{j\omega C + \frac{1}{R + j\omega L}} = \frac{1}{Y(j\omega)}. \qquad (11.36)$$

Winding a coil of $Q \geq 10$ at 1000 kHz does not pose a problem at all; thus from the results derived under 'other types of resonant circuits' discussed earlier,

$$\omega_o = 2\pi \times 10^6 \cong \frac{1}{\sqrt{LC}} \qquad (11.37)$$

and

$$Z(j\omega_o) = Z_o \cong \frac{L}{RC} = RQ^2 \qquad (11.38)$$

The fractional deviation in frequency corresponding to $\omega_2 = 2\pi \times 1050 \times 10^3$ r/s is

$$\delta = \frac{1050 - 1000}{1000} = \frac{50}{1000} = 0.05 \ll 1, \qquad (11.39)$$

so that we can apply Eq. 11.26. Thus

$$\frac{Z(j\omega_2)}{Z_o} = \frac{1}{1 + j0.1Q}. \qquad (11.40)$$

The condition of the problem demands that

$$\left| \frac{V_g}{1 + R_g Y_o} + \frac{V_g}{1 + R_g Y(j\omega_2)} \right| \geq 2. \qquad (11.41)$$

Combining Eqs. 11.40 and 11.41 and simplifying we get

$$\frac{0.1Q}{1 + Z_o/R_g} \geq \sqrt{3}. \qquad (11.42)$$

Let

$$\frac{0.1Q}{1 + Z_o/R_g} = 2. \qquad (11.43)$$

Then

$$\frac{Z_o}{R_g} = .05\,Q - 1. \qquad (11.44)$$

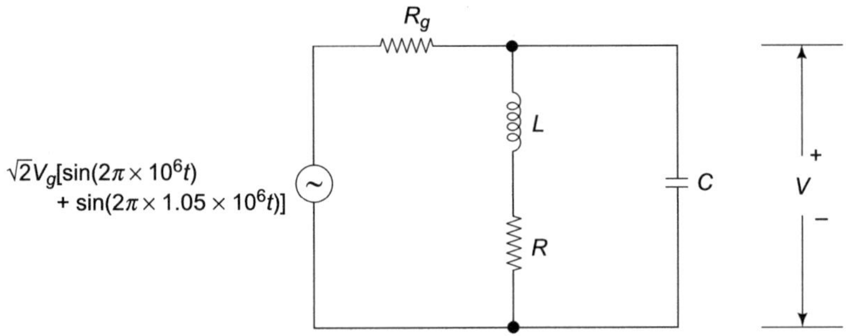

Fig. 11.8 Circuit for Example 1

Thus, we must have $Q > 20$. Let $Q = 40$; then

$$Z_o = R_g = 2000\,\Omega \quad (11.45)$$

From Eq. 11.38, therefore

$$R = Z_o/Q^2 = 2000/16,000 = 1.25\,\Omega \quad (11.46)$$

Also, from $Q =$ $Q = \omega_o L/R$, we have

$$L = \frac{RQ}{\omega_o} = \frac{1.25 \times 40}{2\pi \times 10^6}\text{H} = \frac{25}{\pi}\mu\text{H}. \quad (11.47)$$

Hence, finally, from Eq. 11.37,

$$C = \frac{1}{\omega_o^2 L} = \frac{1}{4\pi^2 \times 10^{12} \times \frac{25}{\pi} \times 10^{-6}}\text{F}$$
$$= \frac{1}{100\pi}\mu\text{F} \quad (11.48)$$

and the design is complete.

Some Problems

If you have understood this chapter, then you should be able to work out the following problems. Try them.

1. Show that for a capacitor, represented by an equivalent circuit consisting of a pure capacitance C in parallel with a resistance R, the Q is given by $Q = \omega CR$.
2. Show that for a series RC circuit, the Q is given by $Q = 1/(\omega CR)$.
3. Show that for a parallel RL circuit, the Q is given by $Q = R/(\omega L)$.
4. With V as the reference phasor, draw the phasor diagrams for the circuit of Fig. 11.4 for $\omega < \omega_o$, $\omega = \omega_o$ and $\omega > \omega_o$ where $\omega_o = 1/\sqrt{LC}$.
5. In a series resonant circuit, derive an expression for the voltage across the capacitor. Find the frequency at which this voltage is a maximum; find also this maximum value.
6. Derive an expression for the bandwidth on the basis of expression Eq. 11.26.
7. Verify Eq. 11.29.
8. Analyze the circuit of Fig. 11.7 and derive the conditions for all-pass resonance.
9. A medium wave broadcast receiver spans the range 570–1560 kHz, with tuning accomplished by a series resonant circuit using an air variable capacitor, ranging from 3 to 500 pF. What value of inductance is needed? At 570 kHz, the capacitor voltage is desired to be 10 times the signal picked up by the aerial. What is the total resistance in the tuned circuit? What will be the signal multiplication factor at 1560 kHz? What are the bandwidths of the circuit when tuned at 570 and 1560 kHz?
10. A signal generator produces a fundamental at 1 kHz of 1 V amplitude and its second and third harmonics at 0.5 and 0.3 V amplitudes respectively. It is required to suppress each harmonic to less than 1% of the fundamental. Design a suitable circuit for the purpose.
11. Determine the frequency of resonance for the circuit of Fig. 11.7 exactly, and the value of the impedance at resonance. Also find the condition for maximum impedance.

12. A 1 V, 1 MHz, 2500 Ω internal resistance source is to deliver maximum power to a load of 1 Ω. Explain how this can be achieved using resonance.

Bibliography

1. F.F. Kuo, *Network Analysis and Synthesis* (Wiley, New York, 1966)

12 The Many Faces of the Single-Tuned Circuit

It is shown that the simple single-tuned circuit is capable of performing a variety of filtering functions and that it can be analyzed graphically for obtaining the relevant performance parameters.

Keywords
Single-turned circuit · Low-pass · High-pass
Band-pass

Figure 12.1 shows the simple single-tuned *RLC* series circuit under consideration. In the usual textbooks, the circuit is analyzed for its behaviour at and near resonance, where resonance implies the in-phase condition of V_1 and I. That it is capable of performing a variety of filtering functions, depending on how the output is chosen, is not generally discussed. Also, except for Kuo [1] who treated one of the configurations graphically, it is hard to find a reference in which graphical construction is used to find the major performance indices of the circuit. In this chapter, we bring out the versatility of the circuit and extend the graphical analysis of [1] to other configurations.

Source: S. C. Dutta Roy, "The Many Faces of the Single Tuned Circuit," *IETE Journal of Education*, vol. 41, pp. 101–104, July-December 2000.

S. C. Dutta Roy, *Circuits, Systems and Signal Processing*,
https://doi.org/10.1007/978-981-10-6919-2_12

Notations: First Things First

In the analysis of the various configurations of Fig. 12.1, we shall adopt the following notations, which are more or less standard, and illustrated in Fig. 12.2.

p_1, p_1^* = complex conjugate poles of the network function
$-\alpha = -R/(2L)$ = real part of either pole
$\beta = \{[1/(LC)] - [R^2/(4L^2)]\}^{1/2}$ = imaginary part of either pole
$\zeta = (R/2)(L/C)^{1/2}$ = damping factor
$\theta = \tan^{-1}(\beta/\alpha) = \cos^{-1}\zeta$
$\omega_n = 1/(LC)^{1/2} = (\alpha^2 + \beta^2)^{1/2}$ = undamped natural frequency
ω_m = frequency of maximum response
$M_1(\omega)$, $M_2(\omega)$ = vectors drawn from the poles to an arbitrary point $j\omega$
ψ = angle between $M_1(\omega)$ and $M_2(\omega)$
$\omega_{2,1}$ = upper and lower 3-dB cutoff frequencies

* denotes complex conjugate

The Possible Configurations

The possible configurations in which the circuit of Fig. 12.1 can be used differ from each other in the location of the output and are shown in Fig. 12.3. Let H_x denote the transfer function V_2/V_1 of

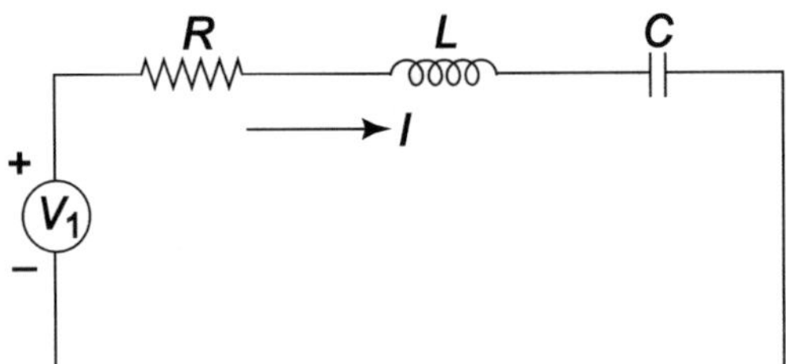

Fig. 12.1 The circuit under consideration

Fig. 12.3, where x = *a, b, c, d, e* or *f*. Then, the variety of transfer functions obtained is as follows.

$$H_a(s) = \omega_n^2/(s^2 + 2\zeta\omega_n s + \omega_n^2) \quad (12.1a)$$

$$H_b(s) = s^2/(s^2 + 2\zeta\omega_n s + \omega_n^2) \quad (12.1b)$$

$$H_c(s) = 2\zeta\omega_n s/(s^2 + 2\zeta\omega_n s + \omega_n^2) \quad (12.1c)$$

$$H_d(s) = (s^2 + \omega_n^2)/(s^2 + 2\zeta\omega_n s + \omega_n^2) \quad (12.1d)$$

$$H_e(s) = (2\zeta\omega_n s + \omega_n^2)/(s^2 + 2\zeta\omega_n s + \omega_n^2) \quad (12.1e)$$

$$H_f(s) = (s^2 + 2\zeta\omega_n s)/(s^2 + 2\zeta\omega_n s + \omega_n^2) \quad (12.1f)$$

The first four transfer functions are basically those of low-pass, high-pass, band-pass and band-stop filters, respectively. However, if ζ is small, H_a, as well as Hb, may be used as band-pass filters. The transfer functions H_e and H_f represent mixed type filters and are not of much interest; hence, they will not be considered any further in this chapter.

We shall analyze each configuration in a conventional manner and follow it up with graphical constructions and interpretations. While only the final results are of interest, the use of graphics gives much more physical meaning than is possible by using routine algebra and calculus.

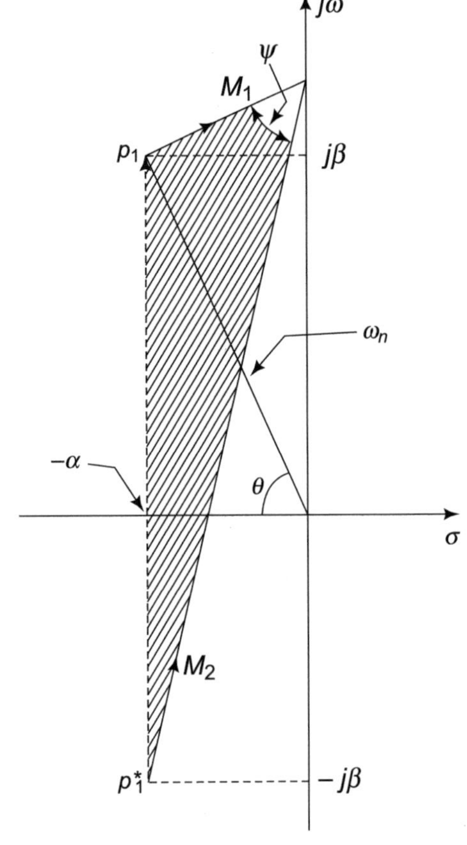

Fig. 12.2 Illustrating the notations used in the chapter

The Low-Pass Configuration

Figure 12.1a represents a low-pass configuration with unity DC response and zero response at infinite frequency. In between these two extremes, the response may be monotonically decreasing or may show a maximum. By putting $s = j\omega$ in (1a) and taking the magnitude, we get

$$|H(j\omega)| = \omega_n^2/\sqrt{(\omega_n^2 - \omega^2)^2 + 4\zeta^2\omega_n^2\omega^2} \quad (12.2)$$

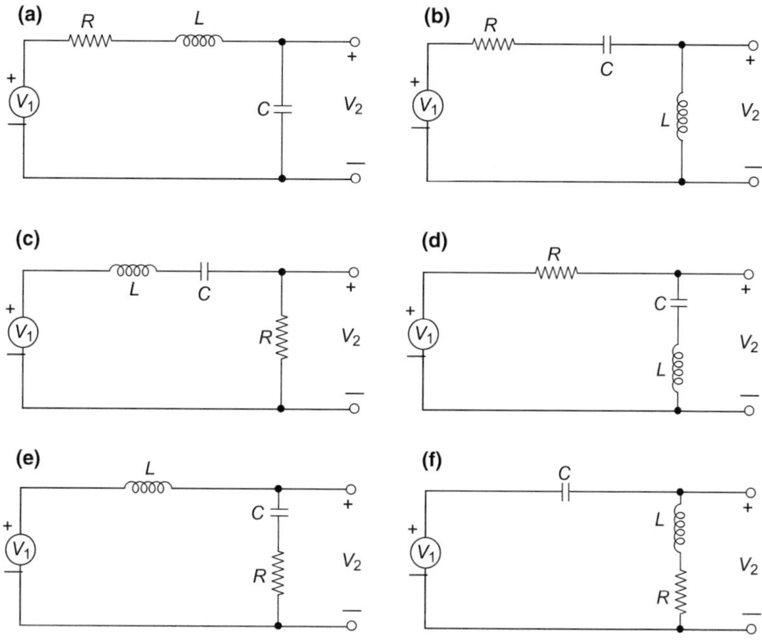

Fig. 12.3 Various possible configurations of the single-tuned circuit

By differentiating Eq. 12.2 with respect to ω and putting the result to zero shows that the maximum response occurs at the frequency ω_{ma}, where

$$\omega_{ma} = \sqrt{\beta^2 - \alpha^2} = \omega_n\sqrt{1 - 2\zeta^2} \qquad (12.3)$$

Obviously, the maximum will exist if $\zeta < 1/\sqrt{2}$, which is equivalent to $\beta > \alpha$ or $\theta > \pi/4$. For $\zeta = \zeta < 1/\sqrt{2}$, the response will be maximally flat (MF), while for $\zeta > 1/\sqrt{2}$, the response will lie below the MF curve, as shown in Fig. 12.4. The maximum response is obtained by combining Eqs. 12.2 and 12.3 and is given by

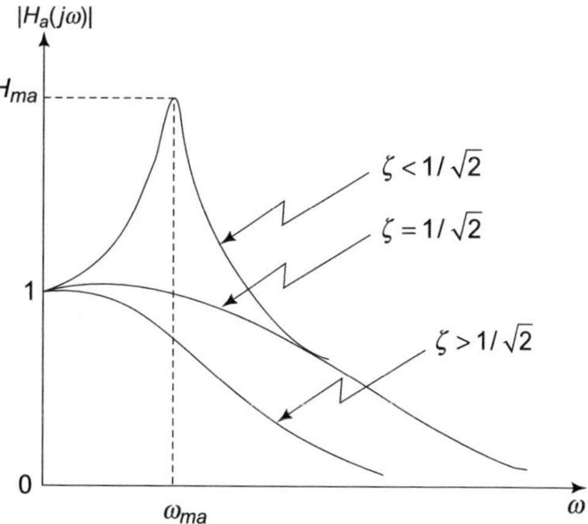

Fig. 12.4 Magnitude response of the low-pass configuration of Fig 12.1 for various values of ζ

$$H_{ma} = 1/(2\zeta\sqrt{1 - \zeta^2}) \qquad (12.4)$$

If H_{ma} is greater than $\sqrt{2}$, which occurs for $\zeta < 0.383$, then there will exist two 3 dB cutoff frequencies ω_{2a} and ω_{1a}, otherwise there will exist only one upper cut off frequency ω_{2a}. These frequencies can be calculated by solving the equation obtained by equating Eq. 12.2 to $H_{ma}/\sqrt{2}$. After a considerable amount of algebra, we get

$$\omega_{2a}, \omega_{1a} = \omega_n\left(1 - 2\zeta^2 \pm 2\zeta\sqrt{1 - \zeta^2}\right)^{1/2} \qquad (12.5)$$

Note that

$$\omega_{2a}, \omega_{1a} = \omega_n^2\sqrt{1 - 8\zeta^2 + 8\zeta^4} \neq \omega_{ma}^2 \quad (12.6)$$

in contrast to what Kuo claims in [1], p. 236.

At this point, it is instructive to bring in Kuo's graphical analysis of the configuration of Fig. 12.1a. Note that the area of the shaded triangle in Fig. 12.2 is given by $\beta\alpha$ as well as (1/2)$|\boldsymbol{M}_1||\boldsymbol{M}_2|$ sin ω so that

$$|M_1||M_2| = 2\beta\alpha/\sin\psi \quad (12.7)$$

Now the magnitude response of H_a can be written as

$$|H_a| = \omega_n^2/|M_1||M_2| = [\omega_n^2/(2\beta\alpha)]\sin\psi \quad (12.8)$$

Thus, the variation of $|H_a|$ with ω is consolidated in the variation of the angle ψ with ω. Now refer to Fig. 12.5, where $p_1 A p_1^* B$ is the so-called peaking circle of radius β, centred at C. Its intersection with the $j\omega$-axis previously occurs at $j\omega_{ma}$ (ω_{ma} being the frequency of maximum response) because the value of ψ at this point is the maximum, equal to $\pi/2$, being the angle subtended by a diameter on the circumference. It is easily shown from the right-angled triangle O-$j\omega_{ma}$-C that ω_{ma} is indeed given by Eq. 12.3.

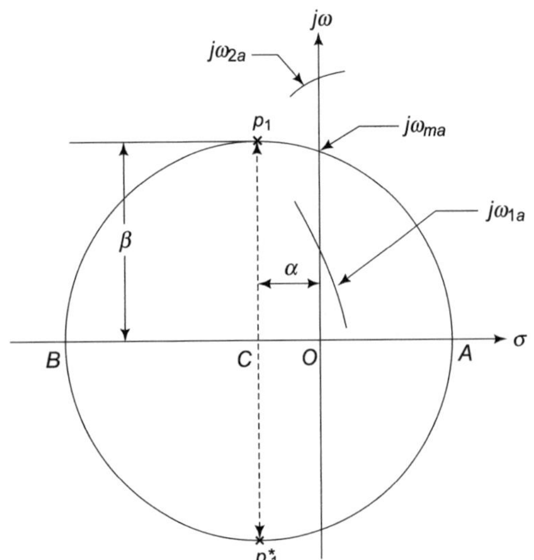

Fig. 12.5 Graphical analysis of the low-pass configuration of Fig. 12.1

Now if an arc is drawn with A as the centre and $Ap_1 = \sqrt{2}\beta$ as the radius, then its intersection with the positive $j\omega$-axis occurs at $j\omega_{2a}$, ω_{2a} being the upper cutoff frequency, because the value of ω there is $\pi/4$, being half of $\pi/2$, the angle subtended at the centre A by the arc from p_1 to p_1^*. It can be verified by considering the right-angled triangle O-$j\omega_{2a}$-A that ω_{2a} is indeed given by Eq. 12.5. Kuo [1], at this point, claimed that the lower cutoff frequency ω_{1a} is given by $\omega_{ma}^2/\omega_{2a}$, which, as we have shown in Eq. 12.6, is not true. However, a graphical construction is also possible for finding ω_{1a}, as shown by Martinez [2]. Draw an arc with B as the centre and Bp_1 as the radius. The point of its intersection with the positive $j\omega$-axis is $j\omega_{1a}$, because the value of ω there is $(\pi/2) + (\pi/4)$ i.e. sin ω is again $1/\sqrt{2}$. From the right-angled triangle O-$j\omega_{1a}$-B, it can now be verified that ω_{1a} is indeed given by Eq. 12.5.

The High-Pass Configuration

The configuration of Fig. 12.1b is the dual of Fig. 12.1a, because its dc response is zero while at infinite frequency, H_b becomes unity. By following the same procedure as in the low-pass case, it can be shown that the maximum response occurs at the frequency ω_{mb}, where

$$\omega_{mb} = \omega_n/\sqrt{1 - 2\zeta^2} \quad (12.9)$$

Comparing Eqs. 12.3 and 12.9, we see that $\omega_{ma}\,\omega_{mb} = \omega_n^2$ i.e. ω_{ma} and ω_{mb} are geometrically symmetrical about ω_n. In fact, ω_{mb} can be obtained by the graphical construction shown in Fig. 12.6. Mark the points $-\omega_{ma}$ and $-\omega_n$, on the negative real axis. Join $-\omega_{ma}$ to $j\omega_n$, thereby creating the angle ω. Construct the same angle ω at $-\omega_n$. The intersection of the new line with the positive $j\omega$-axis gives $j\omega_{mb}$, because

$$\tan\varphi = \omega_n/\omega_{ma} = \omega_{mb}/\omega_n \quad (12.10)$$

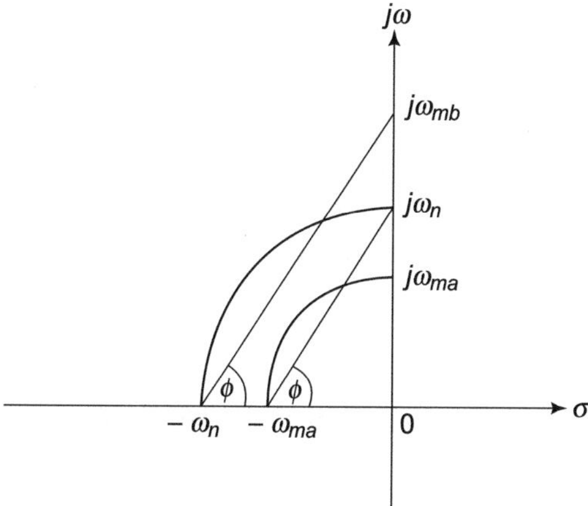

Fig. 12.6 Graphical construction for obtaining ω_{mb} from ω

It is easily shown that H_{mb}, the maximum response of $|H_b(j\omega)|$ is the same as H_{ma}, as given by Eq. 12.4.

Continuing the analysis, the 3 dB cutoff frequencies ω_{2b} and ω_{1b} can be obtained by equating $|H_b(j\omega)|$ to $H_{mb}\sqrt{2}$. After considerable algebra, one obtains the following result:

$$\omega_{2\mathrm{b}}, \omega_{1\mathrm{b}} = \omega_n / \left(1 - 2\zeta^2 \mp 2\zeta\sqrt{1-\zeta^2}\right)^{1/2} \tag{12.11}$$

Comparing Eqs. 12.5 and 12.11, we note that

$$\omega_{2\mathrm{b}}\omega_{1\mathrm{a}} = \omega_{2\mathrm{a}}\omega_{1\mathrm{b}} = \omega_n^2 \tag{12.12}$$

Thus, following the same procedure as depicted in Fig. 12.6, one can find ω_{2b} and ω_{1b} graphically from Fig. 12.5.

The Band-pass Configuration

The configuration of Fig. 12.1c, characterized by the transfer function H_c of (1c), is a true band-pass filter because of its response at dc, as well as infinite frequency, are zero. By writing $H_c(j\omega)$ as

$$H_c(j\omega) = 2\zeta\omega_n \Big/ \left[2\zeta\omega_n + j\omega_n\left(\frac{\omega}{\omega_n} - \frac{\omega_n}{\omega}\right)\right] \tag{12.13}$$

It is clear that the magnitude characteristic will have true geometric symmetry about ω_n, which is also the frequency of maximum response. The maximum response, in this case, is unity, and by the usual procedure, the 3 dB cutoff frequencies are obtained as

$$\omega_{2\mathrm{c}}, \omega_{1\mathrm{c}} = \omega_n\left(\sqrt{1+\zeta^2} \pm \zeta\right) \tag{12.14}$$

Clearly, $\omega_{2\mathrm{c}}, \omega_{1\mathrm{c}} = \omega_n^2$, which is a reflection of the geometric symmetry of the transfer function $H_c(j\omega)$.

A graphical construction for finding ω_{2c} and ω_{1c} is shown in Fig. 12.7. With the origin as the centre and Op_1 $(=\omega_n)$ as the radius, draw the circle $p_1 A p_1^* B$. It cuts the positive $j\omega$-axis at C which is $j\omega_n$. Since OD is $\zeta\omega_n$, the distance DC must be $\omega_n\sqrt{1+\zeta^2}$. With D as the centre and DC as the radius, draw the circle $ECFG$. Now the distance OE is $\omega_n\left(\sqrt{1+\zeta^2}+\zeta\right)$, while the distance OF is $\omega_n\left(\sqrt{1+\zeta^2}-\zeta\right)$. Hence, drawing two arcs with OE and OF as radii will cut the positive $j\omega$-axis at $j_{\omega 2c}$ and $j_{\omega 1c}$, respectively, as shown in Fig. 12.7.

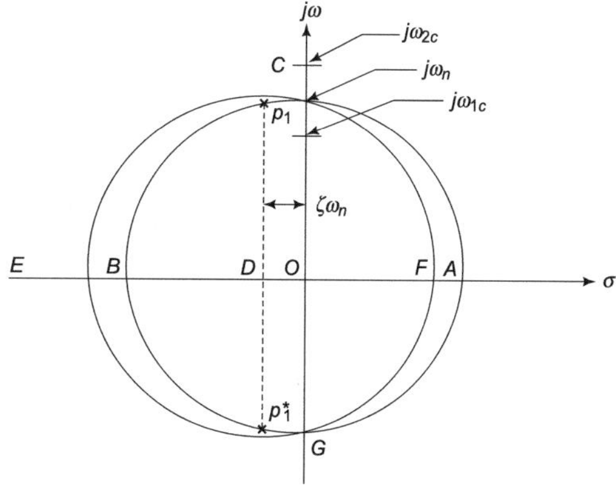

Fig. 12.7 Graphical analysis of the bandpass configuration of Fig. 12.1c

The Band-stop Configuration

By looking at the circuit of Fig. 12.1d or its transfer function H_d, it is clear that it will give null transmission at the frequency ω_n and that the dc and infinite frequency responses would be unity. To determine the 3 dB cutoff frequencies ω_{2d} and ω_{1d}, we let

$$\frac{(\omega^2 - \omega_n^2)^2}{(\omega^2 - \omega_n^2)^2 + 4\zeta^2\omega^2\omega_n^2} = \frac{1}{2} \tag{12.15}$$

and obtain the same values as those given by Eq. 12.14. Hence, the graphical construction of Fig. 12.7 works for this circuit also except that ω_n is now the frequency of rejection and not of maximum response.

Conclusion

We have shown in this chapter, how a relatively simple circuit like that of series *RLC* combination can be used to illustrate various circuit concepts like poles, zeros and their effects on the frequency response; filtering of various kinds; graphical analysis; geometric symmetry, etc. This should be of interest to teachers as well as students of circuit theory.

Problems

P.1. In a single-tuned circuit, if the output is taken across the series combination of *R* and *L*, what kind of frequency response will you obtain? Sketch it.

P.2. If *L* and *C* are in parallel in an *RLC* circuit and the output is taken across *L*, what will be the frequency response? Sketch it.

P.3. You require two poles in the frequency response and are supplied with three reactive elements. Draw an appropriate circuit and sketch its frequency response. Comment on the d.c and infinite frequency responses.

P.4. Same as above except that you require two nulls in the frequency response.

P.5. If you require three nulls, what is the minimum number of reactances needed? Draw an appropriate circuit and comment on the d.c. and infinite frequency responses. Also, comment on the height of the peaks.

References

1. F.F. Kuo, *Network Analysis and Synthesis* (Wiley, 1966)
2. J.R. Martinez, Graphical solution for 3 dB points. Electron. Eng., 48–51 (January 1967)

13 Analyzing the Parallel-T RC Network

Following a review of the various alternative methods available for analyzing the parallel-T RC network, we present yet another conceptually elegant method. This discussion illustrates the famous saying of Ramakrishna Paramhansa: As many religions, as many ways. Don't just grab one method; learn all of them and decide for yourself which one you find to be the simplest.

Keywords
Parallel-T-network · Mesh analysis
Node analysis · Two-port method · Splitting the parallel-T

The parallel-*T* RC network shown in Fig. 13.1a has fascinated many circuits researchers, including me [1–9]. It has many applications, foremost among them being notch filtering [10, 11], active band-pass filtering [12, 13], measurements [14, 15], compensation in control systems [16, 17], FM detection [18], and sine-wave generation [19]. Various methods are available in the literature for analyzing this network, a review of which is given in this chapter. This is followed by yet another method, which is conceptually elegant and is believed to be new.

For illustrating the various methods of analysis, we have used the symmetrical configuration shown in Fig. 13.1b, for simplicity. It should, however, be emphasized that the method of analysis does not depend upon the composition of the individual arms, each of which could as well be a general RLC impedance.

Mesh Analysis

The network of Fig. 13.1b has been redrawn in Fig. 13.2 in order to clearly indicate one choice of four independent meshes. Following standard procedure, the four mesh equations can be written in the matrix form as follows:

Source: S. C. Dutta Roy, "Analyzing the Parallel-T RC Network," *IETE Journal of Education*, vol. 44, pp. 111–116, July–September 2003.

S. C. Dutta Roy, *Circuits, Systems and Signal Processing*,
https://doi.org/10.1007/978-981-10-6919-2_13

$$\begin{bmatrix} R+[1/(2sC)] & -1/(2sC) & -R & 0 \\ -1/(2sC) & R+R_L+[1/(2sC) & -R & -R_L \\ -R & -R & 2R+[2/(sC)] & -1/(sC) \\ 0 & -R_L & -1/(sC) & (R/2)+R_L+[1/(sC)] \end{bmatrix} \begin{bmatrix} I_1 \\ I_2 \\ I_3 \\ I_4 \end{bmatrix} = \begin{bmatrix} V_I \\ 0 \\ 0 \\ 0 \end{bmatrix}. \tag{13.1}$$

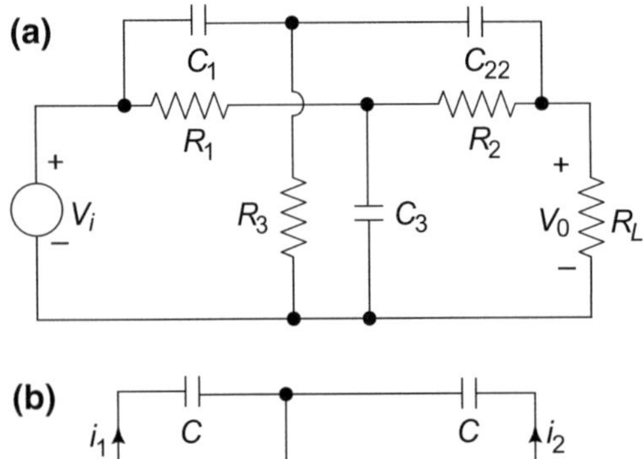

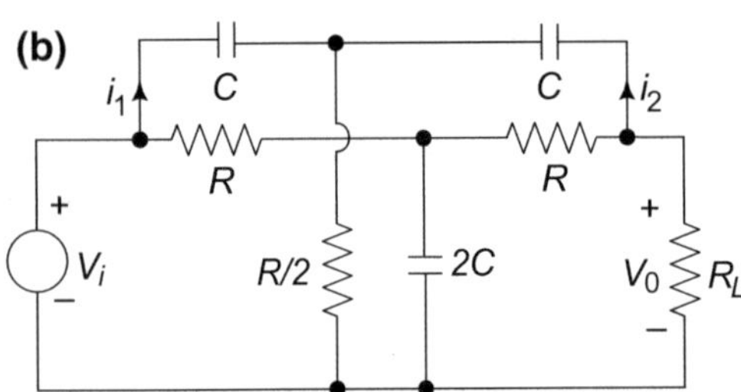

Fig. 13.1 **a** The general parallel-*T*-network; **b** The symmetrical form

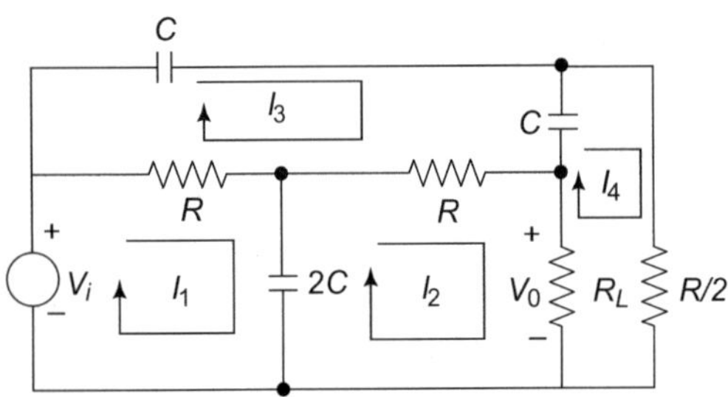

Fig. 13.2 Redrawn form of Fig. 13.lb for mesh analysis

Since

$$V_O = (I_2 - I_4)R_L, \tag{13.2}$$

we need to evaluate I_2 and I_4 only. If Δ denotes the determinant of the 4×4 mesh impedance matrix in Eq. 13.2 and Δ_{ij} its cofactors, then

$$I_2 = V_i\Delta_{12}/\Delta \quad \text{and} \quad I_4 = V_i\Delta_{14}/\Delta \tag{13.3}$$

so that from Eq. 13.2, the voltage transfer function is obtained as

$$\begin{aligned} T = V_O/V_i &= (\Delta_{12} - \Delta_{14})R_L/\Delta \\ &= (M_{14} - M_{12})R_L/\Delta, \end{aligned} \tag{13.4}$$

where M_{ij} is the minor of Δ_{ij} i.e. $\Delta_{ij} = (-1)^{i+j} M_{ij}$. Evaluating Δ and its minors, and simplification of Eq. 13.4 will require pages of calculations. However, if it is done with care, one ends up with the following expression:

$$T = (p^2+1)/[p^2+2(2+r)p+1+2r], \tag{13.5}$$

where we have used the following notations:

$$p = sCR \quad \text{and} \quad r = R/R_L. \tag{13.6}$$

Node Analysis

Refer to Fig. 13.1b again where all node voltages have been identified. The node equations for V_1, V_2 and V_O can be written in the following matrix form:

$$\begin{bmatrix} 2(G+sC) & 0 & -G \\ 0 & 2(G+sC) & -sC \\ -G & -sC & sC+G+G_L \end{bmatrix} \begin{bmatrix} V_1 \\ V_2 \\ V_O \end{bmatrix} = \begin{bmatrix} GV_i \\ sCV_i \\ 0 \end{bmatrix}, \tag{13.7}$$

where

$$G = 1/R \quad \text{and} \quad G_L = 1/R_L. \qquad (13.8)$$

Denoting the determinant of the 3 × 3 matrix on the left of Eq. 13.7 by Δ, we get

$$V_0 = (G\Delta'_{13} + sC\Delta'_{23})/V_i/\Delta' \qquad (13.9)$$

so that

$$T = (GM'_{13} - sCM'_{23})/\Delta'. \qquad (13.10)$$

The evaluation of Δ, M'_{13} and M_{23} is a much easier task than that in mesh analysis. It is not difficult to show that simplifying the resulting expression gives the same result as in Eq. 13.5.

Two-Port Method

The network of Fig. 13.1b is the parallel connection of two *T*-networks, viz. *A*: *R–2C–R* and *B*: *C–R/2–C*, which is terminated in R_L. The *z*-parameters of *T*-networks are as follows:

$$\begin{aligned} z_{11A} &= z_{22A} = R + [1/(2sC)]; \\ z_{12A} &= z_{21A} = 1/(2sC); \\ z_{11B} &= z_{22B} = (R/2) + [1/(sC)]; \\ \text{and } z_{12B} &= z_{21B} = R/2. \end{aligned} \qquad (13.11)$$

The corresponding *y*-parameters can be found from the conversion formulas [20] as

$$\begin{aligned} y_{11A} &= y_{22A} = (2p+1)/[2R(p+1)]; \\ y_{12A} &= y_{21A} = -1/[2R(p+1)]; \\ y_{11B} &= y_{22B} = p(p+2)/[2R(p+1)]; \text{ and} \\ y_{12B} &= y_{21B} = -p^2/[2R(p+1)]. \end{aligned} \qquad (13.12)$$

Thus, the overall *y*-parameters are the following:

$$y_{11} = y_{22} = (p^2 + 4p + 1)/[2R(p+1)] \text{ and } y_{12} = y_{21} = -(p^2+1)/[2R(p+1)]. \qquad (13.13)$$

Finally, the transfer function is

$$\begin{aligned} T &= -y_{21}/(y_{22} + G_L) \\ &= (p^2+1)/[p^2 + 4p + 1 + 2RG_L(p+1)], \end{aligned} \qquad (13.14)$$

which is identical to Eq. 13.5 because $RG_L = r$.

Analysis by Miller's Equivalence

Refer to Fig. 13.1b again. By Miller's theorem, this circuit is equivalent to that shown in Fig. 13.3a

where

$$Y_1 = i_1/V_i \quad \text{and} \quad Y_2 = i_2/V_0. \qquad (13.15)$$

Since Y_1 occurs across the input voltage source, it does not affect the transfer function. We, therefore, have to find Y_2 only. Obviously,

$$i_2 = sC(V_0 - V_2). \qquad (13.16)$$

To find V_2, write the node equation at this node as follows:

$$(2sC + 2G)V_2 = sCV_i + sCV_0 \qquad (13.17)$$

or,

$$V_2 = sC(V_i + V_0)/[2(sC+G)]. \qquad (13.18)$$

Combining Eq. 13.15 with Eqs. 13.16 and 13.18, we get

$$Y_2 = p[(1 - T^{-1})p + 2]/[2R(p+1)]. \qquad (13.19)$$

Although Y_2 involves *T* which we wish to find, one should not be worried. As you would see, we shall find *T* in terms of T^{-1} and then by cross multiplying and simplifying, we shall find an explicit expression for *T*. By applying Thevenin's theorem to the left of the *XX′* line in Fig. 13.3a, we get the equivalent circuit shown in Fig. 13.3b. Thus

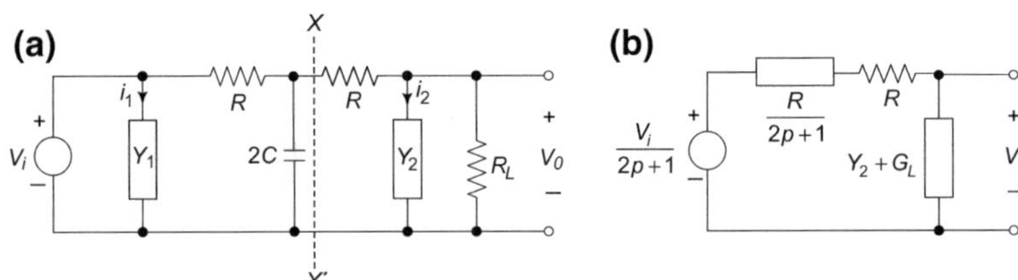

Fig. 13.3 **a** Miller's equivalent of the network of Fig. 13.1b; **b** Equivalent circuit of Fig. 13.3a obtained by using Thevenin's theorem

$$T = V_0/V_i = [1/(2p+1)].\,[1/(Y_2+G_L)]/\{[1/(Y_2+G_L)].\,+R+[R/(2p+1)]\} \tag{13.20}$$

Combining Eqs. 13.20 with 13.19, and simplifying, we get

$$T = 1/\{p[(l-T^{-1})p+2]+2(p+1)RG_L+2p+1\}. \tag{13.21}$$

Now cross multiply and simplify. The final result is the same as Eq. 13.5.

Splitting the T's

Let the *C–R/2–C* *T*-network in Fig. 13.1b be separated at the input side, turned through 180° and be terminated in another voltage source V_i. The result is shown in Fig. 13.4a, which would be completely equivalent to Fig. 13.1b because potentials at all the nodes have been preserved. Now apply Thevenin's theorem to the left of *XX′* and to the right of *YY′* to get the equivalent circuit shown in Fig. 13.4b [8]. Next, write the node equation at the load as follows:

$$\frac{V_0-[V_i/(2p+1)]}{R+[R/(2p+1)]}+\frac{V_0}{R_L}+\frac{V_0-[V_ip/(p+2)]}{(R/p)+[R/(p+2)]} = 0. \tag{13.22}$$

Simplifying Eq. 13.22 and taking the ratio *T* gives the same result as Eq. 13.5.

Yet Another Method

Look at Fig. 13.4a again. Instead of applying Thevenin's theorem, let us apply ladder analysis method [20] starting from R_L and going to the left, and again starting from R_L and going to the right. In order not to clutter Fig. 13.4a, we have redrawn the circuit in Fig. 13.5, where all branch currents have been identified. At node V_0,

$$I_1+I_2 = I_L = V_0G_L. \tag{13.23}$$

Going to the left, we get

$$V_1 = RI_1+V_0, \tag{13.24a}$$

$$I_3 = 2sCV_1 = 2pI_1+2pGV_0, \tag{13.24b}$$

$$I_5 = I_1+I_3 = (2p+1)I_1+2pGV_0 \tag{13.24c}$$

and

$$V_i = RI_5+V_1 = 2(p+l)RI_1+(2p+l)V_0. \tag{13.24d}$$

From the last equation, we have

$$I_1 = [V_i-(2p+l)V_0]/[2(p+1)R]. \tag{13.25}$$

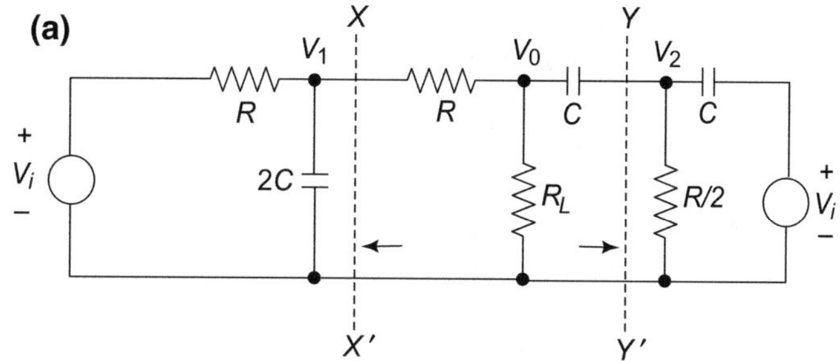

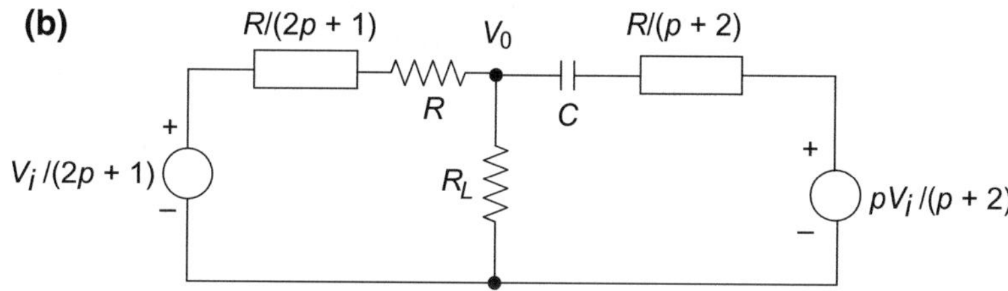

Fig. 13.4 **a** Spread out version of Fig. 13.1b by splitting the two *T's*; **b** Equivalent circuit of Fig. 13.4a obtained by two applications of Thevenin's theorem

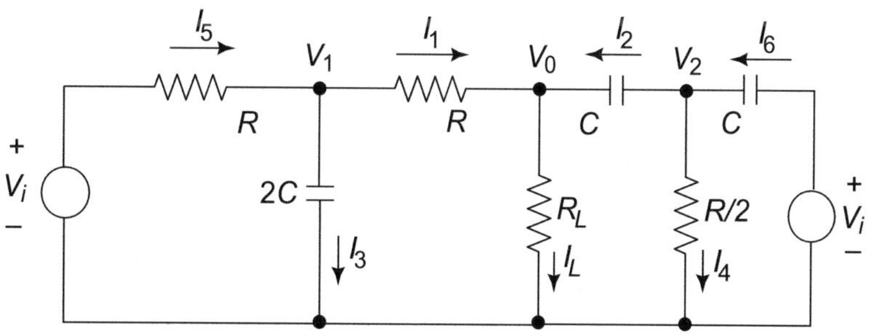

Fig. 13.5 Fig 13.4a circuit redrawn to illustrate the new method

Now start from V_0 and go to the right. This is what we get:

$$V_2 = [I_2/(sC)] + V_0 = (R/p)I_2 + V_0, \quad (13.26a)$$

$$I_4 = 2GV_2 = (2/p)I_2 + 2GV_0, \quad (13.26b)$$

$$I_6 = I_4 + I_2 = [(2/p) + 1]I_2 + 2GV_0 \quad (13.26c)$$

and

$$V_i = [1/(sC)]I_6 + V_2 = (2R/p)[(1/p) + 1]I_2 + [(2/p) + 1]V_0. \quad (13.26d)$$

The last equation gives

$$I_2 = [p^2 V_i - p(p + 2)V_0]/[2(p + 1)R]. \quad (13.27)$$

Now combine Eqs. 13.23, 13.25 and 13.27 to get

$$(p^2 + 1)V_i - (p^2 + 4p + 1)V_0 = 2(p + l)RG_L V_0. \quad (13.28)$$

Simplifying Eq. 13.28 gives the same result as Eq. 13.5.

Conclusion

In this chapter, we have discussed six different methods for analyzing the parallel-*T* RC network. Of these, mesh analysis requires more effort than any other method. The node analysis comes next in terms of computational effort. The efforts needed in the two-port method and the method using Miller's equivalence are comparable and can be bracketed to occupy the joint third position in terms of decreasing computational effort. Splitting the T's is common to the last two methods—one using Thevenin's theorem, and the other using ladder analysis technique. Both are conceptually elegant and require almost the same amount of effort. They, therefore, qualify for the joint fourth position in the list; of these, the last method does not seem to have appeared earlier in the literature and is therefore believed to be new.

Problems

P.1. What kind of transfer function do you get if, in Fig. 13.1a $R_3 \to \infty$ and $C_3 = 0$?

P.2. Same, if in Fig. 13.1b, $R/2 \to \infty$ and $2C = 0$ in the shunt branches?

P.3. Find the two-port parameters of the circuit of P.1 and hence the transfer function.

P.4. Same for the circuit of P.2.

P.5. Apply Miller to P.3 and P.4 and verify that you get the same transfer functions.

References

1. S.C. Dutta Roy, A twin-tuned RC network. Ind. J. Phys. **36**, 369–378 (1962)
2. S.C. Dutta Roy, On the design of parallel-*T* resistance capacitance networks for maximum selectivity. J. Inst. Telecommun. Eng. **8**, 218–233, (1962)
3. S.C. Dutta Roy, Parallel-*T* RC networks: limitations of design equations and shaping the transmission characteristic. Ind. J. Pure Appl. Phys. **1**, 175–181, (1963)
4. S.C. Dutta Roy, The definition of Q of RC networks. Proc. IEEE. **52**, 44, (1964)
5. D.G.O. Morris & S.C. Dutta Roy, Q and selectivity. Proc. IEEE. **53**, 87–89, (1965)
6. S.C. Dutta Roy, Dual input null networks, Proc. IEEE. **55**, 221–222, (1967)
7. S.C. Dutta Roy & N. Choudhury, An application of dual input networks. Proc. IEEE. **58**, 847–848, (1970)
8. S.C. Dutta Roy, A quick method for analyzing parallel ladder networks. Int. J. Elect. Eng. Educ. **13**, 70–75, (1976)
9. S.C. Dutta Roy, Miller's theorem revisited Circ. Syst. Signal Process. **19**, 487–499, (2000)
10. L. Stanton, Theory and applications of the parallel-*T* resistance capacitance frequency selective network. Proc. IRE. **34**, 447–456 (1946)
11. A.E. Hastings, Analysis of the resistance capacitance parallel-*T* network and applications. Proc. IRE. **34**, 126–129 (1946)
12. H. Fleischer, Low frequency feedback amplifiers, in *Vacuum Tube Amplifiers*, ed. by G.E. Valley Jr., H. Wallman, McGrawHill, (1948, Chapter 10)
13. C.K. Battye, A low frequency selective amplifier. J. Sci. Inst. **34**, 263–265 (1957)
14. W.N. Tuttle, Bridged-*T* and parallel-*T* null networks for measurements at rf. Proc. IRE. **28**, 23–30 (1940)
15. K. Posel, Recording of pressure step functions of low amplitude by means of composite dielectric capacitance transducer in parallel-*T* network. Amer. Rocket Soc. J. **21**, 1243–1251 (1961)
16. A.B. Rosenstein, J. Slaughter, Twin T compensation using root locus method. AIEE Trans, Part II (Applications and Industry) **81**, 339–350 (1963)
17. A.C. Barker, A.B. Rosenstein, s-plane synthesis of the symmetrical twin-*T* network. IEEE. Trans. Appl. Indus. **83**, 382–388 (1964)
18. J.R. Tillman, Linear frequency discriminator. Wirel. Eng. **23**, 281–286 (1946)
19. A.P. Bolle, Theory of twin-*T* RC networks and their applications to oscillators. J. Brit. IRE. **13**, 571–587 (1953)
20. F.F. Kuo, *Network Analysis and Synthesis* (John Wiley, 1966, Chapter 9)

14 Design of Parallel-T Resistance–Capacitance Networks For Maximum Selectivity

A simple analysis is presented for obtaining an expression for the transfer function and hence the selectivity, Q_T, of a general parallel-T resistance–capacitance network. The maximum value of Q_T obtainable by a suitable choice of elements is shown to be $\frac{1}{2}$. A design procedure for approaching this maximum value is given.An expression for the selectivity, Q_A, of an amplifier using a general parallel-T resistance–capacitance network in the negative feedback line has been deduced and the advantages of having an increased Q_T explained. Parallel-T *RC* is an important network and you cannot do without it, if you wish to remain in circuit design.An expression has been given for estimating the departure from linearity of the amplitude response characteristic at a particular frequency. This is used to find an optimum value of Q_T for best performance of the network as an F.M. discriminator at low frequencies. It is shown that the required value of Q_T is very near its maximum value.

Source: S. C. Dutta Roy, "Design of Parallel-T Resistance–Capacitance Networks for Maximum Selectivity," *Journal of the Institution of Telecommunication Engineers*, vol. 8, pp. 218–223, September 1962.

S. C. Dutta Roy, *Circuits, Systems and Signal Processing*,
https://doi.org/10.1007/978-981-10-6919-2_14

Keywords
Parallel-T RC network · Selectivity
Selective amplifier

Introduction

In the low-frequency range, an inductance–capacitance-tuned circuit is seldom used as a frequency-selective network because of the following disadvantages: (a) large physical size of the inductor requires space and makes the equipment bulky, (b) an inductor of large value is expensive and (c) the value of Q obtainable is low. A resistance–capacitance network is a better choice for all these considerations. Of all such networks, the parallel-T *RC* network is the most extensively used one.

Much work has been done on the symmetrical configuration of the parallel-T *RC* network and a fairly impressive list of references is available on its theory and applications. The general asymmetrical configuration of the network has received less attention, important work in this line being due to Stanton [1], Wolf [2] and Oono [3]. Stanton [1] has given an expression for the transfer function of a general parallel-T RC network in which the components occur as the ratios

$$\frac{\text{(a series arm resistance or reactance)}}{\text{(total series arm resistance or reactance)}} \tag{14.1}$$

and

$$\frac{\text{(total series arm reactance)}}{\text{(total series arm resistance)}}.$$

A simpler expression is deduced in this chapter by assuming the series arm impedances to be arbitrary multiples of the shunt arm impedances. Using Morris's [4] definition of Q of resistance–capacitance networks, an expression is deduced for the selectivity, Q_T, of the general parallel-T network. It is shown that the maximum value of Q_T is $\frac{1}{2}$, which is in conformity with Wolf's [2] result. A design procedure is then suggested for such networks, which is more general than that given by Wolf [2]. Oono's [3] work is an extension of that of Stanton [1] for the case when the effects of the source and the load impedances are not negligible. Throughout this chapter, however, the source impedance has been assumed to be negligible and the load impedance has been assumed to be infinite.

In its application in an F.M. discriminator [5, 6], it is desired that the amplitude transfer function should have a linear variation with frequency. An expression is given in this chapter for estimating the departure from linearity of the amplitude transfer characteristic at a particular frequency. This is used to find an optimum value of Q_T for best performance of a single parallel-T network in the above application. It is shown that in the frequency range of interest, the required value of Q_T is very near to its maximum value.

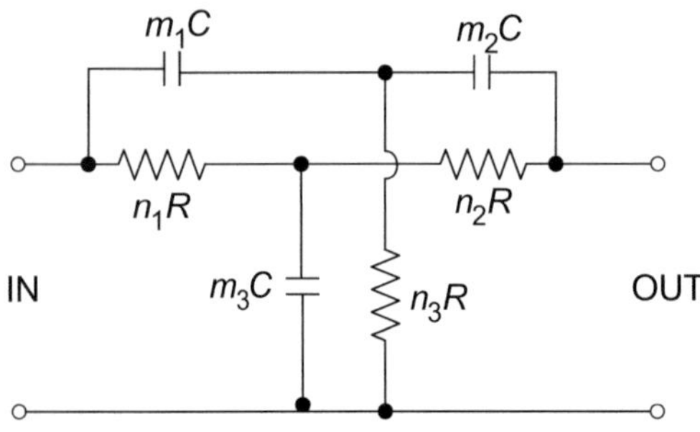

Fig. 14.1 A General Parallel-T network

By itself, the parallel-T RC network behaves as a rejection filter. A resonance characteristic, similar to that of a tuned amplifier, can be obtained by using it as the feedback network of an amplifier which has an odd number of stages. Fleischer [7] has shown that by using a symmetrical configuration of the network, the maximum value of Q obtainable is approximately $G_o/4$, where G_o is the open-loop gain of the amplifier. In this chapter, an expression has been deduced for the selectivity, Q_A, of an amplifier using a general parallel-T network in the negative feedback line and the advantages of having an increased value of Q_T are explained.

Network Configuration and Simplification

A parallel-T RC network with completely arbitrary values of the elements is shown in Fig. 14.1, where R is a resistance parameter, C a capacitance parameter and m_1, m_2, m_3, n_1, n_2 and n_3 are numerical constants. In the conventional symmetrical configuration,

$$m_1 = m_2 = n_1 = n_2 = 1, m_3 = 2/k \quad \text{and} \quad n_3 = 1/(2k),$$

where the parameter k controls the selectivity of the transfer characteristic and has a value of unity for maximum Q equal to $\frac{1}{4}$.

Without any loss of generality, we can assume $m_3 = n_3 = 1$. Each of the two tees in Fig. 14.1 can be converted to an equivalent pi-network. For the two networks in Fig. 14.2 to be equivalent, the elements should be related as follows:

$$Z_A = \Sigma/Z_2, Z_B = \Sigma/Z_3, Z_C = \Sigma/Z_1, \quad (14.1)$$

where $\Sigma = Z_1Z_2 + Z_2Z_3 + Z_3Z_1$. Employing Eq. 14.1 and denoting by subscripts 1 and 2 the equivalent pi-elements of the R-C-R and C-R-C tees respectively, we have

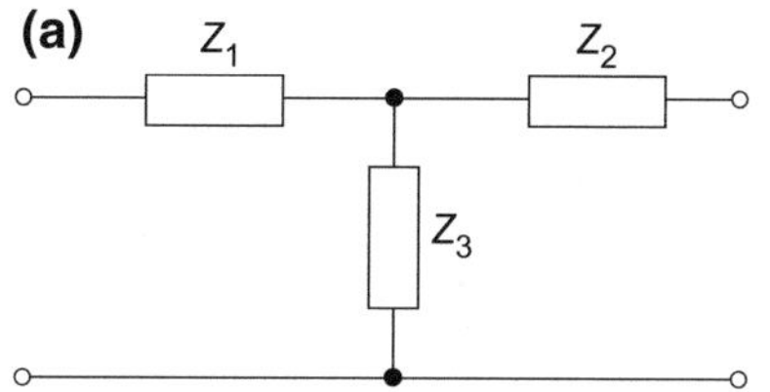

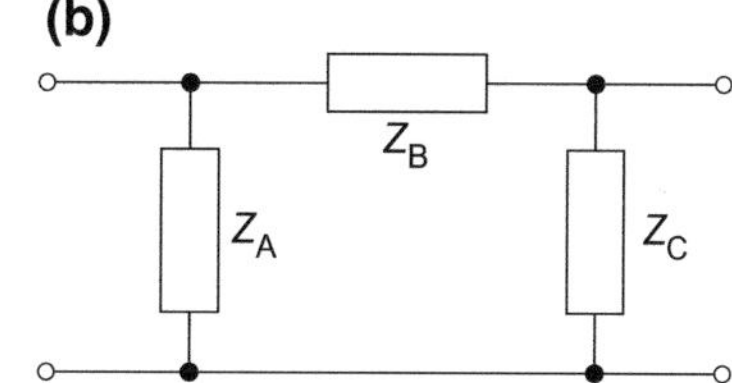

Fig. 14.2 **a** A Tee-network, **b** A Pi-network

$$\left.\begin{aligned} Z_{A1} &= n_1R + (n_1+n_2)/(n_2pC) \\ Z_{B1} &= (n_1+n_2)R + pn_1n_2CR^2 \\ Z_{C1} &= n_2R + (n_1+n_2)/(n_1pC) \\ Z_{A2} &= R(m_1+m_2)/m_1 + 1/(pm_1C) \\ Z_{B2} &= 1/(m_1m_2p^2C^2R) \\ &\quad + (m_1+m_2)/(pCm_1m_2) \\ \text{and} \quad Z_{C2} &= R(m_1+m_2)/m_2 + 1/(pm_2C) \end{aligned}\right\}, \tag{14.2}$$

where $p = j\omega$, ω being the angular frequency. The pi-equivalent of the network of Fig. 14.1 will then have its elements given by

$$Z_A = \frac{Z_{A1}\cdot Z_{A2}}{Z_{A1}+Z_{A2}}, Z_B = \frac{Z_{B1}\cdot Z_{B2}}{Z_{B1}+Z_{B2}}, Z_C = \frac{Z_{C1}\cdot Z_{C2}}{Z_{C1}+Z_{C2}}. \tag{14.3}$$

Null Condition

For zero transmission, since Z_A and Z_C cannot be zero with ordinary circuit elements, we must have $Z_B = \alpha$. At this point, it is convenient to have a look at the elements composing the Z_B arm as shown in Fig. 14.3. We note that this is simply an anti-resonant circuit having infinite impedance at a frequency given by

$$\omega_0^2 = \frac{1}{C^2R^2m_1m_2(n_1+n_2)} = \frac{m_1+m_2}{C^2R^2m_1m_2n_1n_2}.$$

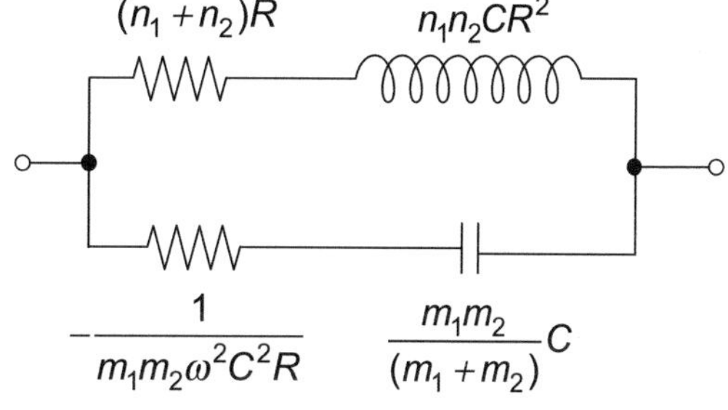

Fig. 14.3 Showing the Z_B arm of the pi-equivalent of the network of Fig. 14.1

Without any loss of generality, we can let $\omega_0^2 = 1/(C^2R^2)$. Then

$$m_1m_2(n_1+n_2) = m_1m_2n_1n_2/(m_1+m_2) = 1. \tag{14.4}$$

Equation 14.4 shows that if ω_0 is fixed, then the number of arbitrary numerical constants is reduced to two only.

Transfer Function

Let $\omega CR = x$; then the rejection frequency is given by $x = 1$ and from Eq. 14.2,

$$\left.\begin{aligned} Z_{B1} &= R\{(n_1+n_2) + jn_1n_2x\} \\ Z_{B2} &= \{-R/(m_1m_2x)\}\{(1/x) + j(m_1+m_2)\} \\ Z_{C1} &= \{R/(n_1x)\}\{n_1n_2x - j(n_1+n_2)\} \\ \text{and} \quad Z_{C2} &= \{R/(m_2x)\}\{(m_1+m_2)x - j\} \end{aligned}\right\}. \tag{14.5}$$

Table 14.1 Examples of design for maximum Q_T

n_1	m_1	Q_T	m_2	n_2
1.0	0.95	0.487	0.026	40.04
1.0	0.90	0.475	0.053	20.11
1.0	0.85	0.459	0.081	13.51
1.0	0.80	0.444	0.111	10.25
1.0	0.75	0.429	0.143	8.33

Combining Eqs. 14.3, 14.4 and 14.5 and simplifying, we get

$$\left.\begin{aligned} Z_B &= R\frac{1+jm_1m_2n_1n_2x}{m_1m_2(1-x^2)} \\ Z_C &= R\frac{m_1m_2n_1n_2x-j}{m_2(1+m_1n_1)x} \end{aligned}\right\}. \tag{14.6}$$

For zero source and infinite load impedances, the transfer function of the network is given by

$$T = Z_C/(Z_B+Z_C).$$

Substituting for Z_B and Z_C from Eq. 14.6 and simplifying, we get

$$T = \frac{1}{1-j(n_1+1/m_1)/(x-1/x)}. \tag{14.7}$$

This expression for the transfer function of a general parallel-T network is much simpler to handle than that of Stanton [1], as it contains only two numerical constants. In Eq. 14.7, the term varying with frequency occurs as $(x - 1/x)$; it thus follows that both the amplitude and phase transfer characteristics of the network will be symmetrical about $x = 1$ when plotted on a log (x) scale.

Selectivity

Equation 14.7 can be written as

$$T = \frac{1-x^2}{1+j(n_1+1/m_1)x-x^2} \tag{14.8}$$

Using Morris's [4] definition of Q of a resistance–capacitance network, we have from Eq. 14.8

$$Q_T = 1/(n_1+1/m_1). \tag{14.9}$$

Thus Q_T can be increased by decreasing n_1 and increasing m_1. The extent to which this can be done depends, however, on m_2 and n_2 also, because these must remain positive as m_1 and n_1 are changed. At this stage, therefore, we require the expressions for m_2 and n_2 in terms of m_1 and n_1. They can be easily obtained from Eq. 14.4 as

$$m_2 = \frac{n_1-m_1}{m_1n_1^2+1} \text{ and } n_2 = \frac{m_1^2n_1+1}{m_1(n_1-m_1)}. \tag{14.10}$$

Thus for m_2 and n_2 to be positive, $(n_1 - m_1)$ must remain positive. Under this restriction, Q_T will have a maximum value of $\frac{1}{2}$ when $m_1 = n_1 = 1$.

Design

At $m_1 = n_1 = 1$, Eq. 14.10 gives $m_2 = 0$ and $n_2 = \alpha$, so that the corresponding arms are effectively open circuited and the output is zero at all frequencies. Even with finite elements of moderate values, however, Q_T can be made to approach this maximum value, as will be evident from the following example. Let $n_1 = 1.0$ and $m_1 = 0.9$; then $Q_T = 0.475$. In the conventional symmetrical case, $Q_T = 0.250$ so that the improvement is as much as 90%. Also from Eq. 14.10, $m_2 = 0.053$ and $n_2 = 20.11$. For a rejection frequency of 1000 c/s., we can choose $C = 0.01$ μF and $R = 16$ KΩ. Then the series resistances required are $n_1R = 16$ KΩ and $n_2R = 321.7$ KΩ and the series capacitances required are $m_1C = 0.009$ μF, and $m_2C = 530$ μμF. Thus elements of reasonable values can be used to approach the maximum selectivity. Table 14.1 shows some typical examples of design for improved Q_T.

Linearity of the Selectivity Curve

Detection of a frequency-modulated signal is usually carried out by first converting it into an amplitude-modulated signal by a device called a discriminator and then applying the A.M. signal to an ordinary A.M. detector. The circuit arrangement of the discriminators used in the high-frequency range may be looked upon as consisting of two channels, each containing an inductance–capacitance circuit. The two LC circuits are tuned to two different frequencies f_1 and f_2 such that $(f_1 \sim f_2)$ is slightly greater than twice the peak deviation and $(f_1 + f_2)/2$ is equal to the carrier frequency of the F.M. wave to be detected. The difference between the rectified outputs of the two channels then varies linearly with frequency in the frequency range of interest. In the low-frequency range, the two tuned circuits are replaced by two parallel-T *RC* networks [5, 6] whose rejection frequencies are chosen in the same manner as f_1 and f_2 in the high-frequency circuit. A single parallel-T network can also be used as a discriminator if it can be so designed that a linear relation exists between the amplitude transfer function ($|T|$) and the frequency (x) in the frequency range of interest. It will be shown that this condition is approximately satisfied when the network selectivity is nearly equal to its maximum value.

From Eqs. 14.7 and 14.9, we can write

$$|T| = \frac{1}{\{1 + 1/(Q_T^2 y^2)\}^{1/2}}$$

or,

$$1/|T|^2 = 1 + 1/(Q_T^2 y^2) \qquad (14.11)$$

where $y = x - 1/x$. Differentiating Eq. 14.11 with respect to y gives

$$\frac{d|T|}{dy} = \frac{|T|^3}{Q_T^2 y^3}. \qquad (14.12)$$

Differentiating again, we get

$$\frac{d^2|T|}{dy^2} = \frac{3|T|^3}{Q_T^2 y^4}\left(\frac{|T|^2}{Q_T^2 y^2} - 1\right).$$

Combining this with Eq. 14.11, we have

$$\frac{d^2|T|}{dy^2} = -\frac{3|T|^5}{Q_T^2 y^4}. \qquad (14.13)$$

Again,

$$\frac{d|T|}{dx} = \frac{d|T|}{dy} \cdot \frac{dy}{dx}$$

$$\therefore \quad \frac{d^2|T|}{dx^2} = \frac{d|T|}{dy} \cdot \frac{d^2y}{dx^2} + \frac{d^2|T|}{dy^2} \cdot \left(\frac{dy}{dx}\right)^2.$$

Substituting the values of $d|T|/dy$ and $d^2|T|/dy^2$ from Eqs. 14.12 and 14.13, we have

$$\frac{d^2|T|}{dx^2} = \frac{|T|^3}{Q_T^2 y^3}\left\{\frac{d^2y}{dx^2} - \frac{3|T|^2}{y}\left(\frac{dy}{dx}\right)^2\right\}. \qquad (14.14)$$

Also,

$$\frac{dy}{dx} = 1 + 1/x^2 \text{ and } \frac{d^2y}{dx^2} = -2/x^3. \qquad (14.15)$$

Combining Eq. 14.14 with Eqs. 14.11 and 14.15 gives

$$\frac{d^2|T|}{dx^2} = -\frac{1}{Q_T^2 y^3 \left(1 + \frac{1}{Q_T^2 y^2}\right)^{3/2}} \times \left\{\frac{2}{x^3} + \frac{3\left(1 + \frac{1}{x^2}\right)^2}{y\left(1 + \frac{1}{Q_T^2 y^2}\right)}\right\}.$$

For a perfectly linear curve, the first differential coefficient is a constant and the second differential coefficient is zero. Thus, the value of $d^2|T|/dx^2$ (neglecting sign) is a measure of the departure from linearity, the least value corresponding to maximum linearity. From the above, we see that $d^2|T|/dx^2$ is a function of both x and Q_T so that for a particular value of Q_T, the linearity varies from point to point.

In the particular application considered, the frequency range of interest is $0 < x < 1$. In this range, y is negative and the expression within the second bracket can be made zero, *i.e.* perfect linearity can be attained at a single frequency by suitably choosing Q_T. If the normalized value of this frequency is denoted by x_0 and $y_0 = x_{0-1}/x_0$, the required value of Q_T is given by

$$Q_T = \left\{\frac{(-1/y_0)}{1.5x_0^3(1+1/x_0^2)+y_0}\right\}^{1/2}. \qquad (14.16)$$

It is natural to suggest that x_0 should be chosen to be somewhere near the centre of the band $0 < x < 1$ so that with the carrier frequency coincident with x_0, a frequency deviation of the order of 50 per cent of the carrier frequency can be detected. Since, however, $|T| \to 1$ as $x \to 0$, there will be a considerable deviation from linearity at very low frequencies. We thus choose x_0 to be nearer to 1 than to 0. Let $x_0 = 0.55$; then the required value of Q_T is 0.485, which is very near to its maximum value. The improvement in linearity as Q_T approaches this value will be evident from Fig. 14.4, where the magnitude of the amplitude transfer function has been plotted in the band $0 \leq x \leq 1$ for various values of Q_T. The curve for $Q_T = 0.495$ is appreciably linear over the range $0.2 < x < 1$.

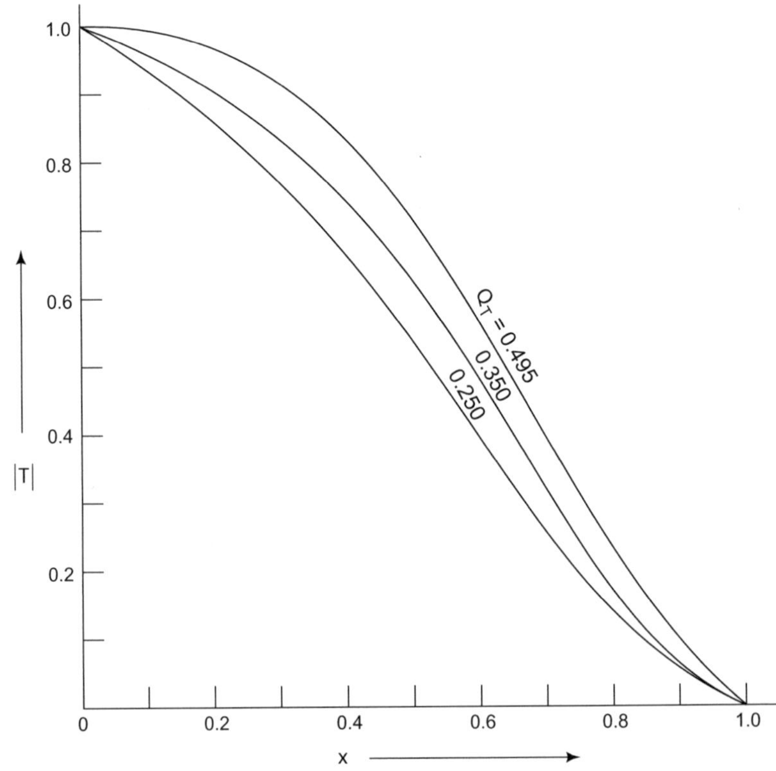

Fig. 14.4 Showing the selectivity curves of the parallel-T network for $Q_T = 0.250$, 0.350 and 0.495

Selectivity of an Amplifier Using the General Parallel-T RC Network in the Negative Feedback Line

In a low-frequency selective amplifier, a parallel-T *RC* network is used in the negative feedback line. If the open-loop gain of the amplifier is G_0, then the gain with feedback is

$$G = G_0/(1 + G_0T).$$

Combining this with Eq. 14.7, we get

$$G = G_0 \frac{1 + \frac{j}{Q_T} \cdot \frac{x}{1-x^2}}{G_0 + 1 + \frac{j}{Q_T} \cdot \frac{x}{1-x^2}}$$

$$\therefore \quad |G| = G_0 \left\{ \frac{1 + \left(\frac{1}{Q_T} \cdot \frac{x}{1-x^2}\right)^2}{(G_0+1)^2 + \left(\frac{1}{Q_T} \cdot \frac{x}{1-x^2}\right)^2} \right\}^{1/2}.$$

The resonant gain is G_0; the gain is 3 dB. below this value at frequencies given by $|G| = 2^{-1/2}\, G_0$ which on simplification reduces to the following:

$$\left(\frac{1}{Q_T} \cdot \frac{x}{1-x^2}\right)^2 = G_0^2 + 2G_0 - 1.$$

The solutions of this equation are

$$x_{1,2} = \frac{1}{2G_0'} \left\{ \left(\frac{1}{Q_T^2} + 4G_0'\right)^{1/2} \mp \frac{1}{Q_T} \right\},$$

where

$$G_0' = (G_0^2 + 2G_0 - 1)^{\frac{1}{2}}.$$

Thus the selectivity of the amplifier is

$$Q_A = 1/(x_2 - x_1) = G_0' Q_T.$$

For $G_0 > 20$, $G_0' \simeq G_0 + 1$ to within an error of less than 0.25% so that

$$Q_A \simeq (G_0 + 1)Q_T.$$

Thus with the network considered previously,

$$Q_A = 0.475(G_0 + 1)$$

while with the conventional symmetrical network,

$$Q_A = 0.25(G_0 + 1).$$

For the same open-loop gain of 50 (say), the values of Q_A in these cases are respectively 24.20 and 12.75 while for the same Q_A of value 12.75, the amplifier with the asymmetrical network need have a gain of 26 only.

Conclusion

In situations where a continuous adjustment of the rejection frequency is desired, a general configuration will, of course, be of limited applicability, as the elements of the same kind are neither equal nor simply related. But for a fixed rejection frequency, a general network with proper asymmetry will definitely be a better choice. Also in its application as an F.M. discriminator in the low-frequency range, a value of Q_T nearly equal to its maximum value is required. Thus, the design procedure given in the chapter will be of much use in these situations.

Problems

P.1. Determine the transfer function of Fig. 14.1 circuit if $m_1 = 0$ and comment on the kind of filtering it can do.
P.2. Same if $m_2 = 0$.
P.3. Same if $m_3 = 0$.
P.4. Same if $n_1 = n_2 = 0$
P.5. Same if $n_3 = \infty$.

Acknowledgments The author is indebted to Prof. J. N. Bhar, D.Sc., F.N.I., and to Dr. A. K. Choudhury, M.Sc., D. Phil., for their kind help and advice in the preparation of this chapter.

References

1. L. Stanton, Theory and applications of parallel-T resistance capacitance frequency selective network. Proc. IRE. **34**, 447 (1946)
2. A. Wolf, Note on a parallel-T resistance capacitance network, *Proc. I.R.E.* **34**, 659 (1946)
3. Y. Oono, Design of parallel-T resistance capacitance network. *Proc. I.R.E.* **43**, 617 (1953)
4. D. Morris, Q as a mathematical parameter. Electron. Eng. 306 (1954)
5. J.R. Tillman, Linear frequency discriminator. Wirel. Engr. **23**, 281 (1946)
6. Paul T. Stine, Parallel-T discriminator design technique. Proc. Natl. Elec. Conf. **IX**, 26 (1950)
7. H. Fleischer, in *Vacuum Tube Amplifiers,* ed. by G.E. Valley Jr. and H. Wallman (McGraw-Hill, 1948), Chap. 10, p. 394

15 Perfect Transformer, Current Discontinuity and Degeneracy

That on connecting a source in the primary circuit of a perfectly coupled transformer, the currents in both the primary and secondary coils may be discontinuous does not appear to have been widely discussed in the literature. In this discussion, we present an analysis of the general circuit and show that in general, the currents will be discontinuous, except for specific combinations of the initial currents in the two coils. Although unity coupling coefficient cannot be realized in practice, a perfectly coupled transformer is a useful concept in circuit analysis and synthesis, and the results presented here should be of interest to students as well as teachers of circuit theory.

Keywords
Perfect transform · Current discontinuity
Degeneracy

Source: S. C. Dutta Roy, "Perfect Transformer, Current Discontinuity and Degeneracy", *IETE Journal of Education*, vol. 43, pp. 135–138, July–September 2002.

It is not common to find an analysis of coupled coils with initial currents in textbooks on circuit theory. Somehow, in the large number of books consulted by the author, it is always assumed that the coils are initially relaxed and imperfectly coupled. The only exception happens to be the book by Kuo [1], where the circuit shown in Fig. 15.1 has been analyzed with due regard to initial conditions. It has been shown that when $M^2 < L_1L_2$, i.e. when the coefficient of coupling $k = M/\sqrt{L_1L_2} < 1$, the currents i_1 and i_2 must be continuous at $t = 0$. On the other hand, if $k = 1$, then for the specific case $i_1(0-) = i_2(0-) = 0$, the currents are discontinuous, with

$$i_1(0+) = VL_2/(R_1L_2 + R_2L_1) \qquad (15.1)$$

and

$$i_2(0+) = -VM/(R_1L_2 + R_2L_1) \qquad (15.2)$$

Kuo is, however, silent on what happens when the coils are not initially relaxed. The specific question is the following: Is $k = 1$ necessary as well as sufficient for the currents to be discontinuous? We show, in this chapter, that this condition is necessary but not sufficient. In other words, even for $k = 1$, the currents may display continuity. First, we demonstrate this through an example. We next consider a more general circuit and analyze it to obtain expressions for $i_1(0+)$ and

S. C. Dutta Roy, *Circuits, Systems and Signal Processing*,
https://doi.org/10.1007/978-981-10-6919-2_15

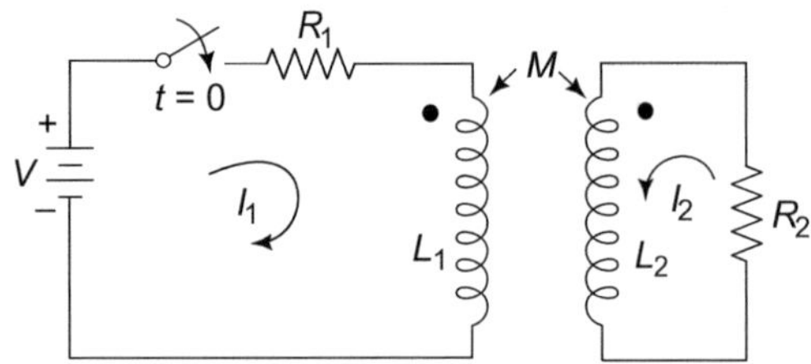

Fig. 15.1 The circuit analyzed by Kuo [1] and used in the example of this chapter with specific values

$i_2(0+)$ in terms of circuit parameters, source value at $t = 0+$, and currents and voltages in the circuit at $t = 0-$. We then derive the condition for current continuity in a perfectly coupled transformer. Finally, we consolidate the main results of the chapter and make some concluding remarks.

An Example

For the sake of completeness and for ready reference, we include, briefly, the analysis and results of Kuo for the circuit shown in Fig. 15.1 in Appendix A. Let

$$\left.\begin{array}{l} L_1 = 4H, L_2 = 1H, M = 2H \\ R_1 = 8\Omega, R_2 = 3\Omega, V = 6V \\ i_1(0-) = 0 \quad \text{and } i(0-) = -1A \end{array}\right\} \tag{15.3}$$

From Eqs. 15.25 and 15.26, then, we get

$$\left.\begin{array}{l} i_1(0+) = 0 \\ i_2(0+) = -1A \end{array}\right\} \tag{15.4}$$

Hence, the currents are continuous despite $k = 1$.

This counterexample is sufficient to prove that $k = 1$ is only a necessary but not a sufficient condition for current discontinuity.

Analysis of the General Circuit

We now consider the general circuit shown in Fig. 15.2 which includes an initially charged capacitor in each loop and, in addition, the voltage source is generalized to $v(t)$ instead of a battery. We assume that $v(t)$ does not contain impulses. The loop equations now become

$$\begin{aligned} v(t) =& R_1 i_1(t) + L_1 i_1'(t) + M i_2'(t) \\ &+ \frac{1}{C_1}\int_{0-}^{t} i_1(t)\,\mathrm{d}t + v_1(0-) \end{aligned} \tag{15.5}$$

$$\begin{aligned} 0 =& M i_1'(t) + R_2 i_2 + L_2 i_2'(t) \\ &+ \frac{1}{C_2}\int_{0-}^{t} i_2(t)\,\mathrm{d}t + v_2(0-) \end{aligned} \tag{15.6}$$

The sum of the last two terms on the right-hand side of Eq. 15.5 represents $v_1(t) = q_1(t)/C_1$, where $q_1(t)$ denotes the charge on C_1. Similarly, $v_2(t) = q_2(t)/C_2$. Integrals of $v_1(t)$, $v_2(t)$ as well as $v(t)$ from $t = 0-$ to $t = 0+$ will be zero because none of them contains impulses. Thus, if we integrate Eqs. 15.5 and 15.6 from $t = 0-$ to $t = 0+$, we get

$$\begin{aligned} 0 = L_1[i_1(0+) \\ - i_1(0-)] + M[i_2(0+) - i_2(0-)] \end{aligned} \tag{15.7}$$

$$0 = M[i_1(0+) - i_1(0-)] + L_2[i_2(0+) - i_2(0-)] \tag{15.8}$$

These are the same as in Kuo's circuit, as given in Eqs. 15.20 and 15.21. Note that Eqs. 15.7 and 15.8 imply that the principle of conservation of flux applies to each coil individually, i.e. the flux in either coil at $t = 0-$ is the same as that at $t = 0+$. Also note that the generalized circuit does not change the conclusion arrived at in Kuo's circuit, viz. that if $k < 1$, then the currents in the two coils must be continuous.

For the case $k = 1$, Eq. 15.6 gives, at $t = 0+$, the following equation:

$$\begin{aligned} 0 =& R_2 i_2(0+) + (M/L_1)[L_1 i_1'(0+) \\ &+ M i_2'(0+)] + v_2(0-), \end{aligned} \tag{15.9}$$

which can be rewritten as

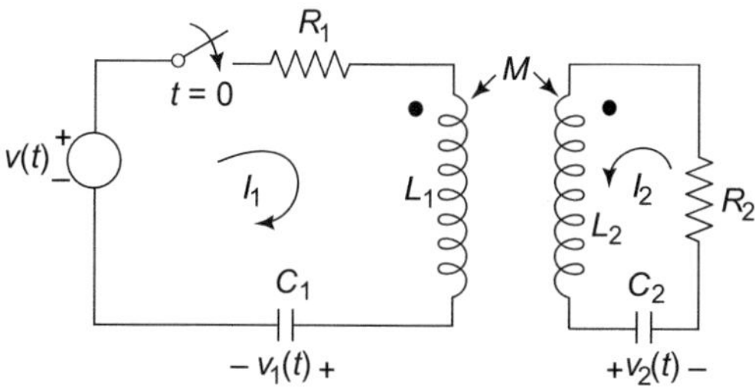

Fig. 15.2 A more general circuit than that shown in Fig. 15.1 with a generalized source $v(t)$, and initially charged capacitors in each loop

$$L_1 i_1'(0+) + M i_2'(0+) = -(L_1/M)[R_2 i_2(0+) + v_2(0-)] \quad (15.10)$$

Now, putting $t = 0+$ in Eq. 15.5 and substituting from Eq. 15.10, we get

$$v(0+) = R_1 i_1'(0+) - (L_1/M)[R_2 i_2(0+) + v_2(0-)] + v_1(0-). \quad (15.11)$$

Combining Eq. 15.11 with 15.7, we get the following two simultaneous equations in $i_1(0+)$ and $i_2(0+)$:

$$i_1(0+) - \frac{R_2 L_1}{R_1 M} i_2(0+) = \frac{v(0+) - v_1(0-) + (L_1/M) v_2(0-)}{R_1} \quad (15.12)$$

$$i_1(0+) + \frac{M}{L_1} i_2(0+) = i_1(0-) + \frac{M}{L_1} i_2(0-) \quad (15.13)$$

Solving Eqs. 15.12 and 15.13 gives, finally,

$$i_1(0+) = \frac{R_2[L_1 i_1(0-) + M i_2(0-)] + L_2[v(0+) - v_1(0-)] + M v_2(0-)}{R_1 L_2 + R_2 L_1} \quad (15.14)$$

$$i_2(0+) = \frac{R_1[L_2 i_2(0-) + M i_1(0-)] + M[v(0+) - v_1(0-)] + L_1 v_2(0-)}{R_1 L_2 + R_2 L_1} \quad (15.15)$$

Condition for Continuity of Currents Under Perfect Coupling

If the currents are to be continuous, then it suffices to equate Eq. 15.14 to $i_1(0-)$ or Eq. 15.15 to $i_2(0-)$ because from Eq. 15.7 or Eq. 15.8, $i_1(0+) = i(0-)$ guarantees that $i_2(0+) = i_2(0-)$, and vice versa. Equating Eq. 15.15 to $i_2(0-)$ gives the following condition for continuity:

$$i_2(0-) = \frac{R_1 M i_1(0-) - M[v(0+) - v_1(0-)] - L_1 v_2(0-)}{R_2 L_1} \quad (15.16)$$

with $i_1(0-)$ arbitrary. In other words, for every $i_1(0-)$, there exists one $i_2(0-)$ for the currents to be continuous and vice versa. For all other combinations of $i_1(0-)$ and $i_2(0-)$, the currents will be discontinuous. It is, of course, implied that other conditions, viz., $v(0+)$, $v_1(0-)$ and $v_2(0-)$, do not change. Should that be the case, it is clear that the relationship between $i_2(0-)$ and $i_1(0-)$, as given by Eq. 15.16, is a straight line with a slope of $R_1M/(R_2L_1)$ and an intercept of

$$-\frac{M[v(0+) - v_1(0-)] + L_1 v_2(0-)}{R_2 L_1} \quad (15.17)$$

on the $i_2(0-)$ axis. For the example considered earlier, the slope is 4/3 while the intercept is −1 A.

Concluding Remarks

We have shown in this chapter that in an imperfectly coupled transformer, the currents in the two coils are always continuous. For perfect coupling, on the other hand, the currents are always discontinuous except for specific combinations of the two initial currents. More specifically, for each initial current in one coil, there exists a particular value of the initial current in the other coil, for which the currents will be continuous. These combinations lie on a straight line, when one current is plotted against the other. These conclusions are valid for any combination of resistors and capacitors in the two loops, with or without initial charges in the capacitors. It is obvious, however, that including another inductor in either or both loops makes the coupling imperfect, and the currents will then be always continuous.

In this context, the following two observations made by Seshu and Balabanian [2] are of interest:

(1) 'If idealized *R, L* and *C* branches, voltage generators, and current generators are arbitrarily connected together, the system may not have the maximum possible order. It is only when such degeneracies are present that discontinuities in inductance currents and capacitance voltages are encountered. No existence theorems have been proved by mathematicians for these cases …' (p. 103).

(2) '… it may be expected that inductance currents will be discontinuous when there are junctions or effective junctions… at which only inductances and current generators are present' (p. 104).

By an effective junction in the second observation, the authors mean 'a junction at which only inductances and current generators would meet if we suitably interchanged series connected two terminal networks or shorted some branches. Thus an effective junction is the same as a *cut set*' (p. 104, footnote).

The first observation regarding degeneracy is clearly demonstrated in the case of coupled coils by the two examples in [1] (pp. 124–126). For $k < 1$ in the circuit shown in Fig. 15.1, the system has two natural frequencies, although it has three inductances L_1, L_2 and M. They are not physically connected at a junction, but in the *equivalent*[1] circuit shown in Fig. 15.3, we do have a junction of $L_1 - M$, M and $L_2 - M$. This is not an 'effective' junction in the sense of [2], but we may call it an 'equivalent' junction. It is no wonder, therefore, that we get the second-order system, instead of the third-order one. An alternative way of justifying the result is to note that we can specify only two initial conditions for the system, the initial current in M being dependent on those in L_1 and L_2. However, despite the degeneracy, there is no discontinuity in the currents!

When $k = 1$, further degeneracy sets in, not because we cannot specify two initial currents, but (in our opinion) because M is completely specified if L_1 and L_2 are specified. As the second example of [2] (pp. 125–126) demonstrates, the system now has only one natural frequency and behaves like the first-order system. *Despite this 'double' degeneracy, however, the currents are not always discontinuous,* as demonstrated in this chapter analytically and by an example. It is clear that a deeper examination of the case is needed to resolve the issue in terms of physical concepts.[2]

Problems

P.1. Suppose R_1 in Fig. 15.1 circuit is shunted by a capacitor C. Investigate the discontinuity in this circuit.

P.2. Same, with C shifted to be across R_2.

P.3. Same, with C shifted to be in series with R_2.

P.4. Same, with C in series with R_1.

[1]This 'equivalent' circuit implies only mathematical equivalence (of the loop equations) but not physical equivalence, because the two coils have no common point.

[2]Notably, in [2], there are no examples or discussions on initial conditions in coupled coils. In the only example in which a coupled coil appears (pp 110–112), inductor junctions are created through additional inductors in each circuit and a current generator in the secondary circuit.

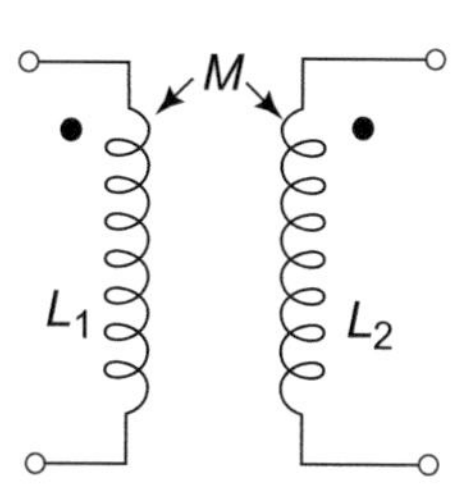

≡

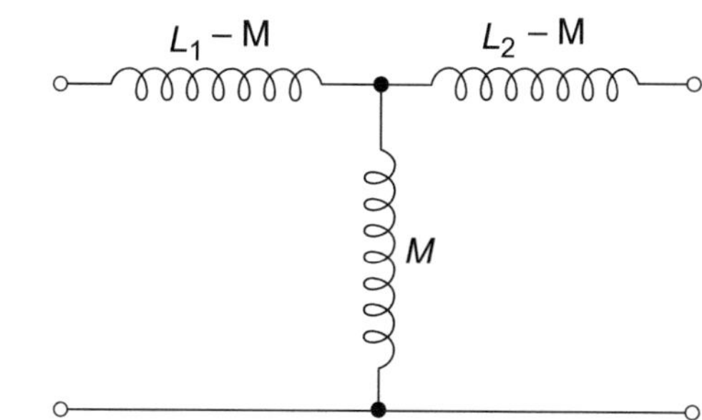

Fig. 15.3 A mathematical equivalent circuit for two coupled coils: coupling is not always good!

P.5. Same, with C_1 in series with R_1, C_2 in series with R_2, and a voltage output taken across R_2.

Appendix

Kuo's analysis and results for the circuit are shown in Fig. 15.1.

The loop equations for the circuit shown in Fig. 15.1 are

$$Vu(t) = L_1 i_1'(t) + R_1 i_1(t) + M i_2'(t) \quad (15.18)$$

$$0 = M i_1'(t) + R_2 i_2(t) + L_2 i_2'(t) \quad (15.19)$$

Integrating Eqs. 15.18 and 15.19 from $t = 0-$ to $t = 0+$, we get

$$L_1[i_l(0+) - i_1(0-)] + M[i_2(0+) - i_2(0-)] = 0 \quad (15.20)$$

$$M[i_1(0+) - i_l(0-)] + L_2[i_2(0+) - i_2(0-)] = 0 \quad (15.21)$$

Combining Eqs. 15.19 and 15.21 gives

$$(L_1 L_2 - M^2)[i_1(0+) - i_1(0-)][i_2(0+) - i_2(0-)] = 0 \quad (15.22)$$

which, along with Eq. 15.20 or 15.21, clearly indicates that if $L_1L_2 > M^2$, i.e. $k < 1$, then the currents are continuous at $t = 0$. On the other hand, if $k = 1$, then they need not be. In fact, in this case, Eq. 15.19 gives at $t = 0+$:

$$R_2 i_2(0+) = -(M/L_1)[L_1 i_1'(0+) + M i_2'(0+)] \quad (15.23)$$

which, substituted in Eq. 15.18 with $t = 0+$, yields

$$V = R_1 i_1(0+) - (L_1/M) R_2 i_2(0+) \quad (15.24)$$

Combining this with Eq. 15.20, one can solve for $i_1(0+)$ and $i_2(0+)$. The results are[3]

$$i_1(0+) = \frac{VL_2 + R_2[L_1 i_1(0-) + M i_2(0-)]}{R_1 L_2 + R_2 L_1} \quad (15.25)$$

$$i_2(0+) = \frac{-VM + R_1[M i_1(0-) + L_2 i_2(0-)]}{R_1 L_2 + R_2 L_1} \quad (15.26)$$

References

1. F.F. Kuo, *Network Analysis and Synthesis* (John Wiley, New York, 1966), pp. 123–126
2. S. Seshu, N. Balabanian, *Linear Network Analysis* (John Wiley, New York, 1963), pp. 101–112

[3]Kuo [1], at this point, assumes $i_1(0-) = i_2(0-)$, presumably, as an example. We give general results in Eqs. 15.25 and 15.26.

16 Analytical Solution to the Problem of Charging a Capacitor Through a Lamp

An analytical solution is presented for the problem of charging a capacitor through a lamp, by assuming a polynomial relationship between the resistance of the lamp and the current flowing through it. The total energy dissipated in the lamp is also easily calculated thereby. An example of an available practical case is used to illustrate the theory.

Keywords
Capacitor charging · Differential equation
Energy

Introduction

The charging of a capacitor from a battery through a resistance is a standard topic in the undergraduate curriculum of Physics or Engineering in the theory as well as laboratory classes. A 2006 paper by Ross and Venugopal [1] (hereafter referred to as RV) deals with an interesting variation of this topic in which the resistor is replaced by a lamp, whose resistance *R* varies with the current *i* flowing through it. They solved the resulting differential equation by applying numerical techniques and found a close fit between these results and the experimental ones. The aim of this chapter is to present an analytical, rather than numerical solution to the problem. For this purpose, we assume a polynomial relationship between $R(i)$ and i. The total energy dissipated in the lamp is also easily calculated thereby. The experimental data of RV are used to illustrate the validity of the theory.

Source: S. C. Dutta Roy, "Analytical Solution to the Problem of Charging a Capacitor through a Lamp," IETE Journal of Education, vol. 47, pp. 145–147, July–September 2006.

The Circuit and the Differential Equation

The circuit under consideration is shown in Fig. 16.1, which obeys the integral equation

$$iR(i) + \frac{1}{C}\int_0^t i\,\mathrm{d}t = V \qquad (16.1)$$

Differentiating Eq. 16.1, we get

$$R\frac{\mathrm{d}i}{\mathrm{d}t} + i\frac{\mathrm{d}R}{\mathrm{d}t} + \frac{i}{C} = 0, \qquad (16.2)$$

where, for brevity, the dependence of *R* on *i* is not shown explicitly. Assuming, as in RV, that the thermal relaxation time of the lamp filament is much less than the time during which the

S. C. Dutta Roy, *Circuits, Systems and Signal Processing*,
https://doi.org/10.1007/978-981-10-6919-2_16

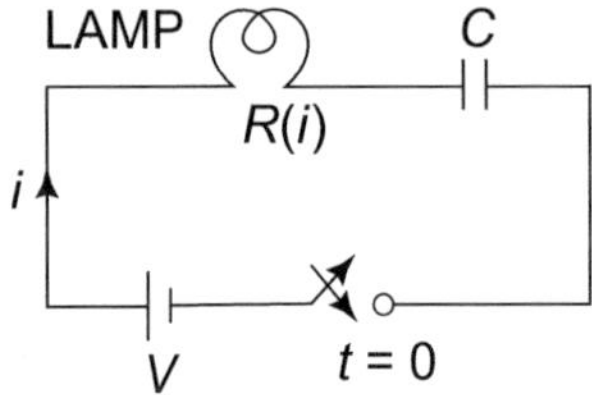

Fig. 16.1 The basic charging circuit

current in the filament changes significantly, we can write

$$\frac{\mathrm{d}R}{\mathrm{d}t} = \frac{\mathrm{d}R}{\mathrm{d}i} \cdot \frac{\mathrm{d}i}{\mathrm{d}t} \tag{16.3}$$

Combining Eqs. 16.2 and 16.3, we get

$$\left(R + i\frac{\mathrm{d}R}{\mathrm{d}i}\right)\frac{\mathrm{d}i}{\mathrm{d}t} + \frac{i}{C} = 0. \tag{16.4}$$

Solution of the Differential Equation

As illustrated in Fig. 3 of RV, the variation of $R(i)$ with i is approximately linear, except at high values of i. In general, we can assume R and i to obey a polynomial relationship of the form

$$R = R_0\left(1 + \sum_{k=1}^{N} a_k i^k\right), \tag{16.5}$$

where N will depend upon the required accuracy. For most practical situations, $N = 2$ or 3 suffices. We shall consider here a third order polynomial, but if required, the treatment can be extended to any order. Let, therefore,

$$R = R_0(1 + a_1 i + a_2 i^2 + a_3 i^3) \tag{16.6}$$

Then

$$\frac{\mathrm{d}R}{\mathrm{d}i} = R_0(a_1 + 2a_2 i + 3a_3 i^2) \tag{16.7}$$

Combining Eq. 16.4 with Eqs. 16.6 and 16.7, we get, on simplification,

$$[(1/i) + 2a_1 + 3a_2 i + 4a_3 i^2]\mathrm{d}i = -\mathrm{d}t/(R_0 C). \tag{16.8}$$

Integrating both sides of Eq. 16.8 gives

$$\begin{aligned} &\ln i + 2a_1 i + (3/2)a_2 i^2 + (4/3)a_3 i^3 \\ &\quad = -[t/(R_0 C)] + K. \end{aligned} \tag{16.9}$$

To evaluate the integration constant K, we note that at $t = 0$, $i = i_0 = V/(R_0)$. Putting this initial condition in Eq. 16.9, we get the value of K as the left hand side of Eq. 16.9 with i replaced by i_0. Finally, therefore, the equation for the current becomes

$$\begin{aligned} t = R_0 C[&\mathrm{In}(i_0/i) + 2a_1(i_0 - i) + (3/2)a_2(i_0^2 \\ &- i^2) + (4/3)a_3(i_0^3 - i^3)]. \end{aligned} \tag{16.10}$$

Equation 16.10 is transcendental in i and for a given t, it has to be solved numerically. A better strategy would be to compute t for various values of i in the range of interest and then to plot the variation of i with t, as we shall do in the example to follow.

Energy Dissipated in the Lamp

The energy dissipated in the lamp is given by

$$E = \int_0^\infty R(i) i^2 \mathrm{d}t \cdot \tag{16.11}$$

Combining Eq. 16.11 with Eq. 16.6, substituting for dt from Eq. 16.8, changing the limits of the integral (from $t = 0$ to $i = i_0$ and $t = \infty$ to $i = 0$) and simplifying, we get

$$E = CR_0^2 \int_0^{i_0} \left[i + 3a_1 i^2 + \left(4a_2 + 2a_1^2\right)i^3 + 5(a_3 + a_1 a_2)i^4 + \left(6a_1 a_3 + 3a_2^2\right)i^5 + 7a_2 a_3 i^6 + 4a_3^2 i^7\right] \mathrm{d}i$$

$$= CR_0^2 \left[\frac{i_0^2}{2} + a_1 i_0^3 + \left(a_2 + \frac{a_1^2}{2}\right) i_0^4 + (a_3 + a_1 a_2) i_0^5 + \left(a_1 a_3 + \frac{a_2^2}{2}\right) i_0^6 + a_2 a_3 i_0^7 + \frac{a_3^2}{2} i_0^8\right] \tag{16.12}$$

A plot of Eq. 16.14 is shown in Fig. 16.2, which, as predicted by RV, is virtually indistinguishable from that given in Fig. 4 of their paper.

Example

We use the experimental data given in RV to illustrate the application of the theory presented here. As mentioned earlier, Fig. 3 of RV shows that the variation of $R(i)$ with i is predominantly linear. By considering the two points (0.03 A, 10 Ω) and (0.07 A, 20 Ω) in this figure, we get

$$R(i) = 2.5(1 + 100i) \tag{16.13}$$

With $C = 0.154$ F and $i_0 = 0.15$ A (as given in Fig. 4 of RV), Eq. 16.10 becomes, for this case,

$$t = 0.385(28.1 - \ln i - 200i). \tag{16.14}$$

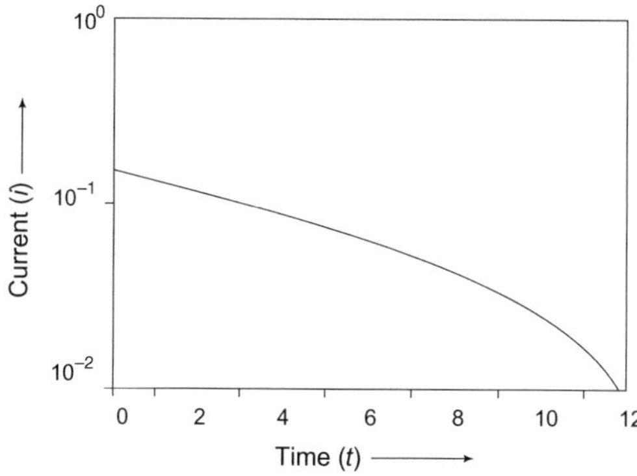

Fig. 16.2 Variation of i with t for the example

The total energy dissipated in the lamp for this case is given by

$$E = CR_0^2 \left[\frac{i_0^2}{2} + a_1 i_0^3 + \frac{a_1^2}{2} i_0^4\right], \tag{16.15}$$

which is calculated as 2.772 J.

Conclusion

It is shown that if the functional dependence of the lamp resistance on current is known in the form of a polynomial relationship, then the charging process of a series capacitor can be analytically determined. It is then also easy to determine the energy dissipated in the lamp during the charging process. It is easily shown that the discharging of a charged capacitor through a lamp also follows Eq. 16.2 and hence the theory presented here also applies to the discharging process.

Problems

P.1. In the circuit of Fig. 16.1, add an inductor L in series. Write the differential equation and solve it.

P.2. Let, in Fig. 16.1, C be shifted to be across the lamp. Obtain the differential equation and solve it.

P.3. Can you solve Eq. 16.10 analytically? After all, it is a cubic equation, and can be solved by Cardan's method. Try it.

P.4. What happens if Eq. 16.6 has another term a_4t^4?

P.5. Repeat the example in the text with an extra term $10i^2$ in Eq. 16.13.

Acknowledgements The author thanks Professor Jayadeva for his help in the preparation of Fig. 16.2.

Reference

1. R. Ross, P. Venugopal, On the problem of (dis) charging a capacitor through a lamp. Am. J. Phys. **74**, 523–525 (2006)

Difference Equations, Z-Transforms and Resistive Ladders

17

It is shown that the semi-infinite and infinite resistive ladder networks composed of identical resistors can be conveniently analyzed by the use of difference equations or z-transforms. Explicit and simple expressions are obtained for the input resistance, node voltages and the resistance between two arbitrary nodes of the network.

Keywords
Infinite networks · Resistive ladders
Difference equations · Z-transforms

Introduction

Difference equations and z-transforms are techniques for dealing with discrete time signals and systems, of which the former is in time domain and the latter is in the frequency domain. Analysis of a purely resistive network does not normally require any tool in the frequency domain, neither does the network process discrete time signals so as to require the use of difference equations. KCL (Kirchoff's Current Law), KVL (Kirchoff's Voltage Law) and Ohm's law should be adequate for dealing with such networks. Yet, there are situations where the use of difference equations and/or frequency domain techniques offers significant advantages over conventional methods. This chapter is concerned with one such situation, viz. a semi-infinite or infinite resistive ladder network.

The semi-infinite resistive ladder network shown in Fig. 17.1 is often posed as a problem [1, 2] to undergraduate students for finding the input resistance $R_i = V_0/I_0$. The solution is easily found by noting that the resistance looking to the right of nodes 1 and ground should also be R_i. Thus,

$$R_i^2 - RR_i - R^2 = 0 \tag{17.1}$$

which gives the quadratic equation

$$R_i^2 - RR_i - R^2 = 0. \tag{17.2}$$

Noting that R_i must be positive, we get

$$R_i = R\left(1+\sqrt{5}\right)/2 = R\Phi, \tag{17.3}$$

where Φ is the so-called 'golden ratio'.

What about the potential v_n at node n, $1 \leq n < \infty$? Solution of this problem appears in [3] in the form of an integral obtained by using the concept of discrete Fourier transform. It will be

Source: S. C. Dutta Roy, "Difference Equations, Z-Transforms and Resistive Ladders," *IETE Journal of Education*, vol. 52, pp. 11–15, January–June 2011.

S. C. Dutta Roy, *Circuits, Systems and Signal Processing*,
https://doi.org/10.1007/978-981-10-6919-2_17

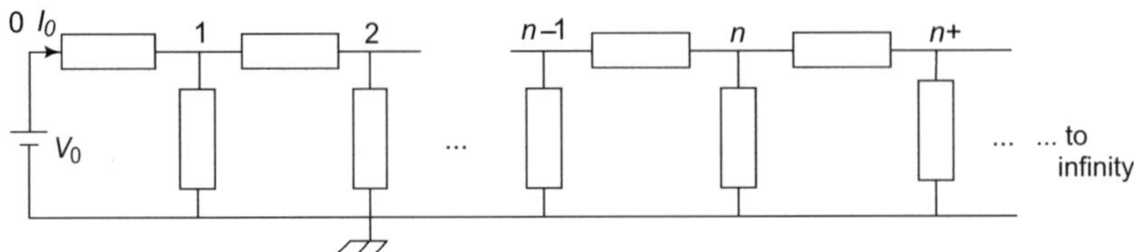

Fig. 17.1 The semi-infinite resistive ladder. Each resistance is of value R

shown in this chapter that the solution can be obtained in a simpler form by using the theory of difference equations or by application of the z-transform technique. In the process, we have also considered the infinite resistive ladder of Fig. 17.2 and have calculated the resistance offered to a battery connected between two arbitrary nodes of this infinite ladder.

Besides [3], there exists a substantial volume of literature on the subject of semi-infinite and infinite resistive ladders. Some of the prominent ones, which are of educational and pedagogic interest, will be reviewed here. Lavatelli [4] considered an infinite balanced ladder i.e. one in which the lower ground line of Fig. 17.2 is replaced by a chain of resistors. He gave a difference equation formulation for the resistance between two arbitrary nodes.

Our treatment here in Part IV has been inspired by his work and follows the same line of analysis. Srinivasan [5] considered the semi-infinite ladder with different values of series and shunt resistors and showed that when they are equal, the input resistance is $R\Phi$, as in Eq. 17.3. He also showed that the successive convergents of the continued fraction form of the input resistance are related to the Fibonacci sequence. As an extension of [5], Thomson [6] analyzed a semi-infinite ladder in which the resistors in the successive sections differ by a factor of b. He showed that by choosing b appropriately, one can obtain the golden ratio, $\sqrt{2}$ and some other irrational numbers in a non-geometric context. Parera-Lopez [7] made some generalizations of [5, 6]. Denardo et al. [8] presented some numerical and laboratory experiments on finite N-section ladders and showed that the convergence of the input resistance to $R\Phi$ is exponential and rapid. For example, for $N \geq 5$, the deviation from $R\Phi$ occurs only in the fourth place of decimal, while for $N \geq 7$, the deviation occurs in fifth place of decimal. Bapeswara Rao [9] related the finite resistance ladder to the effective resistance between the centre and a vertex of an N-sided polygon of resistors.

Besides these papers of pedagogic interest, there have appeared many scholarly papers on infinite networks in IEEE and other professional journals, the most prominent author being Zemanian (see, e.g. [10, 11] and the references cited there). Reference [10] deserves special mention because it is a tutorial paper addressed to undergraduate students in a rather unique and enjoyable style. Zemanian's book [12] gives a comprehensive treatment of the subject with the necessary mathematical rigour.

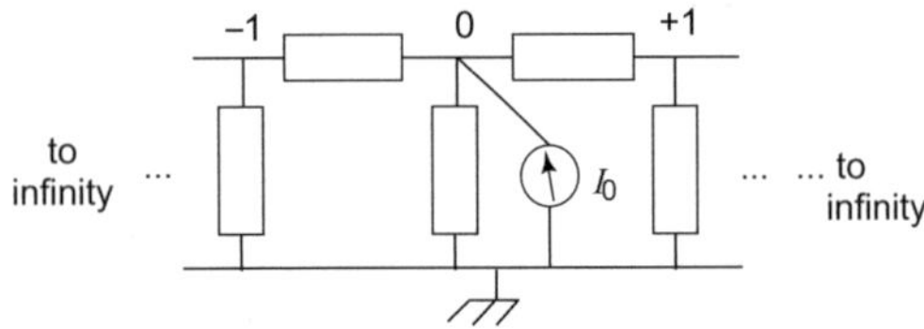

Fig. 17.2 Infinite resistive ladder driven by a current source I_0 at node 0. Each resistance has a value R

Solution by Difference Equation Approach

Consider, in Fig. 17.1, the nodes $n - 1$, n and $n + 1$, $n > 0$. By writing KCL at node n and simplifying, we get

$$3v_n - v_{n-1} - v_{n+1} = 0. \qquad (17.4)$$

This is a difference equation of order 2, and assuming a solution of the form λ^n, we get the characteristic equation

$$\lambda^2 - 3\lambda + 1 = 0. \qquad (17.5)$$

The solution of Eq. 17.5 are

$$\lambda_{1,2} = \left(3 \pm \sqrt{5}\right)/2. \qquad (17.6)$$

Note that $\lambda_1\lambda_2 = 1$; for convenience, we shall call λ_1 as α so that $\lambda_2 = \alpha^{-1}$. Thus, the general solution for v_n is

$$v_n = A\alpha^n + B\alpha^{-n}; \quad \alpha = \left(3 + \sqrt{5}\right)/2. \qquad (17.7)$$

The constants A and B are evaluated from the boundary conditions $v_0 = V_0$ and $v_\infty = 0$, the latter being dictated by physical considerations. The second condition forces A to be zero while the first one makes $B = V_0$. Thus, finally,

$$v_n = V_0\left[\left(3 - \sqrt{5}\right)/2\right]^n. \qquad (17.8)$$

Z-Transform Solution

To apply the z-transform technique [13], it is instructive to consider the infinite ladder of Fig. 17.2, with a current generator I_0 connected between node 0 and ground. Then the difference equation 17.4 is modified to the following:

$$3v_n - v_{n-1} - v_{n+1} = I_0\delta(n), \qquad (17.9)$$

where $\delta(n) = 1$ for $n = 0$ and zero otherwise. Defining the z-transform in the usual manner, i.e.

$$Z[v_n] = V(z) = \sum_{n=-\infty}^{\infty} v_n z^{-n}, \qquad (17.10)$$

we have

$$Z[\delta(n)] = 1, \quad Z[v_{n+1}] = zV(z), \quad Z[v_{n-1}] = z^{-1}V(z). \qquad (17.11)$$

Thus taking the z-transform of both sides of 17.9 and simplifying, we get

$$\frac{V(z)}{I_0R} = \frac{-z^{-1}}{1 - 3z^{-1} + z^{-2}} = \frac{-z^{-1}}{(1 - \alpha z^{-1})(1 - z^{-1}/\alpha)}, \qquad (17.12)$$

where α is the same as that given by 17.7. Expending 17.12 in partial fractions and using the fact that $\alpha - 1/\alpha = \sqrt{5}$, we get

$$\frac{V(z)}{I_0R} = \frac{1}{\sqrt{5}}\left[\frac{1}{1 - z^{-1}/\alpha} - \frac{1}{1 - \alpha z^{-1}}\right]. \qquad (17.13)$$

The pole at $z = \alpha$ is outside the unit circle while that at $z = 1/\alpha$ is inside the unit circle. The physical situation demands that the sequence v_n should decrease on both sides of $n = 0$ and tend to zero when $n \to \infty$. Hence, the first term in 17.13 represents the z-transform of the right-sided sequence $\{v_0, v_1, \ldots \text{ to } \infty\}$ with $|z| < \alpha$ as the region of convergence, while the second term represents the z-transform of the left-sided sequence $\{v_{-1}, v_{-2}, \ldots \text{ to } \infty\}$ with $|z| < \alpha$ as the region of convergence. Thus, the inversion of Eq. 17.13 gives

$$\frac{v_n}{I_0R} = \frac{1}{\sqrt{5}}[\alpha^{-n}u(n) + \alpha^n u(-n-1)], \qquad (17.14)$$

where $u(n)$ is the unit step function, having the value unity for $n \geq 0$ and zero otherwise. More explicitly,

$$v_n = \frac{I_0R}{\sqrt{5}}\left(\frac{3 - \sqrt{5}}{2}\right)^n, \quad n \geq 0; \qquad (17.15)$$

$$v_n = \frac{I_0 R}{\sqrt{5}} \left(\frac{3+\sqrt{5}}{2} \right)^n, \quad n<0 \qquad (17.16)$$

This gives the complete solution for the infinite ladder of Fig. 17.2. The resistance seen by the current generator I_0 is

$$R_\infty = RR_iR_i = R/\sqrt{5} \qquad (17.17)$$

so that $v_0 = I_0 R/\sqrt{5}$; this verifies that Eq. 17.15 gives correct results for the semi-infinite ladder, as derived independently in Eq. 17.8. Also, as expected, $v_n = v_{-n}$, and both tend to zero as $n \to \infty$.

Resistance Between Any Two Arbitrary Nodes of an Infinite Ladder

We now consider another relevant problem in the infinite ladder, viz. that of finding the resistance offered to a source connected between any two arbitrary nodes m and $m + r$. Let the source be a voltage generator V_0 and let a set of $r + 1$ mesh currents be formulated as shown in Fig. 17.3, where the last mesh includes V_0 and the network to the left of node m and that to the right of node $m + r$ have been replaced by an equivalent resistance $R_T = R||R_i = (\sqrt{5} - 1)/2$. Consider the nth mesh, $1 < n < r$. Writing KVL around this mesh gives the equation

$$3i_n - i_{n-1} - i_{n+1} = I_0. \qquad (17.18)$$

The only difference between 17.18 and 17.4 is that the right-hand side in the former is not zero. Hence we shall have a constant term, representing the particular solution of Eq. 17.18, in addition to the solution of the form given by Eq. 17.7. It is easily seen from Eq. 17.18 that this constant term is I_0. Thus the solution of Eq. 17.18 is

$$i_n = A\alpha^n + B\alpha^{-n} + I_0; \quad \alpha = \left(3+\sqrt{5}\right)/2. \qquad (17.19)$$

The constant A and B have to be determined from the boundary conditions that hold at nodes m and $m + r$. Since the network is perfectly symmetrical with respect to an imaginary vertical line at the middle, the voltages at nodes $m + r$ and m are, respectively, $+V_0/2$ and $-V_0/2$. Thus

$$i_1 = i_r = V_0/(2R_T). \qquad (17.20)$$

Combining Eqs. 17.19 and 17.20, we get two simultaneous equations in A and B, the solution of which gives

$$(A, B) = \frac{[V_0/(2R_T)] - I_0}{1 + \alpha^{r-1}} \left(\alpha^{-1}, \alpha^r\right). \qquad (17.21)$$

Thus, finally,

$$i_n = I_0 + \frac{[V_0/(2R_T)] - I_0}{1 + \alpha^{r-1}} \left(\alpha^{n-1} + \alpha^{-n+r}\right). \qquad (17.22)$$

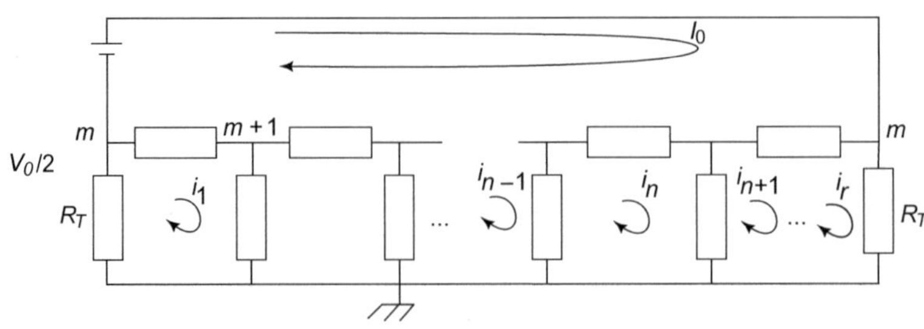

Fig. 17.3 Circuit for determining the resistance between any two arbitrary nodes m and $m + r$. Each unmarked resistance has a value R

Application of KVL around the $(r + 1)$th mesh gives

$$V_0 = \sum_{n=1}^{r} (I_0 - i_n)R. \quad (17.23)$$

Combining Eqs. 17.22 and 17.23 gives

$$V_0 = \frac{-[V_0/(2R_T)] + I_0}{1 + \alpha^{r-1}} \sum_{n=1}^{r} \left(\alpha^{n-1} + \alpha^{-n+r}\right). \quad (17.24)$$

Clearly,

$$\sum_{n=1}^{r} \alpha^{n-1} = \sum_{n=1}^{r} \alpha^{-n+r} = \frac{1-\alpha^r}{1-\alpha}. \quad (17.25)$$

Using Eq. 17.25 in Eq. 17.24 and simplifying, we get

$$R_r = \frac{2R}{\sqrt{5}}\left[1 - \left(\frac{3-\sqrt{5}}{2}\right)^n\right]. \quad (17.26)$$

It is easily verified by direct calculation that Eq. 17.26 give correct results for $r = 1$, 2 and 3, which are, respectively,

$$\begin{aligned} R_1 &= R(1 - 1/\sqrt{5}), \\ R_2 &= R(3 - \sqrt{5}), \\ R_3 &= 8R(1 - 2/\sqrt{5}). \end{aligned} \quad (17.27)$$

For a semi-infinite ladder, the condition of symmetry no longer holds and the appropriate boundary conditions have to be used at both the end meshes 1 and r. For examples, if $m = 1$, then the boundary conditions are $i_1 = i_r = V_0/(R_T + R)$; for $m = 0$, the resistance would be R + (the value for $m = 1$); for $m = 2$, the boundary conditions are $i_1 = i_r = V_0/(R_T + 2/3)$ and so on.

Concluding Discussion

We have used difference equations and z-transform to analyze semi-infinite and infinite resistive ladders. While the former has been used earlier, the use of z-transforms is believed to be new and instructive. The explicit formulas for the node voltages and the resistance between two arbitrary nodes also appear to be new.

Problems

P.1. Suppose in Fig. 17.1, the ladder is terminated at the third node on the right. What impedance does I_0 face? This is easy!

P.2. Suppose, in Fig. 17.2, the current generator is replaced by a voltage generator and the ladder is terminated in node marked + n and $-n$. What current will flow from the generator? This is super-easy!

P.3. Suppose, in Fig. 17.2, each shunt resistors or is replaced by a capacitor C. What is the input impedance? This is not so easy, but not difficult too!

P.4. Same as P.4, but each series resistor is replaced by a capacitor C. What is the input impedance? Same level of difficulty as in P.3.

P.5. Same as P.5, but each shunt resistor is replaced by an inductance L.

Acknowledgments This work was supported by the Indian National Science Academy through the Honorary Scientist scheme.

References

1. E.M. Purcell, in *Electricity and Magnetism, Berkeley Physics Course—Vol. 2*, 2nd edn. (New York, McGraw-Hill, 1985), pp. 167–168
2. F.W. Sears, M.W. Zemansky, in *College Physics, World Students*, 5th edn. (Reading, MA, Addison-Wesley, 1980)
3. R.M. Dimeo, Fourier transform solution to the semi-infinite resistance ladder. American J. Phys. **68**(7), 669–670 (2000)
4. L. Lavatelli, The resistive net and difference equation. American J. Phys. **40**(9), 1246–1257 (1972, September)
5. T.P. Srinivasan, Fibonacci sequence, golden ratio and a network of resistors. American J. Phys. **60**(5), 461–462 (1992)
6. D. Thompson, Resistor networks and irrational numbers. American J. Phys. **65**(1), 88 (1997)

7. J.J. Parera-Lopez, T-iterated electrical networks and numerical sequences. American J. Phys. **65**(5), 437–439 (1997)
8. B. Denardo, J. Earwood, V. Sazonava, Experiments with electrical resistive networks. American J. Phys. **67**(11), 981–986 (1999, November)
9. V.V. Bapeswara Rao, Analysis of doubly excited symmetric ladder networks. American J. Phys. **68**(5), 484–485 (2000)
10. A.H. Zemanian, Infinite electrical networks: a reprise. IEEE Trans. Circuits Sys. **35**(11), 1346–1358 (1988)
11. A.H. Zemanian, Infinite electrical networks. Proc. IEEE **64**(1), 1–17 (1976)
12. A.H. Zemanian, *Transfiniteness for graphs, electrical networks and random walks* (Birkhauser, Boston, MA, 1996)
13. S.K. Mitra, in *Digital Signal Processing—A Computer Based Approach*, 3rd edn, Chapter 6 (New York, McGraw-Hill, 2006)

A Third-Order Driving Point Synthesis Problem

18

Minimal realizations of an interesting third-order impedance function are discussed. The solution, based on an elegant algebraic identity, illustrates several basic concepts of driving point function synthesis.

Keyword

Driving point synthesis · Third-order impedance function

Introduction

Consider the impedance function

$$Z(s) = \frac{(s+a)(s+b)(s+c)}{s(s+a+b+c)}, \qquad (18.1)$$

where a, b and c are arbitrary non-negative real quantities. The problem is to have a minimal realization of Eq. 18.1, i.e. a realization which uses no more than four elements. Why four? Apparently, there are three specifications, namely a, b and c, but then you should realize that the multiplying constant, which in this case is unity, is a hidden specification.

The order of the impedance function, defined as the degree of the numerator or denominator, whichever is higher, being three, we shall naturally require three reactive elements. The fourth element must then be a resistance. Can all the three reactive elements be of the same kind, viz. either inductance or capacitance? Having a pole at the origin ($s = 0$) obviously excludes an all inductor solution, because an RL impedance cannot have such a pole. How about all reactive elements being capacitances? That is, how about an RC realization of Eq. 18.1? Note that Eq. 18.1 has a pole at $s = \infty$ and we know that an RC impedance cannot have such a pole.

Also note that $Z(s)$ poles are at $s = 0$ and $s = -(a + b + c)$ while its zeros are at $s = -a$, $-b$ and $-c$. Since $a + b + c > a$, b as well as c, we conclude that poles and zeros of $Z(s)$ do not alternate. This alternation of poles and zeros is an essential requirement of RC or RL impedances. Hence we conclude that Eq. 18.1 can neither be RL nor RC; if at all realizable, it must be RLC.

Is *Z*(*s*) at All Realizable?

The question that arises at this stage is the following: Is $Z(s)$ at all realizable? It is known that $Z(s)$ will be realizable if it is a positive real function (PRF) [1], i.e. if (i) $Z(s)$ is real for s real

Source: S. C. Dutta Roy, "A Third-Order Driving Point Synthesis Problem," *Students' Journal of the IETE*, vol. 36, pp. 179–183, October–December 1995.

S. C. Dutta Roy, *Circuits, Systems and Signal Processing*, https://doi.org/10.1007/978-981-10-6919-2_18

and (ii) Re $Z(s) \geq 0$ for Re $s \geq 0$. There are many ways of testing for a PRF, but one pre-processing or simplification that should invariably be carried out is to look for poles and zeros on the $j\omega$-axis, including $s = 0$ and $s = \infty$, and to remove them. This step is the testing for a PRF is known as the 'Foster preamble'. In particular, if $Z(s)$ has a pole at the origin ($s = 0$), then one can write

$$Z(s) = \frac{K_0}{s} + Z_1(s), \tag{18.2}$$

where

$$K_0 = sZ(s)|_{s=0}. \tag{18.3}$$

is the residue of $Z(s)$ at the pole at $s = 0$ and $Z_1(s)$ is the remaining function to be tested. The term K_0/s obviously represents a capacitor of value $1/K_0$. Naturally K_0 has to be positive, otherwise no further testing is needed. In fact if $Z(s)$ is PRF, then it can be shown that its residue at all poles on the $j\omega$-axis have to be real and positive, but not necessarily *vice versa*.

If, instead of the origin, $Z(s)$ has a pole at $s = \infty$, then one can write

$$Z(s) = K_\infty s + Z_2(s), \tag{18.4}$$

where

$$K_\infty = \operatorname*{Lim}_{s\to\infty} \frac{Z(s)}{s} \tag{18.5}$$

is the residue of $Z(s)$ at the pole at $s = \infty$ and $Z_2(s)$ is the remaining function to be tested. Here $K_\infty s$ represents an inductor of value K_∞. Finally, if $Z(s)$ has poles at $s = \pm j\omega_1$, then one can write

$$Z(s) = \frac{K_1 s}{s^2 + \omega_1^2} + Z_3(s), \tag{18.6}$$

where

$$K_1 = \left.\frac{(s^2 + \omega_1^2)Z(s)}{s}\right|_{s^2=-\omega_1^2} \tag{18.7}$$

is twice the residue of $Z(s)$ at $s = \pm j\omega_1$, and $Z_3(s)$ is the remaining function to be tested. The first term in Eq. 18.6 represents a parallel connection between an inductance K_1/ω_1^2 and a capacitance $1/K_1$.

If instead of a pole, one finds one or more zeros of $Z(s)$ on the $j\omega$-axis, then one removes them from $Y(s) = 1/Z(s)$ which will have a pole at those points. Here, removal of a pole at $s = 0$, $s = \infty$ and $s = \pm j\omega_1$ corresponds to the removal of an inductance, capacitance and a series connection of inductance and capacitance, respectively, all in parallel with the remaining admittance function to be tested.

It can be shown that if the original function was PR, then so is the remainder function after removal of any pole or zero on the $j\omega$-axis. This, in fact, gives validity of the Foster preamble! But then, how are we simplifying the testing? Note that $Z_1(s)$ of Eq. 18.2 as well as $Z_2(s)$ of Eq. 18.4 will be one order less than $Z(s)$, while $Z_3(s)$ of Eq. 18.6 will have an order reduction by two. Hence, indeed, the remainder functions are simplified.

In the present case of $Z(s)$ given by Eq. 18.1, we have a pole $s = 0$ due to the factor s in the denominator, and also at $s = \infty$ because the degree of the numerator is one greater than that of the denominator (it cannot be more than one or less than one, see [1]). Let us remove them. The residues are, from Eqs. 18.3 and 18.5,

$$K_0 = abc/(a+b+c) \text{ and } K_\infty = 1. \tag{18.8}$$

If we remove the pole at $s = \infty$ first, the remainder function is

$$\begin{aligned} Z_1'(s) &= Z(s) - s \\ &= \frac{(s+a)(s+b)(s+c)}{s(s+a+b+c)} - s \\ &= \frac{s(ab+bc+ca)+abc}{s(s+a+b+c)}. \end{aligned} \tag{18.9}$$

This step, as explained earlier, leads to the partial realization of Fig. 18.1a and reduces the order from three to two. As is obvious from

Eq. 18.9, $Z_1'(s)$ does not (and cannot) have a pole at $s = \infty$, but it retains the pole at $s = 0$ of $Z(s)$ with the same residue. If we now remove this pole from $Z_1'(s)$, we have a remainder function

$$Z_2'(s) = Z_1'(s) - \frac{abc}{s(a+b+c)}. \quad (18.10)$$

On simplification, this reduces to the following:

$$Z_2'(s) = \frac{(a+b+c)(ab+bc+ca) - abc}{(a+b+c)(s+a+b+c)}. \quad (18.11)$$

The partial realization resulting from this step is shown in Fig. 18.1b. Note also that the order of $Z_2'(s)$ is one, which is one less than that of $Z_1'(s)$, as expected.

In order to proceed further with the testing, it is necessary to ensure that the numerator constant of Eq. 18.11 is positive. It is indeed so, because of the algebraic identity

$$(a+b+c)(ab+bc+ca) = (a+b)(b+c)(c+a) + abc, \quad (18.12)$$

which can be easily verified. Thus $Z_2'(s)$ has no obvious defect for positive realness. Not only that, because $Z_2'(s)$ has a zero at $s = \infty$, its reciprocal $Y_2'(s)$ has a pole at $s = \infty$ which can be removed. The corresponding residue is

$$K'_{\infty 2} = \text{Lim}_{s\to\infty} \frac{Y_2'(s)}{s} = \text{Lim}_{s\to\infty} \frac{(a+b+c)(s+a+b+c)}{s(a+b)(b+c)(c+a)} \quad (18.13)$$

$$= \frac{a+b+c}{(a+b)(b+c)(c+a)}, \quad (18.14)$$

where in Eq. 18.13, we have used the identity Eq. 18.12 in conjunction with Eq. 18.11. If we remove this pole from $Y_2'(s)$ which corresponds to a capacitance of value $K'_{\infty 2}$ in parallel, we are left with the following remainder function

$$Y_3'(s) = Y_2'(s) - \frac{(a+b+c)s}{(a+b)(b+c)(c+a)}. \quad (18.15)$$

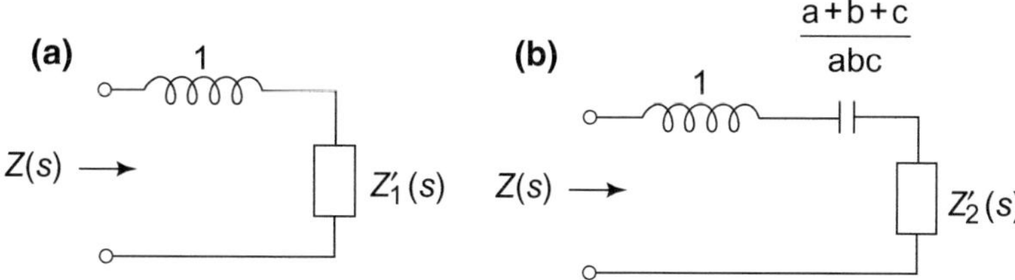

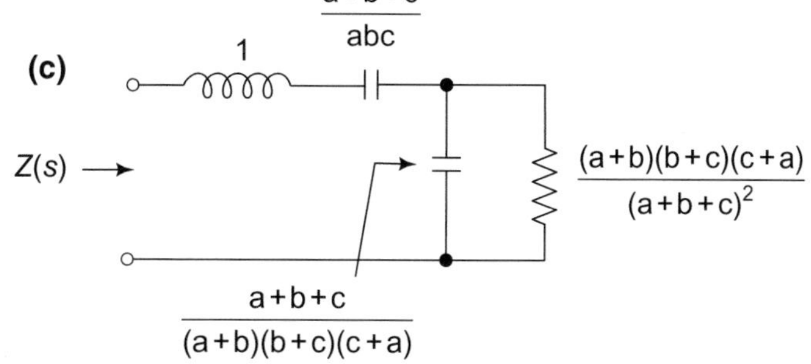

Fig. 18.1 Various steps in the testing of $Z(s)$ for positive realness, leading to a complete realization through Foster preamble only

Simplification of Eq. 18.15 leads to

$$Y_3'(s) = \frac{(a+b+c)^2}{(a+b)(b+c)(c+a)} \quad (18.16)$$

which is a positive constant, equivalent to a resistance of value $(a + b)\ (b + c)\ (c + a)/(a + b + c)^2$. The realization obtained at this stage is shown in Fig. 18.1c, which is, in fact, a complete realization. Nothing is left to test anymore!

We have, therefore, shown that $Z(s)$ is PR and in the process, which involved only the Foster preamble, we have solved the synthesis problem.

Alternative Realization

It is known that solution to a synthesis problem, if it exists, is never unique [1]. Can we, in the present case, find another realization? Let us see.

As in the previous section, let us first remove the pole at $s = \infty$, leaving the remainder $Z_1'(s)$ given by Eq. 18.9. Instead of removing the pole at $s = 0$ from $Z_1'(s)$, note that $Z_1'(s)$ has a zero at $s = \infty$. Let us, therefore, consider the admittance $Y_1'(s) = 1/Z_1'(s)$ and remove its pole at $s = \infty$. The residue is

$$\begin{aligned} K'_{\infty 1} &= \mathrm{Lim}_{s\to\infty} \frac{Y_1'(s)}{s} \\ &= \mathrm{Lim}_{s\to\infty} \frac{s(s+a+b+c)}{s[s(ab+bc+ca)+abc]} \\ &= \frac{1}{ab+bc+ca}. \end{aligned} \quad (18.17)$$

This removal means partial realization through a capacitance of value $K'_{\infty 1}$ in parallel, leaving a remainder $\bar{Y}_2'(s)$, as shown in Fig. 18.2b, where

$$\begin{aligned} \bar{Y}_2(s) &= \frac{s(s+a+b+c)}{s(ab+bc+ca)+abc} - \frac{s}{ab+bc+ca} \\ &= \frac{s[(a+b+c)(ab+bc+ca)-abc]}{(ab+bc+ca)[s(ab+bc+ca)+abc]}. \end{aligned} \quad (18.18)$$

Once again, because of Eq. 18.12, the coefficient of s in the numerator of Eq. 18.18 is positive, and we can re-write $\bar{Y}_2(s)$ as

$$\bar{Y}_2(s) = \frac{s(a+b)(b+c)(c+a)}{(ab+bc+ca)[s(ab+bc+ca)+abc]}. \quad (18.19)$$

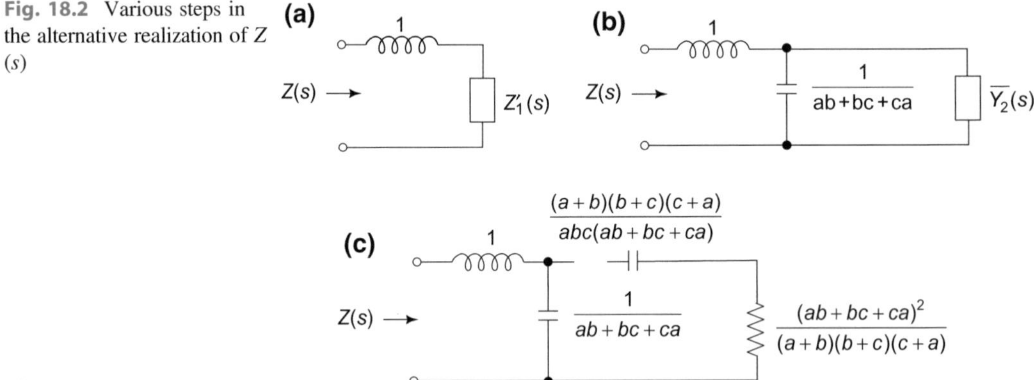

Fig. 18.2 Various steps in the alternative realization of $Z(s)$

$\bar{Y}_2(s)$ has a zero at the origin, which can be removed as the pole of $\bar{Z}_2(s) = 1/\bar{Y}_2(s)$. In fact, we can easily see that

$$\bar{Z}_2(s) = \frac{(ab+bc+ca)^2}{(a+b)(b+c)(c+a)} + \frac{abc(ab+bc+ca)}{s(a+b)(b+c)(c+a)}. \quad (18.20)$$

The second term corresponds to the pole at the origin. Also, observe that this decomposition corresponds to a series combination of a capacitance and resistance value indicated in Fig. 18.2c. The synthesis is complete and as you can see, this is different from the network of Fig. 18.1c.

It is interesting to observe that the alternative realization can be mechanized through continued fraction expansion starting with the highest powers, as follows:

$$Z(s) = \frac{s^3 + s^2(a+b+c) + s(ab+bc+ca) + abc}{s^2 + s(a+b+c)} \quad (18.21)$$

$$= s + 1/\left(\frac{s}{ab+bc+ca} + 1/\left(\left\{\frac{(ab+bc+ca)^2}{(a+b)(b+c)(c+a)} + 1/\left[\frac{s(a+b)(b+c)(c+a)}{abc(ab+bc+ca)}\right]\right\}\right)\right). \quad (18.22)$$

A word of caution must be sounded here. That continued fraction expansion works here is a matter of luck; it may not work in a general RLC case. Even in this case, you may try continued fraction expansion starting with the lowest powers and soon get frustrated!

A Problem for the Student

Can you find out another alternative minimal realization? If this proves tough, try relaxing on the minimal requirement—first with three reactances and more than one resistance and later with more than three reactances and more than one resistance. No more! Isn't life simple?

Acknowledgments Acknowledgement is made to S. Tirtoprodjo who first posed the problem [2] and to S. Erfani et al. who gave the solution of Fig. 18.1, although in a cryptic form [3].

References

1. F.F. Kuo, *Network Analysis and Synthesis* (Wiley, New York, 1966). Chapter 10
2. S. Tirtoprodjo, On the lighter side. IEEE CAS Magazine, **5**(1), 25 (1983, March)
3. S. Erfani et al., On the lighter side—Solution to the march puzzle. IEEE CAS Magazine **5**(2), 22 (1983)

Interference Rejection in a UWB System: An Example of LC Driving Point Synthesis

19

Synthesis of an LC driving point function is one of the initial topics in the study of network synthesis. This chapter gives a practical example of application of such synthesis in the design of a notch filter for interference rejection in an ultra wide-band (UWB) system. The example can used to motivate students to learn network synthesis with all seriousness, and not merely as a matter of academic exercise.

Keywords
LC driving point synthesis · Notch filter
UWB systems · Network synthesis

Introduction

The subject of network synthesis is considered by most students as moderately difficult, mathematical and mainly of academic interest. The underlying reason is that it encounters very few practical applications, except in the design of filters, which is a two-port network, for which extensive tables are available in textbooks [1] and handbooks [2]. In particular, one-port or driving point synthesis, one of the starting topics in the subject, appears to be of little use in practice. This chapter deals with a recent application of LC driving point synthesis, which may be used to enhance the motivation of students to learn the subject with all seriousness.

The example is taken from a 2009 paper [3] dealing with an integrated double-notch filter and implemented with 0.13 μm CMOS technology, for rejection of interference in an ultra wide-band (UWB) system. The problem, translated to network synthesis language, is to design a filter to reject the frequencies around f_1 and f_2 and pass those around f_p, where $f_1 < f_p < f_2$. The authors of [3] set the design values as $\omega_1 = 2\pi$ 2.4×10^9 rad/s, $\omega_2 = 2\pi \times 5.2 \times 10^9$ rad/s and $\omega_p = 2\pi \times 4.8 \times 10^9$ rad/s, where $\omega = 2\pi f$, and suggested and designed the network shown in Fig. 19.1 for this purpose. The current generator and the shunt resistance in Fig. 19.1 represent the equivalent circuit of an amplifier and the LC network has a driving point impedance $Z(s)$, which is required to have series resonance (and hence zero impedance) at ω_1 and ω_2, thus shunting out all the current from the load at these frequencies, and parallel resonance (and hence infinite impedance) at ω_p, thus passing all the current through the load at this frequency. Thus,

Source: S. C. Dutta Roy, "Interference Rejection in a UWB System: An Example of LC Driving Point Synthesis," *IETE Journal of Education*, vol. 50, pp. 55–58, May–August 2009.

S. C. Dutta Roy, *Circuits, Systems and Signal Processing*,
https://doi.org/10.1007/978-981-10-6919-2_19

basically, one requires to design an impedance $Z(s)$ of the form

$$Z(s) = \frac{\left(s^2+\omega_1^2\right)\left(s^2+\omega_2^2\right)}{s\left(s^2+\omega_p^2\right)}, \tag{19.1}$$

where without any loss in generality, the scaling constant is assumed to be unity. In this chapter, we shall treat Eq. 19.1 as the function to be synthesized, and derive the form of Fig. 19.1, as well as the other alternative canonic forms, along with their element values. We shall then compare the various networks on the basis of the required total inductance, total capacitance and grounded and ungrounded capacitors, which are important considerations for integrated circuit implementation.

The Four Canonical Realizations

The network of Fig. 19.1 is easily recognized as the Foster I realization of Eq. 19.1 [4]. As is well known, there are four basic structures for the canonical synthesis of an LC driving point synthesis, viz. Foster I, Foster II, Cauer I and Cauer II [4].

Foster I form is obtained by partial fraction expansion (PFE) of $Z(s)$, given by

$$Z(s) = \frac{K_0}{s} + K_\infty s + \frac{K_p s}{s^2+\omega_p^2}, \tag{19.2}$$

where with reference to Fig. 19.1,

$$K_0 = \omega_1^2\omega_2^2/\omega_p^2 \Rightarrow C_1 = 1/K_0, \tag{19.3a}$$

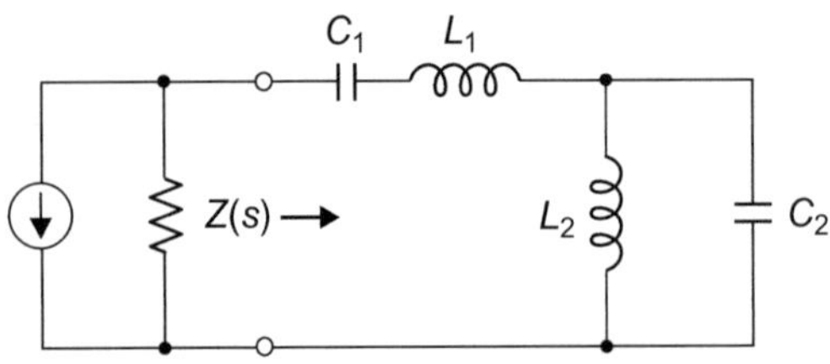

Fig. 19.1 Foster I network connected to a current generator and a shunt resistance, which represent the equivalent circuit of an amplifier

$$K_\infty = 1 = L_1, \tag{19.3b}$$

and

$$K_p = \left(\omega_p^2-\omega_1^2\right)\left(\omega_2^2-\omega_p^2\right)/\omega_p^2 \Rightarrow C_2 = 1/K_p,\quad L_2 = K_p/\omega_p^2. \tag{19.3c}$$

These results are reproduced in Table 19.1, in which the capacitors are given as $C_1\omega_1^2$ and $C_2\omega_1^2$, for later convenience.

Foster II form is obtained by the PFE of $Y(s) = 1/Z(s)$; the results are as follows:

$$Y(s) = \frac{K_1 s}{s^2+\omega_1^2} + \frac{K_2 s}{s^2+\omega_2^2},$$
$$K_1 = \left(\omega_p^2-\omega_1^2\right)/\left(\omega_2^2-\omega_p^2\right), \tag{19.4a}$$
$$K_2 = \left(\omega_2^2-\omega_p^2\right)/\left(\omega_2^2-\omega_1^2\right),$$

$$L_1 = 1/K_1,\quad C_1 = 1/\left(\omega_1^2 L_1\right),\quad L_2 = 1/K_2,\quad C_2 = 1/\left(\omega_2^2 L_2\right). \tag{19.4b}$$

The elements in Eq. 19.4b refer to the network shown in Fig. 19.2, and the values are shown in Table 19.1.

For Cauer I network, we make a continued fraction expansion (CFE) of Eq. 19.1 starting with the highest powers. The quotients of the CFE give the following element values with reference to the structure shown in Fig. 19.3. These values are also shown in Table 19.1.

$$L_1 = 1,\quad C_1 = \frac{1}{\omega_1^2+\omega_2^2-\omega_p^2} = \frac{1}{\omega_3^2},$$

$$L_2 = \frac{\omega_3^4}{\omega_3^2\omega_p^2-\omega_1^2\omega_2^2},\quad C_2 = \frac{\omega_3^2\omega_p^2-\omega_1^2\omega_2^2}{\omega_3^2\omega_2^2\omega_1^2}. \tag{19.5}$$

The Cauer II realization will be of the form shown in Fig. 19.4, and the element values are obtained from the quotients of the CFE of Eq. 19.1 starting with the lowest powers. These

Table 19.1 I Comparison of the four canonical structures

Parameter	Expressions and values of network					
	L_1	L_2	$L_1 + L_2$ (Num. value)	$\omega_1^2 C_1$	$\omega_1^2 C_2$	$\omega_1^2(C_1 + C_2)$ (Num. value)
Foster I	1	$\left(1-\frac{\omega_1^2}{\omega_p^2}\right)\left(\frac{\omega_2^2}{\omega_p^2}-1\right)$ (0.130)	2.130	$\frac{\omega_p^2}{\omega_2^2}$ (0.852)	$\frac{\omega_1^2}{\omega_p^2 L_2}$ (1.923)	2.775
Foster II	$\frac{(\omega_2^2-\omega_1^2)}{(\omega_p^2-\omega_1^2)}$ 0.306)	$\frac{(\omega_2^2-\omega_1^2)}{(\omega_2^2-\omega_p^2)}$ (1.324)	1.630	$1/L_1$ (3.268)	$\frac{\omega_1^2}{\omega_2^2 L_2}$ (0.161)	3.429
Cauer I	1	$\frac{\omega_3^4}{\omega_3^2\omega_p^2-\omega_1^2\omega_2^2}$ (1.378) $\omega_3^2=\omega_1^2+\omega_2^2-\omega_p^2$	2.378	$\frac{\omega_1^2}{\omega_3^2}$ (0.590)	$\frac{\omega_3^2}{\omega_2^2 L_2}$ (0.262)	0.852
Cauer II	$\frac{\omega_4^2}{\omega_p^2}$ (1.130) $\omega_4^2=\omega_1^2\omega_2^2-\frac{\omega_1^2\omega_2^2}{\omega_p^2}$	$\frac{\omega_4^2}{\omega_4^2-\omega_p^2}$ (8.680)	9.810	$\frac{\omega_p^2}{\omega_2^2}$ (0.852)	$\frac{\omega_1^2}{\omega_4^2 L_2}$ (0.025)	0.877

Note The expressions are slightly modified versions of those in the text

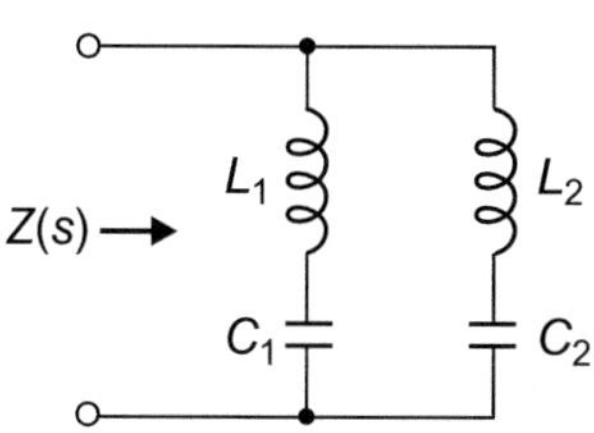

Fig. 19.2 Foster II form of $Z(s)$

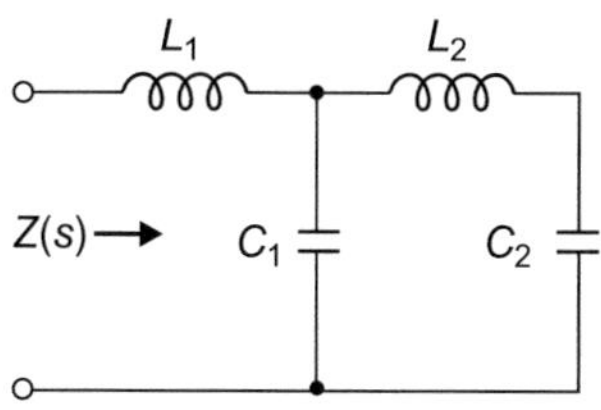

Fig. 19.3 Cauer I form of $Z(s)$

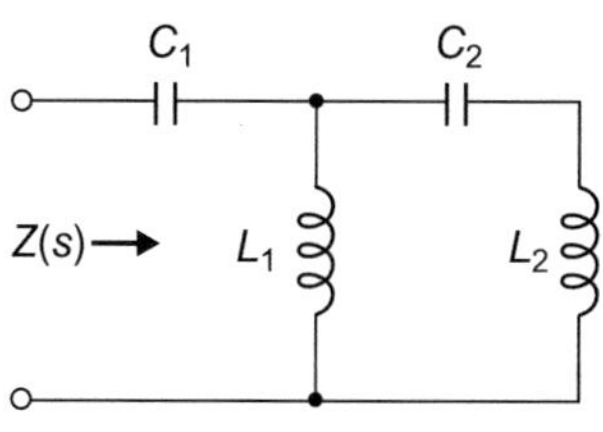

Fig. 19.4 Cauer II form of $Z(s)$

values are given below and are also shown in Table 19.1.

$$L_1 = \omega_4^2/\omega_p^2, \quad \omega_4^2 = \omega_1^2 + \omega_2^2 - \omega_1^2\omega_2^2/\omega_p^2,$$
$$L_2 = \omega_4^2/\left(\omega_4^2 - \omega_p^2\right), \quad C_2 = 1/\left(\omega_4^2 L_2\right), \tag{19.6a}$$

and

$$C_1 = \omega_p^2/\left(\omega_1^2\omega_2^2\right). \tag{19.6b}$$

Comparison

Using the same specifications as given in [3], and mentioned in the Introduction, we have computed the numerical values of the elements for the various structures. These are shown in Table 19.1 inside brackets below the corresponding algebraic expression. Note that no powers of 10 are involved in the expressions for $C_1\omega_1^2$ and $C_2\omega_1^2$ because of multiplication of the capacitors by ω_1^2. Also, for computational convenience, some of the algebraic expressions given in Table 19.1 are also slightly modified versions of the formulas given in the text.

A look at the total capacitance (C_t) and total inductance (L_t) values in Table 19.1 show that Cauer realizations have considerably smaller C_t as compared to the Foster realizations, with Cauer I having the lowest value and Cauer II having a marginal increase over the same. The reverse is the case with respect to L_t with Foster I having the lowest value and Foster II having a marginal increase over it. Another point to be noted is that both Foster II and Cauer I networks have both capacitors connected to ground, which is, in general, a desirable feature in integrated circuits.

Effect of Losses

In practice, all reactive elements are lossy, i.e. all inductors have a series resistance and all capacitors have a shunt conductance. However, the losses in inductors dominate over those in the capacitors. A practical scheme for effective compensation of the losses for the network in Fig. 19.1 has been given in [3] using a single negative resistance realized with active devices. Analysis of the effects of losses on the notch depths and maximum output for the four structures will be a worthwhile project for the students, and a comparison may reveal the superiority of one network over the others.

Problems

P.1. Can you find an alternative network to the C_1, L_1, L_2, C_2 configuration in Fig. 19.1?

P.2. Could we do with third-order impedances for C_1, L_1 combination as will as L_2, C_2 combination in Fig. 19.1. What will this circuit perform as?

P.3. Suppose there are two frequencies which have to be rejected. Draw the necessary circuit configuration.

P.4. Same as P.3 except that three frequencies have to be rejected. Draw an alternation circuit also.

P.5. Draw another alternative circuit for P.4 and compare the two.

Acknowledgements The work was supported by the Indian National Science Academy through their Honorary Scientist scheme. The author acknowledges the help of Dr. Sumantra Dutta Roy for his help in the preparation of the diagrams.

References

1. L. Weinberg, *Network Analysis and Synthesis* (McGrawHill, New York, 1962)
2. W.K. Chen (ed.), in *Passive, Active and Digital Filters, Volume 3 of Handbook of Circuits and Filters* (Boca Raton, CRC Press, 2009)
3. A. Vallese, A. Bevilacqua, C. Sandner, M. Tiebout, A. Gerosa, A. Neviani, Analysis and design of an integrated notch filter for the rejection of interference in UWB systems. IEEE J. Solid-State Circuits **44**, 331–343 (2009)
4. F.F. Kuo, *Network Analysis and Synthesis* (Wiley, New York, 1966). Chapter 11

Low-Order Butterworth Filters: From Magnitude to Transfer Function

20

A simple method is given for obtaining the transfer function of Butterworth filters of orders 1 to 6.

Keywords
Butterworth filters · Transfer functions
Magnitude · Orders of filter · Chebyshev filter

Butterworth Filters

Butterworth filter is the most elegant of all filters in more than one way. The magnitude squared function of a low-pass Butterworth filter, having a normalized 3 dB cutoff frequency of 1 rad/sec, is given by

$$|H_N(j\omega)|^2 = 1/\left(1+\omega^{2N}\right), \tag{20.1}$$

where N is the order of the filter. As is well known, it has a monotonically decaying response in the whole frequency range. It is maximally flat at dc, and the corresponding transfer function $H_N(s)$ has all its poles on the left half of the unit circle centered at $s = 0$, at equal angular intervals of π/N, with none occurring on the $j\omega$-axis. This gives rise to the property that if we write

$$H_N(s) = 1/B_N(s) \tag{20.2}$$

then, $B_N(s)$, the so-called Butterworth polynomial, has symmetrical coefficients, i.e. $B_N(s)$ is of the form

$$B_N(s) = 1 + b_1 s + b_2 s^2 + \cdots + b_2 s^{N-2} + b_1 s^{N-1} + s^N \tag{20.3}$$

In standard textbooks (see, e.g. [1–3]), the procedure prescribed for finding $B_N(s)$ is to locate all the roots of

$$B_N(s)B_N(-s) = 1 + \left(-s^2\right)^N \tag{20.4}$$

and then to take the left half-plane ones for $B_N(s)$, i.e.

$$B_N(s) = \prod_{k=1}^{N}(s - s_k); \quad \mathrm{Re}\, s_k < 0, \quad k = 1 \text{ to } N \tag{20.5}$$

Source: S. C. Dutta Roy, "Low Order Butterworth Filters: From Magnitude to Transfer Function," *Journal of the IETE*, vol. 37, pp. 221–225, October–December 1996.

S. C. Dutta Roy, *Circuits, Systems and Signal Processing*,
https://doi.org/10.1007/978-981-10-6919-2_20

It may be noted that Eq. 20.4 is simply the denominator of Eq. 20.1 with ω^2 replaced by $-s^2$, a process known as 'analytic continuation'. An explicit formula for s_k is [3]:

$$s_k = -\sin[(2k-1)\pi/(2N)] + j\cos[(2k-1)\pi/(2N)],\quad k = 1 \text{ to } N \tag{20.6}$$

Obviously, these roots occur in symmetry with respect to the real as well as the imaginary axes. It is also clear that if N is odd, then there is a root of $B_N(s)$ at $s = -1$, i.e. $(s + 1)$ is a factor of $B_N(s)$; the other $(N - 1)$ roots will be complex and will occur in conjugate pairs. Consequently, one can write

$$B_N(s) = \begin{cases} \prod_{k=1}^{N/2} \{s^2 + 2s\sin[(2k-1)\pi/(2N)] + 1\}, N \text{ even} \\ (s+1) \prod_{k=1}^{(N-1)/2} \{s^2 - 2s\sin[(2k-1)\pi/(2N)] + 1\}, N \text{ odd} \end{cases} \tag{20.7}$$

When N is large, one can use Eq. 20.7 for finding $B_N(s)$ in quickest possibly way. However, when N is low, an alternative trick can be applied, and that is the subject matter of this chapter.

Basis of the Alternative Method

Since the roots of $B_N(s)$ are strictly in the left half of the s-plane, $B_N(s)$ qualifies as a strict Hurwitz polynomial. We can write

$$B_N(s) = (1 + b_2s^2 + \cdots) + (b_1s + b_3s^3 + \cdots) = m_N(s) + n_N(s), \tag{20.8}$$

where m_N and n_N are the even and odd parts of $B_N(s)$. By definition

$$m_N(-s) = m_N(s) \text{ and } n_N(-s) = -n_N(s), \tag{20.9}$$

so that

$$B_N(-s) = m_N(s) - n_N(s) \tag{20.10}$$

Combining Eqs. 20.4, 20.8 and 20.9, we get

$$m_N^2 - n_N^2 = 1 + (-1)^N s^{2N} \tag{20.11}$$

or,

$$(1 + b_2s^2 + \cdots)^2 - (b_1s + b_3s^3 + \cdots)^2 = 1 + (-1)^N s^{2N} \tag{20.12}$$

Obviously, the constant terms and the coefficients of s^{2N} on both sides of Eq. 20.12 are identically equal. The coefficients of $s^2, s^4, \ldots, s^{2N-2}$ must then each be zero. This, combined with the symmetry of the coefficients, makes it easy to find $B_N(s)$ for low orders.

A further simplification can be obtained for the odd-order case. It follows from Eq. 20.11 that if N is odd, then $(1 - s^2)$ will be a factor of $B_N(s)\,B_N(-s)$, which contributes $(1 + s)$ to $B_N(s)$ and $(1 - s)$ to $B_N(-s)$ as factors. The other factor of $B_N(s)\,B_N(-s)$ will be $(1 + s^2 + s^4 + \cdots + s^{2N-2})$. Hence, the procedure simplifies to that of finding the coefficients of the polynomial

$$C_{N-1}(s) = 1 + c_1s + c_2s^2 + \cdots + c_2s^{N-3} + c_1s^{N-2} + s^{N-1}, \tag{20.13}$$

where

$$C_{N-1}(s)C_{N-1}(-s) = 1 + s^2 + s^4 + \cdots + s^{2N-2} \tag{20.14}$$

Notice that $C_{N-1}(s)$ is also written with coefficient symmetry. This follows from the fact that all real polynomial factors of a polynomial with symmetrical coefficients retain the symmetry property. From Eqs. 20.13 and 20.14, we have

$$\left(1+c_2s^2+\cdots\right)^2-\left(c_1s+c_3s^3+\cdots\right)^2 = 1+s^2+s^4+\cdots+s^{2N-2} \tag{20.15}$$

Again, the coefficients of s^0 and s^{2N-2} are identical on both sides, while the coefficients of $s^2, s^4, \ldots, s^{2N-4}$, each equated to unity, will form a set of equations for determining the c_i coefficients.

We shall now apply the method to all the possible low orders; in the process, the difficulties encountered for high orders will become obvious. It will also be clear that one cannot apply the method blindly, because of the occurrence of multiple solutions for the coefficients, and the consequent need for identifying the correct solutions. An obvious constraint is that all coefficients must be positive, but, as will be shown, this alone is not enough.

Application to Low Orders

First-Order Case

For $N = 1$, we have from Eq. 20.11,

$$m_1^2 - n_1^2 = 1 - s^2 \tag{20.16}$$

Consequently,

$$m_1 = 1 \text{ and } n_1 = s \tag{20.17}$$

so that

$$B_1(s) = 1+s \tag{20.18}$$

This is, of course, a trivial case.

Second-Order Case

For $N = 2$, we have from Eq. 20.3,

$$B_2(s) = 1+b_1s+s^2 \tag{20.19}$$

and hence, from Eq. 20.12,

$$\left(1+s^2\right)^2-b_1^2s^2 = 1+s^4 \tag{20.20}$$

Equating the coefficients of s^2 on both sides gives

$$2-b_1^2 = 0 \tag{20.21}$$

or

$$b_1 = \sqrt{2} \tag{20.22}$$

Hence,

$$B_2(s) = 1+\sqrt{2}s+s^2 \tag{20.23}$$

Third-Order Case

As already pointed out, $(1 + s)$ will be a factor of $B_3(s)$, so that

$$B_3(s) = (1+s)C_2(s) \tag{20.24}$$

where

$$C_2(s) = 1+c_1s+s^2 \tag{20.25}$$

From Eq. 20.15, then, we get

$$\left(1+s^2\right)^2-c_1^2s^2 = 1+s^2+s^4 \tag{20.26}$$

Equating the coefficients of s^2 on both sides gives

$$2-c_1^2 = 1 \tag{20.27}$$

or,

$$c_1 = 1 \tag{20.28}$$

so that

$$B_3(s) = (1+s)(1+s+s^2) \tag{20.29}$$

$$= 1 + 2s + 2s^2 + s^3 \tag{20.30}$$

Fourth-Order Case

For $N = 4$, we have from Eq. 20.8 and the symmetry of coefficients,

$$B_4(s) = 1 + b_1 s + b_2 s^2 + b_1 s^3 + s^4 \tag{20.31}$$

and from Eq. 20.12, we get

$$(1 + b_2 s^2 + s^4)^2 - (b_1 s + b_1 s^3)^2 = 1 + s^8 \tag{20.32}$$

Equating the coefficients of s^6 and s^4 on both sides of Eq. 20.32 gives rise to the following set of nonlinear equations:

$$2b_2 = b_1^2 \tag{20.33}$$

and

$$b_2^2 + 2 = 2b_1^2 \tag{20.34}$$

Note that because of symmetry, equating the coefficients of s^2 on both sides of Eq. 20.32 gives the same result as Eq. 20.33. Combining Eqs. 20.33 and 20.34 gives

$$b_2^2 - 4b_2 + 2 = 0 \tag{20.35}$$

Solving this quadratic, we get the following two solutions for b_2:

$$b_2 = 2 \pm \sqrt{2} \tag{20.36}$$

Consequently, from Eq. 20.33

$$b_1 = \sqrt{2\left(2 \pm \sqrt{2}\right)} \tag{20.37}$$

Both of these values of b_2 and b_1 are positive and are candidates for belonging to $B_4(s)$. Which ones are admissible? To test this, we bring in the strict Hurwitz character of $B_4(s)$, which demands that the ratio $m_4(s)/n_4(s)$ should be an LC driving point function of the fourth order. Let us try

$$b_2 = 2 - \sqrt{2} \text{ and } b_1 = \sqrt{2\left(2 - \sqrt{2}\right)}$$

We can write

$$\frac{m_4(s)}{n_4(s)} = \frac{s^4 + (2 - \sqrt{2})s^2 + 1}{\sqrt{2(2 - \sqrt{2})}(s^3 + s)},$$

$$= \frac{1}{\sqrt{2(2 - \sqrt{2})}}\left[\frac{(1 - \sqrt{2})s^2 + 1}{s^3 + s} + s\right], \tag{20.38}$$

Since $(1 - \sqrt{2})$ is negative, Eq. 20.38 cannot qualify as an LC driving point function. Hence, the acceptable solutions are the ones with positive signs in Eqs. 20.36 and 20.37. We therefore get

$$B_4(s) = 1 + \sqrt{2\left(2+\sqrt{2}\right)}s + \left(2+\sqrt{2}\right)s^2 + \sqrt{2\left(2+\sqrt{2}\right)}s^3 + s^4 \tag{20.39}$$

$$= 1 + 2.6131\,s + 3.4142s^2 + 2.6131\,s^3 + s^4 \tag{20.40}$$

This case illustrates the difficulty we encounter with multiple solutions. Finding the acceptable solution requires a Hurwitz test, which, in this particular case, has not proved to be difficult. Note that for the Hurwitz test, one can also use continued fraction expansion, rather than partial fraction expansion.

Fifth-Order Case

As in the third-order case, $(1 + s)$ is a factor of $B_5(s)$ and we can write

$$B_5(s) = (1+s)C_4(s), \quad (20.41)$$

$$= (1+s)\left(1+c_1 s+c_2 s^2+c_1 s^3+s^4\right), \quad (20.42)$$

where

$$\left(1+c_2 s^2+s^4\right)^2-\left(c_1 s+c_1 s^3\right)^2 = 1+s^2+s^4+s^6+s^8 \quad (20.43)$$

Equating the coefficients of s^2 and s^4 on both sides gives

$$2c_2 - c_1^2 = 1 \quad (20.44)$$

and

$$c_2^2+2-2c_1^2 = 1 \quad (20.45)$$

Combining Eqs. 20.44 and 20.45 give the following quadratic equation in c_2:

$$c_2^2-4c_2+3 = 0 \quad (20.46)$$

It is easy to see that the solutions for c_2 are

$$c_2 = 3, 1 \quad (20.47)$$

Correspondingly, from Eq. 20.44,

$$c_1 = \sqrt{2c_2-1} = \sqrt{5}, 1 \quad (20.48)$$

Try $c_2 = c_1 = 1$; then

$$C_4(s) = 1+s+s^2+s^3+s^4 \quad (20.49)$$

The ratio of even to odd parts of $C_4(s)$ is

$$\frac{s^4+s^2+1}{s^3+s} = s+\frac{1}{s^3+s}, \quad (20.50)$$

which is obviously not an LC driving point function. Hence, the acceptable solutions are: $c_2 = 3$ and $c_1 = \sqrt{5}$ giving

$$B_5(s) = (1+s)\left(1+\sqrt{5}s+\sqrt{3}s^2+\sqrt{5}s^3+s^4\right) \quad (20.51)$$

$$=(1+(\sqrt{5}+1)s+(\sqrt{5}+3)s^2+(\sqrt{5}+3)s^3 +(\sqrt{5}+1)s^4+s^5), \quad (20.52)$$

$$=1+3.2361\,s+5.2361\,s^2+5.2361\,s^3 +3.2361\,s^4+s^5 \quad (20.53)$$

Sixth-Order Case

For $N = 6$,

$$B_6(s) = 1+b_1 s+b_2 s^2+b_3 s^3+b_2 s^4+b_1 s^5+s^6 \quad (20.54)$$

and

$$\left(1+b_2 s^2+b_2 s^4+s^6\right)^2-\left(b_1 s+b_3 s^3+b_1 s^5\right)^2 = 1+s^{12} \quad (20.55)$$

Equating the coefficients of s^2, s^4 and s^6 on both sides give the following equations:

$$2b_2^2 = b_1^2, \quad (20.56)$$

$$2b_2+b_2^2 = 2b_1 b_3, \quad (20.57)$$

and

$$2+2b_2^2 = 2b_1^2+b_3^2 \quad (20.58)$$

From Eq. 20.56, we get

$$b_1 = \sqrt{2b_2} \quad (20.59)$$

and from Eqs. 20.58 and 20.59, we get

$$b_3 = \sqrt{2}|1-b_2|, \quad (20.60)$$

where the magnitude sign is included to indicate that b_3 must be positive. Combining Eqs. 20.57,

20.59 and 20.60 give the following cubic equation in b_2:

$$b_2^3 - 12b_2^2 + 36b_2 - 16 = 0 \tag{20.61}$$

A cubic equation is analytically solvable [4], although not as easily as the quadratic equation. For the general cubic equation

$$ax^3 + bx^2 + cx + d = 0 \tag{20.62}$$

all the three roots are real if $b^2 - 3ac > 0$, while if $b^2 - 3ac < 0$, then there is only one real root, the other two being complex conjugates of each other. As can be easily verified, the second case is valid for Eq. 20.61; hence, our job is simply to find the real root of Eq. 20.61. Trial and error seems to be the best policy at this stage, and after a few trials, we get

$$b_2 = 7.4641 \tag{20.63}$$

as a reasonably good solution. The corresponding values of the other two coefficients are obtained from Eqs. 20.59 and 20.60 as

$$b_1 = 3.8637 \text{ and } b_3 = 9.1416 \tag{20.64}$$

Hence, finally,

$$\begin{aligned} B_6(s) =& 1 + 3.8637\, s + 7.4671\, s^2 + 9.1416\, s^3 \\ &+ 7.4671\, s^4 + 3.8637\, s^5 + s^6 \end{aligned} \tag{20.65}$$

Seventh-Order Case

Since the order is odd, we have

$$B_7(s) = (1 + s)C_6(s), \tag{20.66}$$

$$= (1 + s)\left(1 + c_1 s + c_2 s^2 + c_3 s^3 + c_2 s^4 + c_1 s^5 + s^6\right), \tag{20.67}$$

where

$$\begin{aligned} &\left(1 + c_2 s^2 + c_2 s^4 + s^6\right)^2 - \left(c_1 s + c_3 s^3 + c_1 s^5\right)^2 \\ &= 1 + s^2 + s^4 + s^6 + s^8 + s^{10} \end{aligned} \tag{20.68}$$

Equating the coefficients of like powers of s on both sides gives

$$2c_2 - c_1^2 = 1, \tag{20.69}$$

$$c_2^2 + 2c_2 - 2c_1 c_3 = 1, \tag{20.70}$$

and

$$2 + 2c_2^2 - c_3^2 - 2c_1^2 = 1 \tag{20.71}$$

From Eq. 20.69, we get

$$c_1 = \sqrt{2c_2 - 1} \tag{20.72}$$

while Eq. 20.71 gives

$$c_3 = \sqrt{3 - 4c_2 + 2c_2^2} \tag{20.73}$$

Substituting these values in Eq. 20.70, we get

$$c_2^2 + 2c_2 - 1 = 2\sqrt{(2c_2 - 1)(3 - 4c_2 + 2c_2^2)} \tag{20.74}$$

Squaring both sides and simplifying, we get the following quartic equation in c_2:

$$c_2^4 - 12c_2^3 + 42c_2^2 - 44c_2 + 13 = 0 \tag{20.75}$$

A cubic equation was bad enough; this is worse! Fortunately, however, a quartic equation can also be solved analytically [4]; however, the effort involved does not justify proceeding further. Using Eq. 20.7 appears to be a much better proposition, not just for $N = 7$, but for all higher orders.

Application to Chebyshev Filters

What about filters other than Butterworth? For example, Chebyshev? Does the procedure presented here offer any simplicity? Let us examine.

For the Chebyshev low-pass filter, the normalized magnitude squared function is given by

$$|H_N(j\omega)|^2 = 1/\left[1+\varepsilon^2 T_N^2(\omega)\right], \qquad (20.76)$$

where T_N is the Chebyshev polynomial of the first kind, of order N. For the first few orders, we have

$$\begin{aligned} T_1(\omega) &= \omega \\ T_2(\omega) &= 2\omega^2 \\ T_3(\omega) &= 4\omega^3 - 3\omega - 1 \\ T_4(\omega) &= 8\omega^4 - 8\omega^2 + 1 \end{aligned} \qquad (20.77)$$

Let

$$H_N(s) = 1/D_N(s) = 1/[m_N(s) + n_N(s)], \qquad (20.78)$$

$$= 1/\left(b_0 + b_1 s + b_2 s^2 + \cdots + b_N s^N\right), \qquad (20.79)$$

where the usual notations have been used. Note that neither b_0 nor b_N is fixed, as in the Butterworth case; there is no coefficient symmetry either. The equation for finding the coefficients is

$$\begin{aligned} &\left(b_0 + b_2 s^2 + \cdots\right)^2 - \left(b_1 s + b_3 s^3 + \cdots\right)^2 \\ &\quad = 1 + \varepsilon^2 T_N^2(s/j) \end{aligned} \qquad (20.80)$$

Although the right-hand side of Eq. 20.80 involves j, in actual practice, $C_N^2(\omega)$ shall involve only $\omega^2, \omega^4, \ldots$ so that ω^2 has to be replaced by $-s^2$; hence, only real coefficients would be encountered.

For the first-order case, we have

$$b_0^2 - b_1^2 s^2 = 1 - \varepsilon^2 s^2 \qquad (20.81)$$

Thus,

$$b_0 = 1 \text{ and } b_1 = \varepsilon \qquad (20.82)$$

and

$$D_1(s) = 1 + \varepsilon s \qquad (20.83)$$

For the second-order case, we get

$$\left(b_0 + b_2 s^2\right)^2 - b_1^2 s^2 = 1 + \varepsilon^2\left(2\omega^2 - 1\right)^2\Big|_{\omega^2=-s^2} \qquad (20.84)$$

On simplification, Eq. 20.84 becomes

$$\begin{aligned} &b_0^2 + \left(2b_2 - b_1^2\right)s^2 + b_2^2 s^4 \\ &\quad = \left(1 + \varepsilon^2\right) + 4\varepsilon^2 s^2 + 4\varepsilon^2 s^4 \end{aligned} \qquad (20.85)$$

Equating the coefficients of like powers of s gives

$$b_0 = \sqrt{1+\varepsilon^2} \qquad (20.86)$$

$$b_2 = 2\varepsilon \qquad (20.87)$$

and

$$2b_0 b_2 - b_1^2 = 4\varepsilon^2 \qquad (20.88)$$

Combining Eqs. 20.86–20.88, we get

$$b_1 = 2\sqrt{\varepsilon\left(\sqrt{1+\varepsilon^2} - \varepsilon\right)} \qquad (20.89)$$

Thus,

$$D_2(s) = \sqrt{1+\varepsilon^2} + 2\sqrt{\varepsilon\left(\sqrt{1+\varepsilon^2} - \varepsilon\right)}\,s + 2\varepsilon s^2 \qquad (20.90)$$

Now consider the third-order case, for which we get, from Eqs. 20.77 and 20.80.

$$\begin{aligned} \left(b_0 + b_2 s^2\right)^2 - \left(b_1 s + b_3 s^3\right)^2 &= 1 + \varepsilon^2\left(4\omega^3 - 3\omega\right)^2\Big|_{\omega^2=-s^2} \\ &= 1 + \varepsilon^2(-s)^2\left(-4s^2 - 3\right)^2 \\ &= 1 - 9\varepsilon^2 s^2 - 24\varepsilon^2 s^4 - 16\varepsilon^2 s^6 \end{aligned} \qquad (20.91)$$

Equating the coefficients of like powers of s on both sides, we get

$$b_0 = 1, \quad b_3 = 4\varepsilon \tag{20.92}$$

$$2b_2 - b_1^2 = -9\varepsilon^2 \tag{20.93}$$

and

$$b_2^2 - 8\varepsilon b_1 = -24\varepsilon^2, \tag{20.94}$$

where, in Eqs. 20.93 and 20.94, the values given in Eq. 20.92 have been utilized. Combining Eqs. 20.93 and 20.94 give the following cubic equation for b_2:

$$b_2^3 + 48\varepsilon^2 b_2 - 128\varepsilon^2 = 0 \tag{20.95}$$

According to the theory of cubic equations [4], this also has only one real root, which obviously depends on ε.

One should, at this point, be convinced that the applicability of the technique presented in this chapter: to the Chebyshev filter, or for that matter, to any other kind of filter would be limited. Even for the Butterworth case, the limit appears to be set by the sixth order.

Conclusion

A simple method is presented for finding the Butterworth polynomials of orders one to six, and its limitations for higher orders of Butterworth filter and other types of filters, even of a lower order, have been pointed out.

Problems

P.1. Without finding poles and zeroes, can you formulate a procedure and give the equations for finding an Nth-order Butterworth polynomial? No, no, I am not asking you to solve these equations, because that will be too must to ask for. With the kind of training I have given to you, I believe you should be able to do it.

P.2. For the third-order case, find the zeroes of the third-order Butterworth polynomial. Do not bring poles and zeroes into the scene. They pollute and hamper you intellectual development! Of course, you substitute the values of the coefficients from the text.

P.3. Same as P.1 for order = 4.

P.4. Same as P.2 for order = 5.

P.5. Same as P.2 for order = 7.

References

1. M.E. Van Valkenburg, *Introduction to Modern Network Synthesis* (Wiley, New York, 1964)
2. N. Balabanian, in *Network Synthesis* (Englewood Cliffs, NJ, Prentice Hall, Inc, 1958)
3. S. Karni, *Network Theory: analysis and Synthesis* (Allyn and Bacon Inc, Boston, 1966)
4. S. Neumark, *Solution of Cubic and Quartic Equations* (Pergamon Press, London, 1965)

21 Band-Pass/Band-Stop Filter Design by Frequency Transformation

Given the specifications of a band-pass filter (BPF) or a band-stop filter (BSF), the same can be translated to those of a normalized low-pass filter (LPF) by frequency transformation. Once the latter is designed, one can realize the BPF/BSF by using the same transformation in a reverse manner. The process of translation to the normalized LPF is usually not explained in details in standard textbooks, and in some of them, the process has even been wrongly stated or illustrated. This chapter clarifies this important step in BPF/BSF design.

Keywords
Band-Pass · Band-Stop · Frequency transformation

Introduction

As is well known the normalized LPF response of Fig. 21.1a can be transformed to the BPF response of Fig. 21.1b by the transformation

$$S = \frac{\omega_0}{B}\left(\frac{s}{\omega_0} + \frac{\omega_0}{s}\right), \tag{21.1}$$

where $S = \Sigma + j\Omega$ is the LPF complex frequency variable, $s = \sigma + j\omega$ is the BPF complex frequency variable,

$$B = \omega_{p2} - \omega_{p1} \tag{21.2}$$

is the bandwidth and

$$\omega_0 = \sqrt{\omega_{p1}\omega_{p2}} = \sqrt{\omega_{s1}\omega_{s2}} \tag{21.3}$$

is the centre frequency of the BPF response, which is geometrically symmetrical about ω_0. Similarly, the BSF response of Fig. 21.1c can be obtained from the LPF response of Fig. 21.1a through the transformation.

$$S = 1/\left[\frac{\omega_0}{B}\left(\frac{s}{\omega_0} + \frac{\omega_0}{s}\right)\right], \tag{21.4}$$

where, again Eqs. 21.2 and 21.3 are valid, but B does not have the interpretation of bandwidth. As in the BPF case, the BSF characteristic is also geometrically symmetrical about ω_0.

Given Fig. 21.1a, it is easy to obtain the characteristics of Fig. 21.1b or Fig. 21.1c by using Eqs. 21.1 or 21.4 as the case may be, but given a BPF or BSF response, how does one go to the normalized LPF response? In particular, how does one find the edge of the stop-band Ω_s,

Source: S. C. Dutta Roy, "Band-Pass/Band-Stop Filter Design by Frequency Transformation," *IETE Journal of Education*, vol. 45, pp. 145–149, July–September 2004.

S. C. Dutta Roy, *Circuits, Systems and Signal Processing*,
https://doi.org/10.1007/978-981-10-6919-2_21

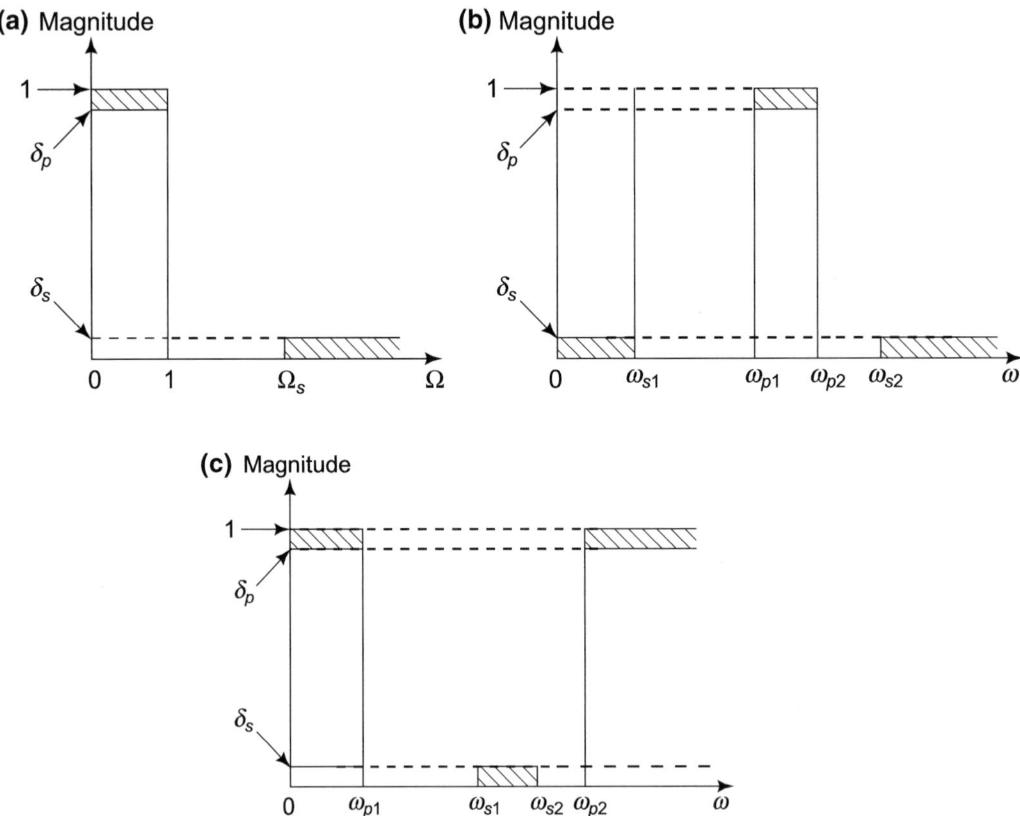

Fig. 21.1 **a** Normalized LPF characteristic. **b** BPF response obtained through Eq. 21.1. **c** BSF response obtained through Eq. 21.4

in Fig. 21.1a? Also, obviously $\sqrt{\omega_{p1}\omega_{p2}}$ may not be equal to $\sqrt{\omega_{s1}\omega_{s2}}$ in the given specifications. How does one proceed? These questions are either not answered or not adequately explained in textbooks [1–3]. Some textbooks [4, 5] in fact have given wrong answers/illustrations (see Appendix). The purpose of this chapter is to clarify these important points.

Band-Pass Case

Let the BPF specifications be:

$1 \leq \text{magnitude} \leq \delta_p$, for $\omega_{p1} \leq \omega \leq {}_{p2}$, and

$0 \leq \text{magnitude} \leq \delta_s$, for $0 \leq \omega \leq \omega'_{s1}$ and $\omega'_{s2} \leq \omega \leq \infty$.

If $\omega_{p1}\omega_{p2} = \omega'_{s1}\omega'_{s2}$then no modification of the characteristics is necessary. However, if this is not the case, then two cases may arise:

I. $\omega_{p1}\omega_{p2} < \omega'_{s1}\omega'_{s2}$
II. $\omega_{p1}\omega_{p2} > \omega'_{s1}\omega'_{s2}$

In case I, reduce ω'_{s2} to $\omega_{s2} = \omega_{p1}\omega_{p2}/\omega'_{s1}$ and rename ω'_{s1} as ω_{s1}, as shown in Fig. 21.2a. In case II, increase ω'_{s1} to $\omega_{s1} = \omega_{p1}\omega_{p2}/\omega'_{s2}$ and rename ω'_{s2} as ω_{s2}, as shown in Fig. 21.2b. In both cases adjustments have been made to guarantee geometric symmetry, thus facilitating the application of Eq. 21.1.

Now comes the question of finding Ω_s. Note that $S = \pm j\Omega_s$ should correspond to $s = \pm j\omega_{s1}$ as well as $s = \pm j\omega_{s2}$, where the signs may or may

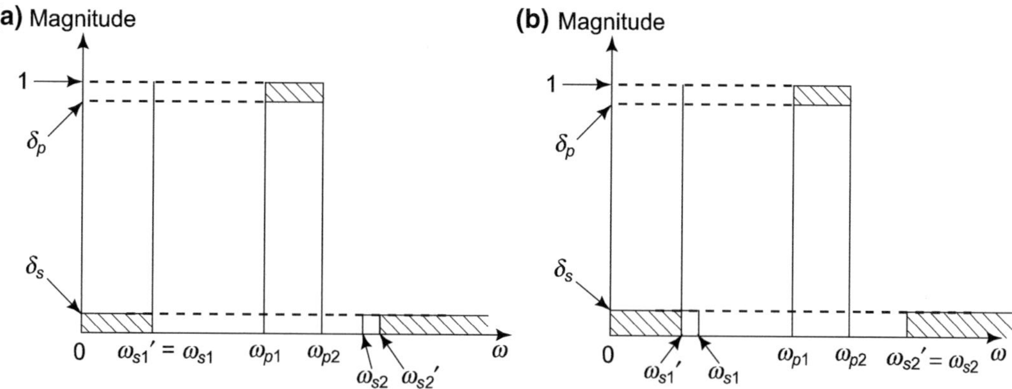

Fig. 21.2 Adjustments in given BPF response to ensure that $\omega_{p1}\omega_{p2} = \omega_{s1}\omega_{s2}$

not correspond to each other. Putting $S = j\Omega_s$ and $s = \pm j\omega_{s1}$ in Eq. 21.1 and simplifying, we get

$$\Omega_s = \frac{\omega_{s2} - \omega_{s1}}{\omega_{p2} - \omega_{p1}} \tag{21.5}$$

Since a positive value of Ω_s has been obtained, the correspondence of $S = j\Omega_s$ to $s = \pm j\omega_{s1}$ is validated. Similarly, one can show that $S = j\Omega_s$ also corresponds to $s = \pm j\omega_{s2}$ and that $S = -j\Omega_s$ corresponds to $s = \pm j\omega_{s1}$ as well as $s = \pm j\omega_{s2}$.

Band-Stop Case

The adjustments needed in the band-stop case are illustrated in Fig. 21.3. It is easily shown that, here

$$\Omega_s = \frac{\omega_{p2} - \omega_{p1}}{\omega_{s2} - \omega_{s1}} \tag{21.6}$$

Example

As an example, let us design a maximally flat BSF to satisfy the specifications shown in Fig. 21.4a, where magnitude refers to that of the transfer function V_2/I_1 of the network shown in Fig. 21.4b. Since here $f'_{s1}f'_{s2} = 6\ (\text{kHz})^2 > f_{p1}f_{p2} = 5\ (\text{kHz})^2$, f denoting $\omega/2\pi$, we adjust f'_{s1} to $f_{s1} = 5/3$ kHz, and set $f_{s2} = f'_{s2} = 3\text{kHz}$. Thus, by Eq. 21.6, we get

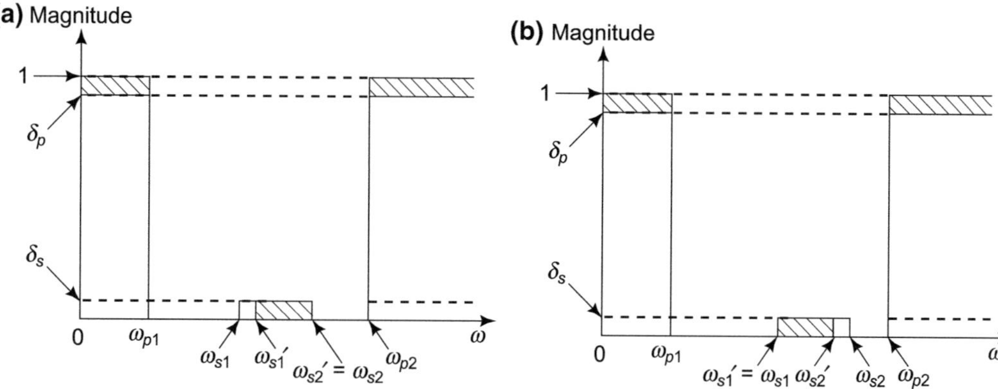

Fig. 21.3 Adjustments in given BSF response to ensure that $\omega_{p1}\omega_{p2} = \omega_{s1}\omega_{s2}$. **a** $\omega_{p1}\omega_{p2} < \omega'_{s1}\omega'_{s2}$; $\omega_{s1} = \omega_{p1}\omega_{p2}/\omega'_{s2}$; $\omega'_{s2} = \omega_{s2}$; **b** $\omega_{p1}\omega_{p2} < \omega'_{s1}\omega'_{s2}$; $\omega_{s2} = \omega_{p1}\omega_{p2}/\omega'_{s2}$; $\omega'_{s1} = \omega_{s1}$

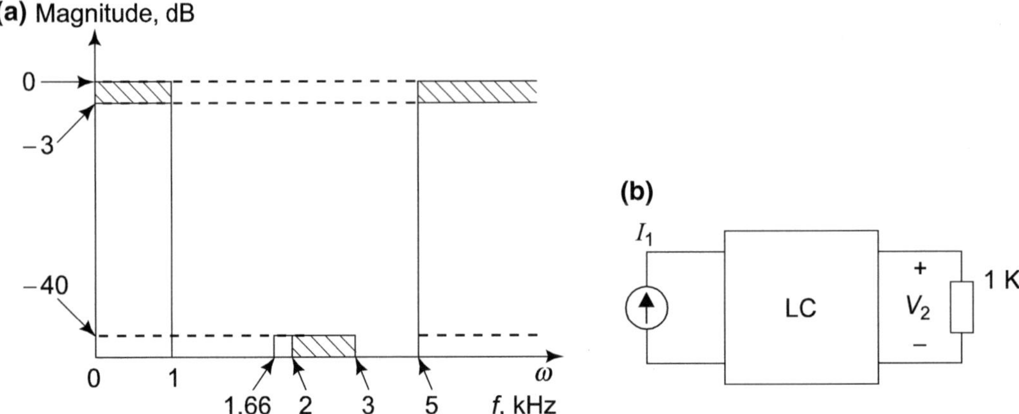

Fig. 21.4 a BSF specification, b desired network

$$\Omega_s = \frac{f_{p2} - f_{p1}}{f_{s2} - f_{s1}} = 2.25 \tag{21.7}$$

Thus, the normalized LPF to be designed has the specifications shown in Fig. 21.5 with a network of the form of Fig. 21.4b but with a terminating resistance of 1 ohm. Since a maximally flat design is needed, the order required is given by

$$N \geq \frac{\log_{10}\sqrt{10^4 - 1}}{\log_{10} 2.25} = 5.67887 \tag{21.8}$$

Thus, a sixth-order Butterworth filter is needed. Taking values from standard Tables [1], we get the complete normalized LPF shown in Fig. 21.6. To convert it to a de-normalized BSF with a termination of 1 K, the following replacements are to be made:

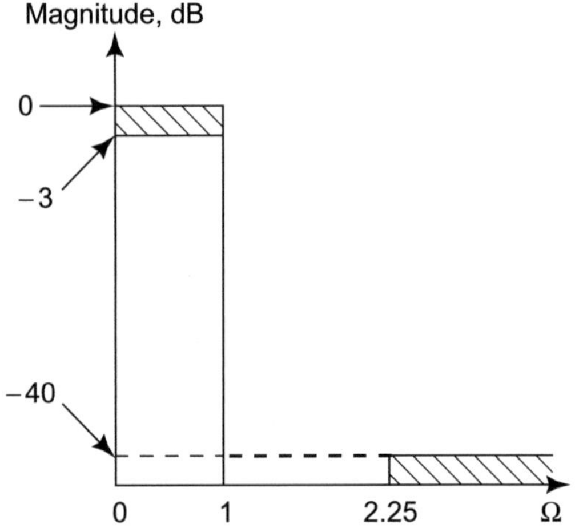

Fig. 21.5 Characteristics of the normalized LPF corresponding to the adjusted BSF response of Fig. 21.4a

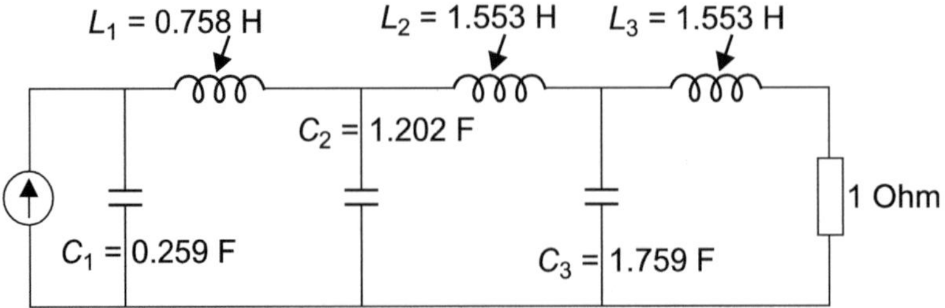

Fig. 21.6 Normalized Butterworth filter satisfying the specifications of Fig. 21.5

(1) each inductance L_i by a parallel combination of inductance RL_iB/ω_0^2 and capacitance $1/(RL_iB)$;
(2) each capacitance C_i by a series combination of inductance $R/(C_iB)$ and capacitance $C_iB/(\omega_0^2R)$; and
(3) resistance 1 ohms by R.

where $R = 1000$ ohms, $\omega_0 = 2\pi\sqrt{5} \times 10^3$ radians/sec and $B = 8\pi \times 10^3$ radians/sec.

Concluding Comments

This chapter attempts to supply the important steps needed in designing a BPF/BSF through frequency transformation. As noted in the introduction, these steps are either wrongly stated/illustrated, as discussed in the Appendix, or not explained in details in standard textbooks.

Problems

P.1. Determine the low-pass transfer function corresponding to a BPF having the same specification as those given in the example in the text.
P.2. Design the BPF corresponding to the above.
P.3. Design the LPF corresponding to P.1.
P.4. What will happen if geometric symmetry is ignored in the BPF design?
P.5. Same as above except that the design is to be for a BSF.

Appendix

Temes and Lapatra [4] recommend adjusting either of the pass-band edges to obtain geometric symmetry. This is obviously not advisable because admitting part of the transition band into the pass-band allows undesirable frequencies to be passed, along with the noise at these frequencies, thus deteriorating the signal to noise ratio. On the other hand, adjusting the stop-band by pushing some of the transition band into the stop-band not only attenuates undesired frequencies to a greater extent, but also improves the signal to noise ratio.

Karni [5], in an example to illustrate the design of a BPF, computes the stop-band edge of the normalized LPF as the ratio of the upper stop-band edge of the BPF to the bandwidth. This is obviously wrong!

References

1. F.F. Kuo, in *Network Analysis and Synthesis* (Wiley, 1996)
2. H. Ruston, J. Bordogna, in *Electric Networks: functions, Filters, Analysis* (McGraw-Hill, 1966)
3. M.E. Van Valkenburg, in *Introduction to Modem Network Synthesis* (Wiley, 1964)
4. G.C. Temes, J.W. Lapatra, in *Introduction to Circuit Synthesis and Design* (McGraw-Hill, 1977), pp. 556–557
5. S. Karni, in *Network Theory–Analysis and Synthesis* (Allyn and Bacon, 1966), p. 379

22 Optimum Passive Differentiators

A general, nth order, the transfer function (TF) is derived, whose time-domain response approximates optimally that of an ideal differentiator, optimality criterion chosen being the maximization of the first n derivatives of the ramp response at $t = 0^+$. It is shown that transformerless, passive, unbalanced realizability is ensured for $n < 3$, but for $n > 3$, the TF is unstable. For $n = 3$, the TF is not realizable, however, near optimum results can be obtained by perturbation of the pole locations. Optimum TFs are also derived for the additional constraint of inductorless realizability. It is shown that TFs for $n \geq 2$ are not realizable. For all n, however, near optimum results can be achieved by small perturbations of the pole locations; this is illustrated in this chapter for $n = 2$. Network realizations, for a variety of cases, are also given.

Keywords
Differentiators · Networks · Optimization

This chapter is complementary to [1], which deals with optimum passive integrators. Following, a parallel approach, optimum differentiators of order n have been suggested here, the optimality criterion being the maximum possible values for the first n derivatives of the ramp response at $t = 0^+$.

We show that transformerless RLC unbalanced realization is possible only for $n < 3$, and that for $n \geq 3$, the optimum transfer function (TF) is unstable. Near optimum results can be achieved for $n = 3$ by perturbation of the pole locations. RLC realizations for $n \geq 2$ give, in general, a damped oscillatory output around the ideal differentiated value. However, one can reduce the amplitude of these oscillations such that the output is within a prescribed limit of tolerance. In this chapter, we assume the tolerance as ±5% of the ideal differentiated value.

We will also derive an nth-order TF with an additional constraint of RC realizability, and show that for $n > 2$, optimal RC realizations are not possible; however, near optimum results can be achieved by small perturbations of the pole locations for *all n*.

S. C. Dutta Roy, *Circuits, Systems and Signal Processing*,
https://doi.org/10.1007/978-981-10-6919-2_22

Network realizations are given for the following cases: $n = 3$, suboptimal, RLC; $n = 2$, optimal RLC; $n = 2$, suboptimal RLC (Oscillations limited to ±5% of the ideal differentiated value); $n = 2$, suboptimal, RC.

Optimal Transfer Function and Its Realizability

The ideal differentiator has a normalized TF

$$H_0(s) = s \tag{22.1}$$

so that when the input voltage $v_{in}(t)$ is $t\,u(t)$, i.e. a ramp function, the output voltage $v_{out}(t) = u(t)$, a unit step function.

The TF given by Eq. 22.1 is not realizable and most common approximation used is

$$H_1(s) = sT/(sT + 1). \tag{22.2}$$

which can be realized by an RC or an RL network, shown in Figs. 22.1a, b, respectively. When driven by $v_{in}(t) = t\,u(t)$, the output is

$$v_1(t) = T(1 - e^{-t/T})u(t). \tag{22.3}$$

$v_1(t)$ rises exponentially to the true (ideal) value of differentiation (=$T\,u(t)$), with a time constant T, and reaches the ideal value only at $t = \infty$. However, in practice, the time taken for the output to reach ±10% of the ideal value determines the usefulness of a differentiator. We shall choose a tighter tolerance of ±5% to compare the performance of various differentiators in this chapter.

The approximation will improve with decreasing T, in Eq. 22.3, but it also reduces the output level. We, therefore, derive an optimal TF, of general order n, which does not suffer from this disadvantage. Let, the required nth-order TF be

$$H_n(s) = \frac{a_n s^n + a_{n-1}s^{n-1} + \cdots + a_1 s + a_0}{s^n + b_{n-1}s^{n-1} + \cdots + b_1 s + b_0}. \tag{22.4a}$$

With an input $v_{in}(t) = t\,u(t)$, the output will be

$$v_n(t) = L^{-1}V_n(s) = L^{-1}\left[(1/s^2)H_n(s)\right]. \tag{22.4b}$$

The optimality criteria chosen are:

(i) $$v_n(t)|_{t=\infty} = 1 \tag{22.5a}$$

(ii) maximum possible $v_n^{(1)}(0^+), v_n^{(2)}(0^+), v_n^{(3)}(0^+)\ldots, v_n^{(n)}(0^+)$.
(22.5b)

where $v_n^{(i)}(t)$. denotes the ith derivative of $v_n(t)$, and

(iii) highest possible order zero, at $s = 0$, of

$$L[1 - v_n(t)] \tag{22.5c}$$

Criterion (ii) is chosen to minimise the rise time of $v_n(t)$; the reason for the choice of criterion (iii) will be discussed later. We shall, in the sequel, impose the condition of a grounded transformerless network as part of realization constraints. Combining Eqs. 22.4a, 22.4b and 22.5a with the final value theorem gives

$$a_0 = 0 \quad \text{and}\; a_1 = b_0 \tag{22.6}$$

Using Eq. 22.6, $H_n(s)$ and $v_n(s)$ become

$$H_n(s) = \frac{a_n s^n + a_{n-1}s^{n-1} + \cdots + a_2 s^2 + b_0 s}{s^n + b_{n-1}s^{n-1} + \cdots + b_2 s^2 + b_1 s + b_0} \tag{22.7a}$$

and

$$V_n(s) = \frac{1}{s^2}\left[\frac{a_n s^n + a_{n-1}s^{n-1} + \cdots + a_2 s^2 + b_0 s}{s^n + b_{n-1}s^{n-1} + \cdots + b_2 s^2 + b_1 s + b_0}\right] \tag{22.7b}$$

By the initial value theorem, we have

$$v_n(0) = \lim_{s\to\infty} s\,V_n(s) = 0 \tag{22.7c}$$

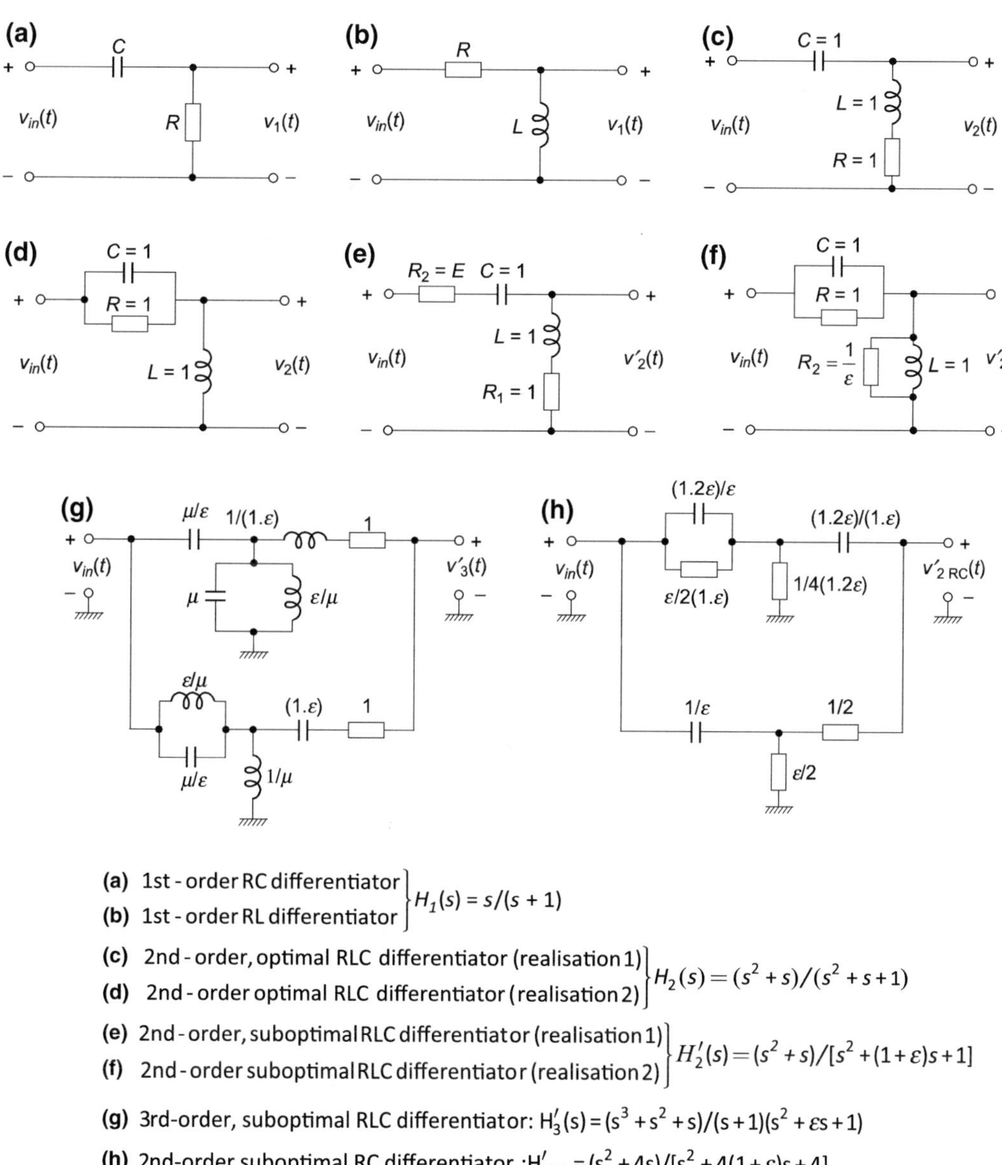

Fig. 22.1 Differentiator Networks

If we let $V_{n,1}(s) \triangleq Lv_n^{(i)}(t)$, then by the differentiation theorem of Laplace transform,

$$V_{n,i}(s) = sV_{n,i-1}(s) - v_n^{i-1}(0), \quad i \geq 1 \quad (22.8)$$

Thus,

$$V_{n,1}(s) - sV_n(s) - v_n(0) = \frac{1}{s}\left[\frac{a_n s^n + a_{n-1}s^{n-1} + \cdots + a_2 s^2 + b_0 s}{s^n + b_{n-1}s^{n-1} + \cdots + b_2 s^2 + b_1 s + b_0}\right] \quad (22.9a)$$

and

$$v_n^{(1)}(0) = \lim_{s\to\infty} s\, V_{n,1}(s) = a_n \quad (22.9b)$$

From criterion Eq. 22.5b, and the Fialkow–Gerst condition, we note that $v_n^{(i)}(0) = a_n = 1$. Thus Eqs. 22.7a and 22.9a become, respectively,

$$H_n(s) = \frac{s^n + a_{n-1}s^{n-1} + \cdots + a_2 s^2 + b_0 s}{s^n + b_{n-1}s^{n-1} + \cdots + b_2 s^2 + b_1 s + b_0} \quad (22.10a)$$

$$V_{n,1}(s) = \frac{1}{s}\left[\frac{s^n + a_{n-1}s^{n-1} + \cdots + a_2 s^2 + b_0 s}{s^n + b_{n-1}s^{n-1} + \cdots + b_2 s^2 + b_1 s + b_0}\right] \quad (22.10b)$$

Again we have, from Eqs. 22.8 and 22.10b

$$V_{n,2}(s) = (a_{n-1} - b_{n-1})s^{n-1} + \frac{(a_{n-2} - b_{n-2})s^{n-2} + \cdots + (a_2 - b_2)s^2 + (b_0 - b_1)s - b_0}{s^n + b_{n-1}s^{n-1} + \cdots + b_2 s^2 + b_1 s + b_0} \quad (22.11a)$$

so that

$$v_n^{(2)}(0) = \lim_{s\to\infty} s\, V_{n,2}(s) = a_{n-1} - b_{n-1} \quad (22.11b)$$

To maximize $v_n^{(2)}(0)$, (criterion Eq. 22.5b), under the Fialkow–Gerst condition, $a_{n-1} \leq b_{n-1}$, we have to choose

$$a_{n-1} = b_{n-1} \quad (22.12)$$

which gives

$$v_n^{(2)}(0) = 0 \quad (22.13)$$

From Eqs. 22.11a and 22.12, we get

$$V_{n,2}(s) = \frac{(a_{n-2} - b_{n-2})s^{n-2} + \cdots + (a_2 - b_2)s^2 + (b_0 - b_1)s - b_0}{s^n + b_{n-1}s^{n-1} + \cdots + b_2 s^2 + b_1 s + b_0} \quad (22.14)$$

Again, using Eq. 22.8, we have

$$V_{n,3}(s) = sV_{n-2}(s) - v_n^{(2)}(0) = \frac{(a_{n-2} - b_{n-2})s^{n-1} + \cdots + (a_2 - b_2)s^3 + (b_0 - b_1)s^2 - b_0 s}{s^n + b_{n-1}s^{n-1} + \cdots + b_2 s^2 + b_1 s + b_0} \quad (22.15a)$$

and

$$v_n^{(3)}(0) = \lim_{s\to\infty} s\, V_{n,3}(s) = a_{n-2} - b_{n-2} \quad (22.15b)$$

By arguments similar to those already used, the maximum value of $v_n^{(3)}(0)$ is obtained when

$$a_{n-2} = b_{n-2} \quad (22.16)$$

for which

$$v_n^{(3)}(0) = 0 \quad (22.17)$$

This yields

$$V_{n,3}(s) = \frac{(a_{n-3} - b_{n-3})s^{n-2} + \cdots + (a_2 - b_2)s^3 + (b_0 - b_1)s^2 - b_0 s}{s^n + b_{n-1}s^{n-1} + \cdots + b_2 s^2 + b_1 s + b_0} \quad (22.18)$$

Proceeding in this manner, till we maximize $v_n^{(n)}(0)$ we get

$$a_1 = b_i, i = 1, 2, \ldots, (n-1) \quad (22.19)$$

Combining these with Eq. 22.6 gives

$$H_n(s) = \frac{s^n + b_{n-1}s^{n-1} + \cdots + b_2 s^2 + b_0 s}{s^n + b_{n-1}s^{n-1} + \cdots + b_2 s^2 + b_0 s + b_0} \quad (22.20a)$$

and

$$V_n(s) = \frac{1}{s}\left[\frac{s^{n-1} + b_{n-1}s^{n-2} + \cdots + b_2 s + b_0}{s^n + b_{n-1}s^{n-1} + \cdots + b_2 s^2 + b_0 s + b_0}\right] \quad (22.20b)$$

We may also write $H_n(s)$ of Eq. 22.20a as

$$H_n(s) = \frac{N_n(s)}{D_n(s)} = \frac{N_n(s)}{N_n(s) + D_n(0)}, \quad (22.21)$$

where $N_n(s)$ and $D_n(s)$ denote, respectively, the numerator and the denominator polynomials of Eq. 22.20a.

Let

$$q(t) \underline{\underline{\Delta}} (1 - v_n(t))\, u(t) \quad (22.22)$$

denote the deviation of $v_n(t)$ from the ideal output $u(t)$. Had $v_n(t)$ been the ideal output; $q(t)$ as well as $Q(s) = Lq(t)$ would be zero. Since this is not the case in practice, we impose condition Eq. 22.5c, i.e. $Q(s)$ should have a zero of the highest possible order at $s = 0$. Now

$$Q(s) = \frac{1}{s} - V_n(s) \quad (22.23)$$

Substituting for $V_n(s)$ from Eq. 22.20b and simplifying, it is easy to observe that $Q(s)$ will have a zero of the highest order ($=n - 1$), at $s = 0$ if

$$b_{n-1} = b_{n-2} = \cdots = b_2 = b_0 = 1 \quad (22.24)$$

and finally, the optimum TF is

$$H_n(s) = \frac{s^n + s^{n-1} + s^{n-2} + \cdots + s^2 + s}{s^n + s^{n-1} + s^{n-2} + \cdots + s^2 + s + 1} \quad (22.25)$$

$$= \frac{s(s^n - 1)}{s^{n+1} - 1} \quad (22.26)$$

The poles of $H_n(s)$ are located on the unit circle, at

$$S_r = c^{j2pr/(n+1)} \times r = 1, 2, \ldots, n \quad (22.27)$$

where $r = 0$ is excluded because $s = 1$ is a pole as well as a zero. For stability, the poles should be in the left half of the s-plane, i.e.

$$\pi/2 \le 2\pi r/(n+1) \le 3\pi/2 \quad (22.28)$$

Equation 22.28 is violated for $n = 4$. The TF for $n = 3$ is

$$H_3(s) = \frac{s^3 + s^2 + s}{s^3 + s^2 + s + 1} \quad (22.29)$$

It can be easily shown that the poles of $H_3(s)$, at $s = \pm j$, do not have purely imaginary residues, hence $H_3(s)$ cannot be realized [3]. Had $H_3(s)$ been realizable, the ramp response would have been given by

$$\begin{aligned} v_s(t) &= L^{-1}\left[(1/s^2)H_3(s)\right] \\ &= \left[1 - \frac{1}{2}e^{-t} - \left(\frac{1}{2}\right)^{\frac{1}{2}}\cos(t + \pi/4)\right]u(t) \end{aligned} \quad (22.30)$$

A plot of $v_3(t)$ is shown in Fig. 22.2 (curve *f*). Clearly, $v_3(t)$ is of little use due to the undamped oscillations.

We thus conclude that the only optimum, passive, grounded and transformerless network realizable approximations of Eq. 22.1 are

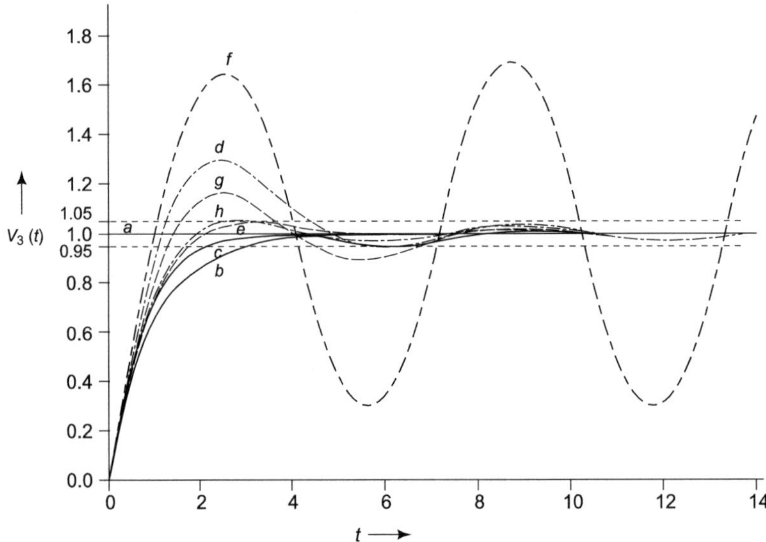

Fig. 22.2 Ramp responses of various differentiators with final value normalised to unity. *a* Ideal case, *b* first-order RC, *c* second-order, suboptimal RC (ε = 0.01), *d* second-order, optimal RLC, *e* second-order, suboptimal RLC (ε = 0.601), *f* third-order, optimal RLC (unrealizable), *g* third-order, suboptimal RLC (ε = 0.5), *h* third-order, suboptimal RLC (ε = 0.71)

$$H_1(s) = s/(s+1) \tag{22.31}$$

$$H_2(s) = (s^2+s)/(s^2+s+1) \tag{22.32}$$

Equation 22.31 is the same as Eq. 22.2 with T normalized to unity.

Second-order Optimal and Suboptimal Differentiators

Dividing the numerator and denominator of Eq. 22.32 by s, two simple network realizations of second-order passive differentiators, as shown in Figs. 22.1c, d are obtained. Realization of Fig. 22.1c may be preferred to that of Fig. 22.1d since the effect of losses in the inductor L can be fully compensated by appropriate reduction in the series resistance R. The ramp response of the second-order differentiator is given by

$$\begin{aligned} v_s(t) &= L^{-1}\left[\left(1/s^2\right)H_2(s)\right] \\ &= \left\{1 - \left(2/\sqrt{3}\right)e^{-t/2}\cos\left[\left(\sqrt{3/2}\right)t + \pi/6\right]\right\} \\ &\quad u(t) \end{aligned} \tag{22.33}$$

and is plotted in Fig. 22.2, curve *d*.

The damped oscillations exhibited by $v_2(t)$ reduce the utility of $H_2(s)$ vis-a-vis $H_1(s)$. However, by shifting the poles of $H_2(s)$, we may bring down the amplitude of the oscillations to achieve the desired tolerance limits of the output. If we take

$$H_2'(s) = (s^2+s)/\left[s^2+(1+\varepsilon)s+1\right], \text{where} 0<\varepsilon<1 \tag{22.34a}$$

then

$$v_2(t) = \left[1 - \frac{2}{(3+\varepsilon)^{1/2}} e^{-(1+\varepsilon)t/2} \cos(\beta t + \alpha)\right] u(t), \quad (22.34b)$$

Where

$$\alpha = \tan^{-1}\left(\frac{1-\varepsilon}{3+\varepsilon}\right)^{\frac{1}{2}} \text{ and } \beta = (3 - 2\varepsilon - \varepsilon^2)^{\frac{1}{2}}/2 \quad (22.34c)$$

It may be seen that damping increases as we increase ε from 0 to 1. In particular, $H_2'(s)\big|_{\varepsilon=0} = H_2(s)$ and $v_2'(t)\big|_{\varepsilon=0} = v_2(t)$. Critical damping is achieved for $\varepsilon = 1$ [4], and $v_2'(t)|_{\varepsilon=1} = 1 = v_1(t)$. Thus for critical damping, $v_2'(t)$ coincides with $v_1(t)$ and the rise time of $v_2'(t)$ is maximal. The rise time of $v_2'(t)$ decreases with the decrease of damping (i.e. with the decrease of ε). The optimum value of ε, such that $v_2'(t)$ may reach the ideal differentiated value, within a tolerance of $\pm 5\%$, in the shortest possible time is found to be $\varepsilon = 0.601$, and the response under this condition is shown by curve e in Fig. 22.2.

Third-order Suboptimal Passive Differentiator

The third-order TF given by Eq. 22.29 is not realizable due to its poles at $s = \pm j$. We may, however, realize a network by shifting the poles slightly to the left in the s-plane. The TF will no longer remain optimal, and we call this a suboptimal realization. The suboptimal TF will be

$$H_3'(s) = \frac{s(s^2+s+1)}{(s+1)(s^2+\varepsilon s+1)}, \text{ where } 0 < \varepsilon < 1 \quad (22.35)$$

The ramp response of Eq. 22.35 is given by

$$v_3'(t) = \left[1 - \frac{e^{-t}}{2-\varepsilon} - \frac{(1-\varepsilon)}{\beta(2-\varepsilon)^{\frac{1}{2}}} e^{-\varepsilon t/2} \cos(\beta t + \alpha)\right] u(t), \quad (22.36a)$$

where

$$\alpha = \tan^{-1}\left(\frac{2-\varepsilon}{2+\varepsilon}\right)^{\frac{1}{2}} \text{ and } \beta = (1 - \varepsilon^2/4)^{\frac{1}{2}} \quad (22.36b)$$

$H_2'(t)$ can be realized using the Fialkow-Gerst technique [3], one such realization being shown in Fig. 22.1g.

As is clear from Eq. 22.36a, $v_3'(t)$ also gives damped oscillations around the ideal differentiated value. However, we can decrease the oscillations by increasing ε. In particular, $\varepsilon = 1$ gives critical damping [4], i.e. no oscillations and $v_3'(t)|_{\varepsilon=1} = 1 = v_1(t)$. The optimum value of ε so that $v_3'(t)$ may reach the ideal differentiated value within a tolerance of $\pm 5\%$, in the shortest possible time, is found to be $\varepsilon = 0.71$. Curves g and h in Fig. 22.2. show $v_3'(t)$ for $\varepsilon = 0.5$ and for $\varepsilon = 0.71$, respectively.

Optimal RC Differentiators

From Eq. 22.21, one can write the nth-order differentiator TF as

$$H_n(s) = \frac{N_n(s)}{N_n(s) + b_o} \quad (22.37)$$

For RC realizability, the roots of the denominator polynomial should be distinct and located on the negative real axis of the s-plane, i.e.

$$N_n(s) + b_o = \prod_{i=1}^{n} (s + \sigma_i), \quad (22.38)$$

where

$$0 < \sigma_1 < \sigma_2 < \cdots < \sigma_n \quad (22.39)$$

Equating the constant terms and the coefficients of s on both sides of Eq. 22.38, we get

$$b_o = \prod_{i=1}^{n} \sigma_i \quad (22.40)$$

and

$$b_o = \sum_{j=1}^{n} \frac{1}{\sigma_j} \left(\prod_{i=1}^{n} \sigma_i \right) \quad (22.41)$$

Combining Eqs. 22.40 and 22.41 gives

$$\sum_{i=1}^{n} 1/\sigma_i = 1 \quad (22.42)$$

For optimum results, we must maximize b_o, as shown in the appendix. Following, a procedure similar to one used in Section 5 of [1]; maximization of b_o yields

$$\sigma_1 = \sigma_2 = \sigma_3 = \cdots = \sigma_n = n \quad (22.43)$$

and

$$b_o = n^n \quad (22.44)$$

Thus nth-order optimal RC differentiator is

$$H_{n_{\mathrm{RC}}}(s) = \frac{(s+n)^n - n^n}{(s+n)^n} \quad (22.45)$$

Suboptimal RC Differentiator

$H_{n_{\mathrm{RC}}}(s)$ is not realizable for $n \geq 2$ as the TFs have multiple poles at $s = -n$ for $n \geq 2$. To make the poles distinct, we perturb them from location $-n$, so that $H_{n_{\mathrm{RC}}}(s)$ becomes realizable for *all n*. The same methodology as suggested in Section 6 of [1] can be followed; then the second-order suboptimal TF becomes

$$H'_{2_{\mathrm{RC}}}(s) = \frac{s^2 + 4s}{s^2 + 4(1+\varepsilon)s + 4} \quad (22.46)$$

which gives, for a ramp input

$$v'_{2_{\mathrm{RC}}}(t) = \left\{ 1 - \frac{1}{2\beta} e^{-\alpha t} \left[(\alpha+\beta-1)e^{\beta t} - (\alpha-\beta-1)e^{-\beta t} \right] \right\} u(t), \quad (22.47a)$$

where

$$\alpha = 2(1+\varepsilon) \quad \text{and}\ \beta = 2(2\varepsilon + \varepsilon^2)^{1/2} \quad (22.47b)$$

Also,

$$v'_{2_{\mathrm{RC}}}(t)\big|_{\varepsilon=0} \triangleq v_{2_{\mathrm{RC}}}(t) = \left[1 - (1+t)e^{-2t}\right] u(t) \quad (22.48)$$

A plot of $v'_{2_{\mathrm{RC}}}(t)$, for $\varepsilon = 0.01$, is shown in Fig. 22.2. (curve *c*). As the plots for $v_{2_{\mathrm{RC}}}(t)$ and $v'_{2_{\mathrm{RC}}}(t)$ do not differ by more than 1% in the time range shown, these are not shown separately in Fig. 22.2.

A realization of $H'_{2_{\mathrm{RC}}}(s)$, using the F-G technique [3] is shown in Fig. 22.1h.

Conclusion

The problem of obtaining an optimum approximation to the ideal differentiator by passive, transformerless, unbalanced network has been investigated. The following conclusions have been arrived at:

(a) RLC, optimal differentiators are not realizable for order $n \geq 3$; however, suboptimal RLC differentiators, for $n = 3$ can be realized by pole perturbation technique.
(b) RC, optimal differentiators are not realizable for order $n \geq 2$; however, suboptimal RC differentiators for *all n* can be realized by pole perturbation.
(c) RC differentiators, of all orders, reach the ideal value of differentiation only at $t = \infty$.
(d) Although the response of optimal RLC differentiators approaches the ideal value faster than that of RC differentiators, the former exhibit damped/undamped oscillations. This restricts the use of optimal RLC differentiators. However, the amplitude of oscillations can be limited to *any* desired tolerance by pole perturbation, such that the RLC differentiators give better performance than the RC differentiators. We have chosen a tolerance of

Table 22.1 Values of τ_5, the normalised time taken by various differentiators to give output voltage within ±5% of the ideal output voltage

Case	Ideal case	RC, optimal $n = 1$	RC, suboptimal $n = 2$	RLC, suboptimal $n = 2$	RLC, suboptimal $n = 3$
TF of the differentiators	$H_0(s) = s$ (unrealisable)	$H_1(s) = \frac{s}{s+1}$	$H'_{2\text{RC}}(s) = \frac{s^2+4s}{s^2+4(1+\varepsilon)s+4}$ $\varepsilon = 0.01$	$H'_2(s) = \frac{s^2+s}{s^2+\varepsilon s+1}$ $\varepsilon = 0.601$	$H'_3(s) = \frac{s^3+s^2+s}{(s+1)(s^2+\varepsilon s+1)}$ $\varepsilon = 0.71$
τ_5	0.0	1.0	0.7	0.587	0.576

±5% of the ideal differentiated value and the results obtained are given in Table 22.1.

A variety of differentiators using operational amplifiers are known. But the constraints of the active device viz. offset voltages and currents, finite gain-bandwidth product, finite dynamic range and slew rate limiting, etc. lead to further problems. The proposed optimum passive differentiators, which are free from aforesaid limitations, may be successfully employed in areas, where the frequency spectrum of the signal is relatively wide or where simple and reliable circuits with minimum power consumption are a necessity. A few such applications are the homing devices of an underwater torpedo, (where the dynamic range requirement is large of the order of 80 dB) and guidance system of a long-range guided missile (where the signal spectrum is wide, up to about 60 MHz).

Problems

P.1. Apply a square pulse of duration 1 s and height 1 V to a first-order differentiator. Find the output v_O/H and sketch it.

P.2. Determine the transfer function of a sub-optimal differentiator of order 4 and obtain the output for a ramp function.

P.3. Obtain the transfer function for an optimal RC differentiator of order 2 and find its support for a unit step input.

P.4. Same as P.3 except that the input is a ramp function.

P.5. Same as P.3 except that the input is an impulse function.

Appendix

In this section, we examine the nature of b_o and substantiate the assertions made in Section 5 that higher the value of b_o, better is the approximation. For the first-order case, the TF and the corresponding ramp response are, respectively,

$$H_1(s) = \frac{b_o s}{s + b_o} \quad (22.49)$$

and

$$v_1(t) = (1 - e^{-b_o t})u(t) \quad (22.50)$$

Clearly, higher the b_o, closer is $v_1(t)$ to $u(t)$ which is the ideal ramp response. Maximum value of b_o can be unity in $H_1(s)$ [F-G conditions of realizability of $H_1(s)$].

For the second-order case, the TF and the ramp response are, respectively

$$H_2(s) = \frac{s^2 + b_o s}{s^2 + b_o s + b_o} \quad (22.51)$$

and

$$v_2(t) = \left[1 - \frac{e^{-b_o t/2}}{(4 - b_o)^{1/2}} \cos\left(\omega_o t - \tan^{-1}\frac{b_o}{2\omega_o}\right)\right] u(t), \quad (22.52)$$

where

$$\omega_o = \left(b_o - b_o^2/4\right)^{1/2} \quad (22.53)$$

As $\left|\cos\left(\omega_o t - \tan^{-1}\frac{b_o}{2\omega_o}\right)\right| \le 1$, smaller the value $\frac{e^{-b_o t/2}}{(4-b_o)^{1/2}}$, closer is $v_2(t)$ to $u(t)$. Increase of b_o decreases $e^{-b_o t/2}$ faster (i.e. exponentially) than $(4 - b_o)^{1/2}$. Thus higher the b_o, smaller is the value of $\frac{e^{-b_o t/2}}{(4-b_o)^{1/2}}$ and consequently closer is the $v_2(t)$ to the ideal value $u(t)$.

For the general case, a semi-rigorous argument can be forwarded as follows. As

$$H_n(s) = \frac{s^n + b_{n-1}s^{n-1} + \cdots + b_2 s^2 + b_o s}{s^n + b_{n-1}s^{n-1} + \cdots + b_2 s^2 + b_1 s + b_o} \quad (22.54)$$

and

$$V_n(s) = \frac{1}{s}\left[\frac{s^{n-1} + b_{n-1}s^{n-2} + \cdots + b_3 s^2 + b_2 s + b_o}{s^n + b_{n-1}s^{n-1} + \cdots + b_2 s^2 + b_1 s + b_o}\right] \quad (22.55)$$

$$\triangleq \frac{1}{s} G(s) \quad (22.56)$$

where

$$G(s) = \frac{s^{n-1} + b_{n-1}s^{n-2} + \cdots + b_3 s^2 + b_2 s + b_o}{s^n + b_{n-1}s^{n-1} + \cdots + b_2 s^2 + b_1 s + b_o} \quad (22.57a)$$

$$= \frac{1 + (b_2/b_o)s + (b_3/b_o)s^2 + \cdots}{1 + (b_1/b_o)s + (b_2/b_o)s^2 + \cdots} \quad (22.57b)$$

Equation 22.56 shows that $v_n(t) = L^{-1} V_n(s)$ can be interpreted as the unit step response of the low-pass function $G(s)$. Equation 22.55, together with the initial and final value theorems of Laplace transforms shows that $v_n(t)$ rises from a value zero at $t = 0$ to unity at $t = \infty$. To enable us make $v_n(t)$ achieve unity value in as short a time as possible, we must choose b_0 such that the rise time τ_r, of $v_n(t)$ is as small as possible. Using Elmore's formula [2], with the assumption that the plot of $v_n(t)$ is monotonic, (whereby Elmore's formula can be applied), we get

$$\tau_r = \frac{1}{b_o}\left\{2\pi\left[\left(b_1^2 - b_2^2\right) - 2b_o(b_2 - b_3)\right]\right\}^{\frac{1}{2}} \quad (22.58)$$

τ_r, decreases monotonically with the increase of b_o. Thus b_o should be as large as possible. The assumption of $v_n(t)$ being monotonic has implications as mentioned in the Appendix of [1].

References

1. S.C. Dutta Roy, Optimum passive integrators, in *IEE Proceedings*, part G, (vol. 130, No. 5, pp. 196–200), Oct 1983
2. W.C. Elmore, Transient response of damped linear network with particular regard to wide band amplifiers. J. Appl. Phys. **19**, 55–63 (1948)
3. N. Balabanian, *Network Synthesis* (Prentice Hall, 1958)
4. M.E. Van Valkenburg, *Network analysis* (Prentice Hall of India, 1983)

Part III
Active Circuits

Passive circuits have their own limitations and can do very little when amplification, oscillation and other essential function of practical circuits can offer. Part III therefore concentrates on active circuits, which are combinations of passive circuits with active devices. Vacuum tubes are the things of the past and are seldom used, except in broadcast applications. We therefore treat circuits with transistors and operational amplifiers as the active devices. Amplifier fundamentals are presented in Chap. 23; this material was the first broadcast in India on actual educational materials and was done from studios of Space Application Centre at Ahmedabad, under the 'Teacher in the Sky' experiment of IETE. Judged by the positive feedback from students, it was a great success.

Again, that appearances can be deceptive occurs in active circuits also. This is illustrated in Chap. 24 with the BJT biasing circuit as an example. BJT biasing is dealt with, comprehensively, in Chap. 25, and it is proved that bias stability is the best in the four resistor circuit. A high-frequency transistor stage, consisting of emitter feedback, is analysed in detail in Chap. 26, using the hybrid equivalent circuit of the transistor, which was carefully avoided till then in most textbooks. Transistor Wien Bridge oscillator is treated comprehensively in Chap. 27, where various circuits and their merits and demerits are enumerated. In contrast to the hybrid parameter, I used the h-parameter equivalent circuit of the transistor because the former was not known till then.

The usual analysis of the oscillator, as given in textbooks, is to use the Nyquist criterion $A\beta=1$. In Chap. 28, I formulate several simpler, in fact much simpler, methods for doing the same, without the difficulty of identifying A and β, which is not easy even for experienced researchers. The triangular to sine wave converter is discussed in Chap. 29 with step-by-step logical analysis. The Wilson current mirror, presented in Chap. 30, is a versatile circuit and is used as an essential component in various analog ICs. The dynamic resistance is calculated easily. That completes our journey through active circuits. I hope it will be a smooth one, without getting lost in the rather complicated equivalent circuits.

Amplifier Fundamentals

23

This chapter presents the fundamentals of a bipolar junction transistor amplifier and includes the following aspects: choice of Q point, classes of operation, incremental equivalent circuit, frequency response, cascading, broadbanding and pulse testing. The emphasis is on understanding the fundamental, rather than rigorous analysis or elaborate design procedure.

Keywords
Amplifier · Transistor characteristics
CE configuration · Biasing · Hybrid Π equivalent circuit

The term 'amplifier' stands for any device which amplifies or magnifies a weak signal so as to make it detectable and useful. An amplifier is perhaps the most important electronic circuit and was the motivation or the leading reason behind the invention of the triode and the transistor. An amplifier is also a part of our daily life. The Radio, the Television and the Stereo are common examples of electronic equipment where the amplifier is an essential component. The Public Address system used at large gatherings, like political rallies and music concerts, is another very common example. Under this topic, we shall discuss the essential features of an amplifier along with the analysis of a typical circuit.

In order to introduce the subject, consider a typical single-transistor amplifier circuit, shown in Fig. 23.1, in which the transistor is connected in the common emitter configuration. The phrase 'common emitter', incidentally, implies that the emitter terminal is common between the input and the output. In the circuit of Fig. 23.1, we have shown an n-p-n transistor, whose dc collector-to-emitter voltage, V_{CE}, is determined by the supply voltage V_{CC}, the collector resistance R_C and the emitter resistance R_E through the equation

$$V_{CE} = V_{CC} - I_C R_C - I_E R_E \tag{23.1}$$

I_C and I_E are the dc collector and emitter currents, which are, of course, approximately equal, because the DC base current I_B is much smaller than I_C ($I_B = I_C/\beta$, $\beta \sim 50$). I_B is determined by the relation (see Fig. 23.2)

$$V_{CC}\frac{R_2}{R_1 + R_2} = I_B(R_1||R_2) + V_{BE} + I_E R_E, \tag{23.2}$$

where V_{BE} is the dc base-to-emitter voltage and is of the order of 0.7 V for a silicon transistor.

S.C Dutta Roy, "Amplifier Fundamentals," *Students' Journal of the IETE*, vol. 35, pp. 143–150, July–December 1994.

S. C. Dutta Roy, *Circuits, Systems and Signal Processing*,
https://doi.org/10.1007/978-981-10-6919-2_23

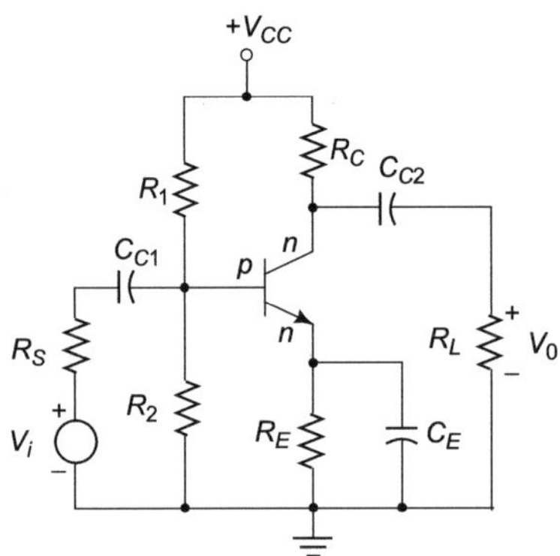

Fig. 23.1 A single-stage CE amplifier

I_C and V_{CE} determine the dc operating point or the so-called Q point of the transistor. More will be said about this later.

C_{C1} and C_{C2} in Fig. 23.1 are coupling capacitors, their values being so chosen that they act as short circuits at the signal frequency. Had C_{C1} not been there (short-circuited), R_2 would have been virtually short-circuited ($R_1 \ll R_2$, usually) and the transistor could not have been biased properly. Similarly, if C_{C2} is not there (short-circuited), the DC through R_C would have found a path through the load resistor R'_L, so that the Q point would shift with variations in the load. The capacitor C_E across R_E is chosen to be large so that at the signal frequency, R_E is virtually short-circuited. If C_E is not there (opened), then the signal voltage developed across R_E would cause a negative feedback, because the actual signal input to the transistor would have been the voltage across R_2 minus the voltage across R_E. The gain is thereby reduced to approximately—$R_L/R_E, R_L = R_C||R'_L$.

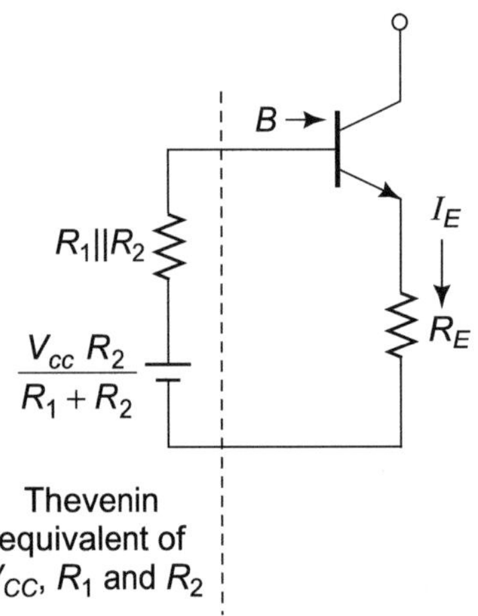

Fig. 23.2 DC equivalent of the base-emitter circuit

Now let us come back to the question of choosing the Q point. Figure 23.3 shows the transistor characteristics along with a plot of Eq. 23.1 given by the line ABC. The usable region of the characteristic is bounded by the maximum ratings of the transistor viz $V_{CE\max}$, $I_{C\max}$, the maximum collector dissipation ($V_{CE}I_C$), $P_{D\max}$, along with the saturation line near $V_{CE} = 0$, and the cutoff line $i_B = 0$. If V_{CE} and I_C are such that the Q point is near B, then signal excursions like that shown will lead to faithful variations in the collector current, as shown on the left side of the figure. This is the linear range of operation and goes under the name of Class A operation in the literature. Obviously, Class A operation cannot extend to a point close to A during the positive excursions of i_B, or to a point close to C during the negative excursions of i_B, because of the distortion due to saturation and cutoff, respectively. If the Q point is the same as the point marked V_{CC} in Fig. 23.3, then current will flow in the collector circuit only during the positive excursions of the signal. No current will flow during the negative excursions (this condition can be achieved by opening R_1 in Fig. 23.1). Obviously, the circuit will act as a half-wave rectifier. In order to reproduce the positive, as well as the negative portions of the signal, we need another transistor to take care of the negative half. These two transistors are configured in the so-called push–pull operation, as shown in Fig. 23.4, which, incidentally, uses a complementary symmetric pair of transistors (n-p-n and p-n-p) in a transformerless configuration. The usual arrangement of push–pull uses a centre-tapped transformer at the input and another such transformer at the output. The important point is that the transistors in Fig. 23.4 operate under what is defined as the Class B condition. Naturally, the Class B condition

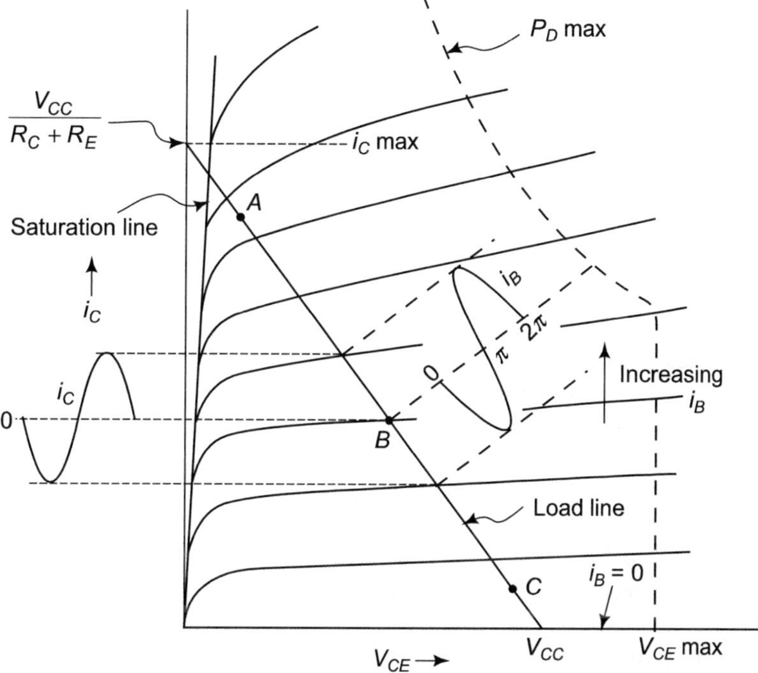

Fig. 23.3 Transistor characteristics, showing load line and the bounds of operation

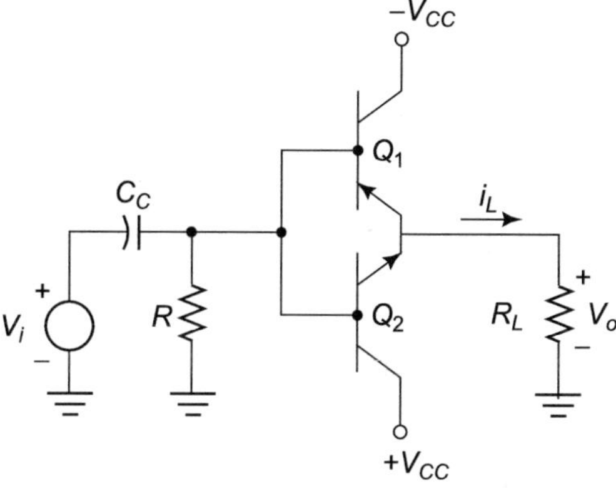

Fig. 23.4 Complementary symmetry push–pull class B amplifier

produces more distortion than the Class A operation. Nevertheless, Class B is invariably used in low-frequency (audio) large signal or power amplifiers, because of a drastic improvement in efficiency (from a maximum of 25% in Class A to a maximum of 78.5% in Class B) arising out of the reduced (to zero, ideally) power dissipation under the quiescent or no-signal condition. A further improvement in efficiency is possible if the transistor is biased beyond cutoff under quiescent conditions; the collector current will then flow for only a part of the positive half cycle (assuming an n-p-n transistor). This, of course, leads to excessive distortion, which can be avoided by using a parallel resonant circuit in the collector, tuned to the signal frequency. Then the voltage developed across the tuned circuit will mostly consist of the signal component. This is called the Class C mode of operation, and can be used to achieve practical efficiencies of about 85%.

We shall now confine our attention to small signal Class A operation so that the analysis is possible through an incremental or ac equivalent circuit. The adjective ‘incremental’ refers to the condition that signal components of current and voltage in the transistor are small perturbations to the total current and voltage respectively. Under this assumption, the amplifier behaves as a linear system and DC and AC analyses can be done separately, the latter being carried out with the equivalent circuit.

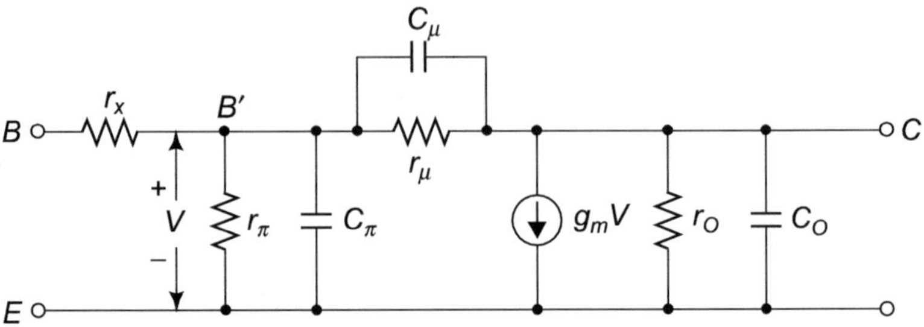

Fig. 23.5 Hybrid-π equivalent circuit of transistor

We shall now consider the circuit of Fig. 23.1 again, and carry out its analysis. Although various equivalent circuits have been proposed in the literature, the most versatile one viz the hybrid π, as shown in Fig. 23.5, shall be used here. We have used somewhat simplified notation as compared to what is used in most textbooks. While *B*, *E* and *C* stand for the base, emitter and collector respectively, *B′* is used to denote the internal base. The difference between *B* and *B′* is the occurrence of the 'base-spreading' resistor, r_x which is usually an order of magnitude smaller than r_π, the base-emitter resistor. Typically, $r_x = 100\ \Omega$, $r_\pi = 1$ K, $r = 4$ MΩ, $r_o = 80$ K, $C_\pi = 100$ pF, $C = 3$ pF and $C_o = 1$ pF, while g_m is determined by the quiescent collector current I_C according to the relation

$$g_m = [I_C(\text{in mA})/26]\ \text{mhos}$$

Except at high frequencies, or in rigorous analysis at other frequencies, we can ignore r_x. Also, notice that r_o will come across the load which is normally much smaller than 80 K; hence it can be ignored, along with C_o which is basically the stray capacitance. Normally, *r* can also be ignored in comparison to the impedance of *C* at the frequencies at which it counts. Hence, *r* also qualifies to be omitted from consideration, thus leading to the much simplified form of Fig. 23.6.

First, we consider midband frequencies at which the effects of all capacitances—internal (C_π, *C*) as well as external (C_{C1}, C_{C2} and C_E) can be ignored, the former acting as open circuits while the latter act as short circuits. Then, the equivalent circuit of Fig. 23.1 becomes that shown in Fig. 23.7, where the source resistance

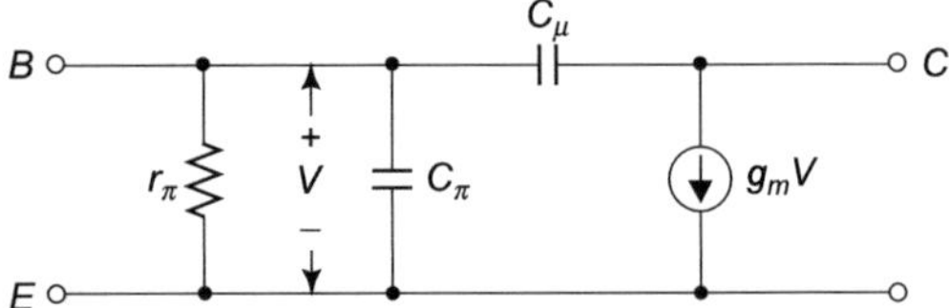

Fig. 23.6 Simplified hybrid-π equivalent circuit of the transistor

has also been ignored. By inspection, you see that the gain is

$$A_o = \frac{v_o}{v_\text{i}} = -g_m R_L \tag{23.3}$$

Next, consider the low-frequency response for which we ignore internal capacitances but not the external ones. There are three such capacitances viz C_{C1}, C_{C2} and C_E, and it becomes rather involved to consider the effects of all three simultaneously. We, therefore, consider them one by one. Suppose C_E, $C_{C2} \to \infty$; then the equivalent circuit becomes that shown in Fig. 23.8. With $R_\pi = R_1 \| R_2 \| r_\pi$, we have the low-frequency gain

$$A_L(s) = \frac{v_o}{v_\text{i}} = \frac{v_o}{V} \cdot \frac{V}{v_\text{i}},$$

$$= -g_m R_L \frac{R_\pi}{R_\pi + \frac{1}{sC_{C1}}}, \tag{23.4}$$

Putting $s = j\omega$ and $R_\pi\ C_{C1} = 1/\omega_1$ we get

$$A_L(j\omega) = \frac{A_\text{o}}{1 - j\frac{\omega_1}{\omega}} \tag{23.5}$$

This shows that with increasing frequency, the gain rises from zero to the midband value A_o and

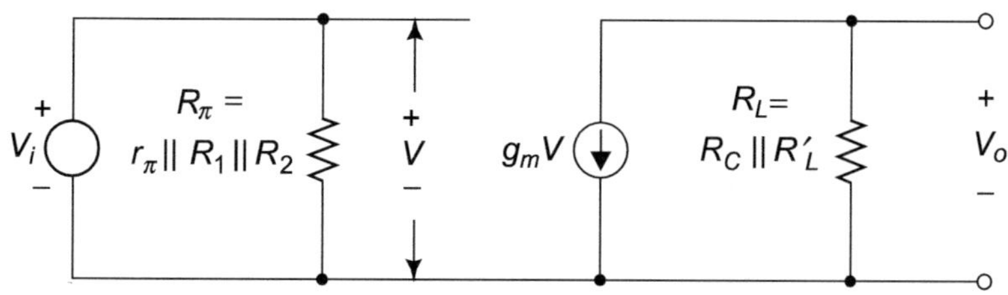

Fig. 23.7 Equivalent circuit of Fig. 23.1 at midband frequencies

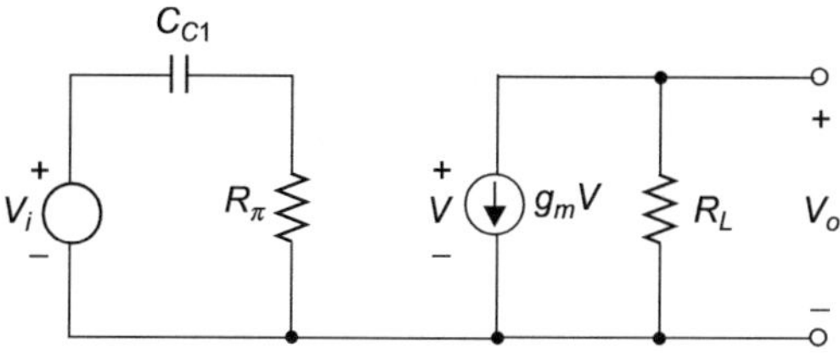

Fig. 23.8 Low-frequency equivalent circuit with C_E, $C_{C2} \to \infty$

reaches the value $A_o/\sqrt{2}$ when $\omega = \omega_1$. In terms of decibels, this is equivalent to saying that the gain reaches 3 dB below the midband value at $\omega = \omega_1$. Hence, ω_1 is called the low-frequency cutoff point.

In a similar manner, we can calculate the effect of C_{C2} with C_{C1}, $C_E \to \infty$ as giving rise to an expression of the form Eq. 23.5 with ω_1 replaced by $\omega_1' = 1/\left[C_{C2}(R_C + R_L')\right]$. The effect of C_E with C_{C1}, $C_{C2} \to \infty$ is a bit more involved, and it can be shown that the gain is proportional to

$$\frac{1 - j\omega_z/\omega}{1 - j\omega_1''/\omega}, \tag{23.6}$$

where

$$\omega_z = \frac{1}{(C_E R_E)} \tag{23.7}$$

and

$$\omega_1'' = \frac{1}{R_E C_E}\left[1 + \left(g_m + \frac{1}{r_\pi}\right)R_E\right], \tag{23.8}$$

$$\approx \frac{g_m}{C_E} \tag{23.9}$$

The form of Eq. 23.9 arises because

$$g_m r_\pi = \beta (= h_{fe}) \gg 1$$

Usually, $\omega_1'' \gg \omega_z$.

The question now arises: how to determine the low-frequency 3 dB cutoff (ω_L) when none of the three capacitances can be considered as short-circuits. A guideline for the designer is that one of the capacitances should be used to control ω_L while the other two should be so chosen that the critical frequencies due to them are an order of magnitude less than the desired ω_L. For example, with $C_{C1} \to \infty$, $C_E = 200$ μF, $C_{C2} = 10$ μF, $R_E = 100\ \Omega$, $R_c = R'_L = 2$ K, $R_\pi \approx r_\pi = 1$ K and $\beta = g_m r_\pi = 100$, the gain is of the form

$$\frac{(\text{constant})(1 - j\omega/\omega_z)}{(1 - j\omega/\omega_1')(1 - j\omega/\omega_1'')}, \tag{23.10}$$

where ω_z, = 50 r/s, ω'_1 = 25 r/s and $\omega_1'' = 509.1$ r/s. The value of ω_L is then determined by ω_1'' and is given by $f_L = 509.1/(2\pi) = 81.02$ Hz.

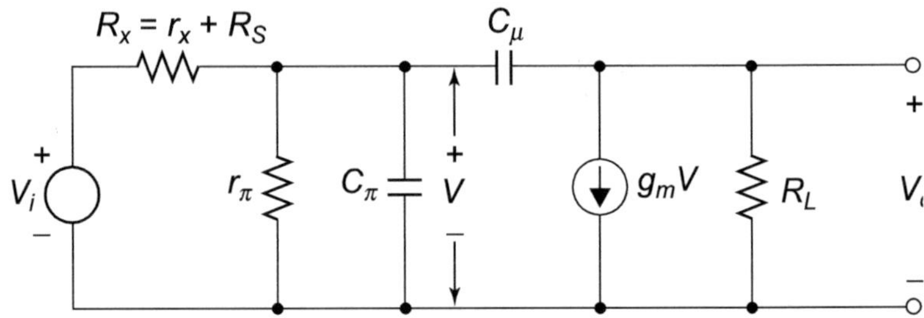

Fig. 23.9 High frequency equivalent circuit of the amplifier of Fig. 23.1

Finally, we consider the high-frequency response of the amplifier, for which the equivalent circuit is shown in Fig. 23.9. Notice that we have no longer ignored the effect of r_π or of the source internal resistance R_s. The reason is that the small capacitor C_μ reflects at the input (across C_π) as a much larger capacitance

$$C_M = C_\mu(1 + g_m R_L), \tag{23.11}$$

which is approximately the midband gain times C_μ. This is known as the 'Miller effect'.

The total capacitance

$$C_T = C_\pi + C_M \tag{23.12}$$

may have a reactance which is comparable to $R_x = r_x + R_s$ at high frequencies. With Miller effect taken into account, the equivalent circuit simplifies to that shown in Fig. 23.10, where we have ignored $R_1 \parallel R_2$ in comparison to r_π. Obviously, the gain is given by

$$\begin{aligned} A_H(s) &= \frac{v_o}{V}\frac{V}{v_i} = \frac{-g_m R_L(r_\pi \parallel 1/sC_T)}{R_x + (r_\pi \parallel 1/sC_T)} \\ &= \frac{-g_m R_L r_\pi}{sC_T r_\pi R_x + R_x + r_\pi} \\ &= \frac{-g_m R_L r_\pi}{R_x + r_\pi}\frac{1}{1 + sC_T(r_\pi \parallel R_x)} \end{aligned} \tag{23.13}$$

The effect of R_x, as expected, is to reduce the midband gain by the factor $r_\pi/(R_x + r_\pi)$. Also putting $s = j\omega$, and denoting $C_T(r_\pi \parallel R_x)$ by $1/\omega_2$, we see that the HF 3-dB cutoff is given by

$$\omega_2 = \frac{1}{C_T(r_\pi \parallel R_x)} \tag{23.14}$$

To get an idea of ω_2, let, for a typical transistor amplifier,

$$\begin{aligned} &R_S = 900\,\Omega, r_x = 100\,\Omega.\ r_\pi = 1\,\mathrm{K}, C_\mu = 4\,\mathrm{pF} \\ &C_\pi = 31\,\mathrm{pF}, g_m = 58\,\mathrm{m\text{-}mhos} \quad \text{and}\, R_L = 2\,\mathrm{K} \end{aligned} \tag{23.15}$$

Then

$$\begin{aligned} C_T &= 31 + 4 \times (1 + 58 \times 10^{-3} \times 2 \times 10^3) \\ &\approx 500\,\mathrm{pF} \end{aligned} \tag{23.16}$$

$$r_\pi \parallel R_x = 1\,\mathrm{K} \parallel 1\,\mathrm{K} = 500\,\Omega \tag{23.17}$$

and

$$\begin{aligned} f_2 &= \frac{\omega_2}{2\pi} = \frac{1}{2\pi \times 500 \times 10^{-12} \times 500}\,\mathrm{Hz} \\ &= 636\,\mathrm{kHz} \end{aligned} \tag{23.18}$$

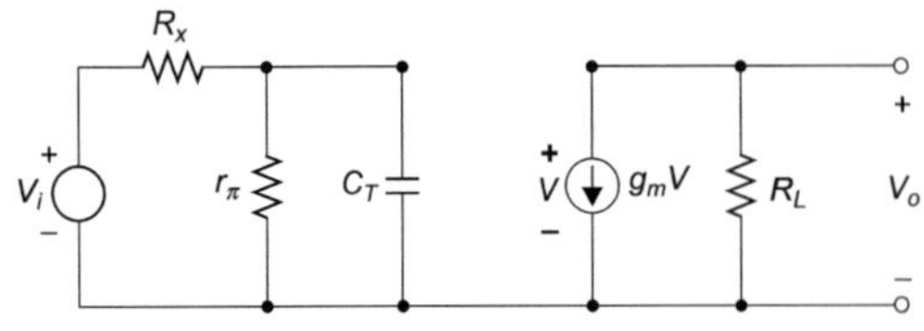

Fig. 23.10 Simplified equivalent of Fig. 23.9 using Miller effect

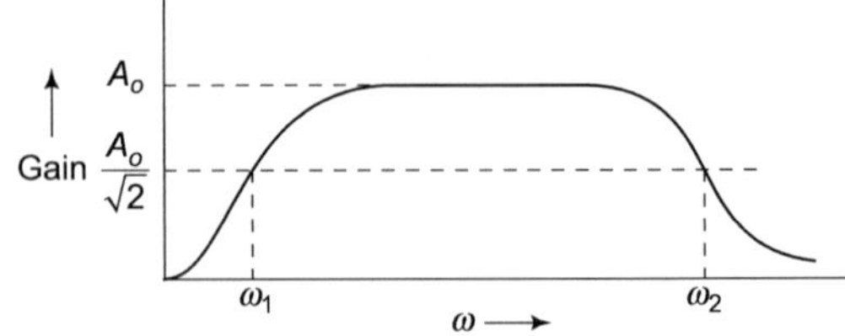

Fig. 23.11 Frequency response of the gain of circuit of Fig. 23.1

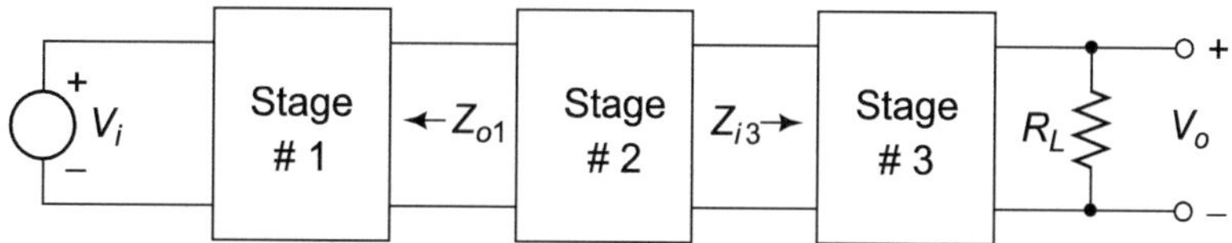

Fig. 23.12 A cascade of three stages

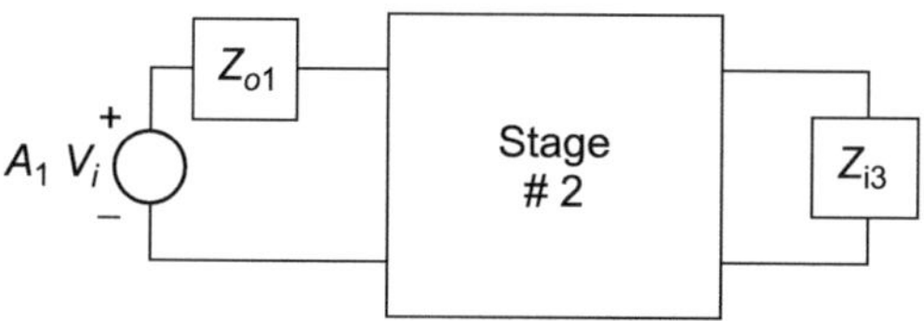

Fig. 23.13 Showing how stages interact with each other

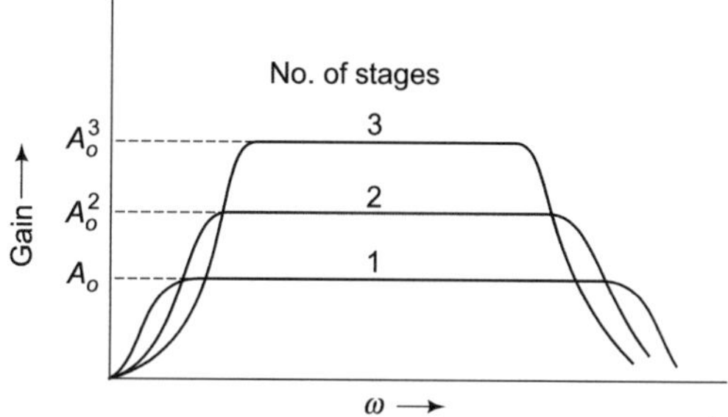

Fig. 23.14 Effect of cascading a number of stages

As we can see from the analysis, the gain of the typical amplifier of Fig. 23.1, called an RC coupled amplifier for obvious reasons, will have a band-pass characteristic as shown in Fig. 23.11.

The gain of a single-stage amplifier may not be adequate for the specific application. Hence, one uses multistage amplifiers by cascading several stages as shown in Fig. 23.12. Analysis of such multistage amplifiers proceeds stage by stage by considering the Thevenin equivalent of the previous stage as constituting the source, while the succeeding stage constitutes the load. This is illustrated in Fig. 23.13 for analysis of Stage#2 of the circuit in Fig. 23.12. In general, therefore, cascading does not lead to a multiplication of gains.

Suppose that we have succeeded in cascading a number of identical non-interacting stages, each having a midband gain A_o and low-and high-frequency 3-dB cutoff at ω_1 and ω_2 respectively. The term 'non-interacting' here means that each stage has an input impedance which is much higher than the output impedance of the previous stage, i.e. $Z_{i,\ n+1} \gg Z_{0,\ n}$. Under this condition, what happens to the overall ω_1 and ω_2? This is illustrated qualitatively in Fig. 23.14. Obviously, ω_1 increase and ω_2 decreases, i.e. the overall bandwidth decreases while the gain increases. It can be easily shown that when n such stages are cascaded, the overall parameters are given by

$$A_{on} = A_o^n \tag{23.19}$$

$$\omega_{1n} = \omega_1 / \sqrt{2^{1/n} - 1} \tag{23.20}$$

$$\omega_{2n} = \omega_2 \sqrt{2^{1/n} - 1} \tag{23.21}$$

In a given amplifier, how does one increase the ω_2 and decrease ω_1? A number of such compensation techniques are available and we shall discuss, qualitatively, one example of each.

A simple philosophy of increasing ω_2 is to use a load whose impedance increases with frequency, so that the fall in gain due to the factor $1 + j\omega/\omega_2$ in the denominator can be partially compensated. Such a load is shown in Fig. 23.15.

A similar philosophy can be applied to compensate for low-frequency fall off, by using a load which increases with decreasing frequency. One such load is shown in Fig. 23.16.

We conclude this discussion by pointing out that in order to test a given amplifier for its low- and high-frequency responses, it is convenient to use a pulse as the input. As shown in Fig. 23.17, the response to a pulse will be a gradually rising waveform, which after reaching a maximum, does not stay there, but sags a little before settling down to the zero value. The rise time,

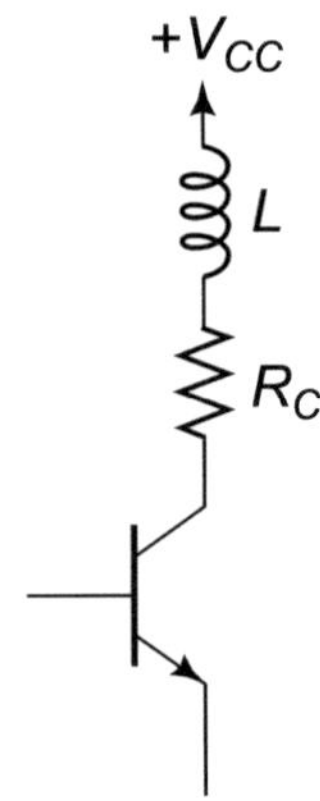

Fig. 23.15 HF compensation

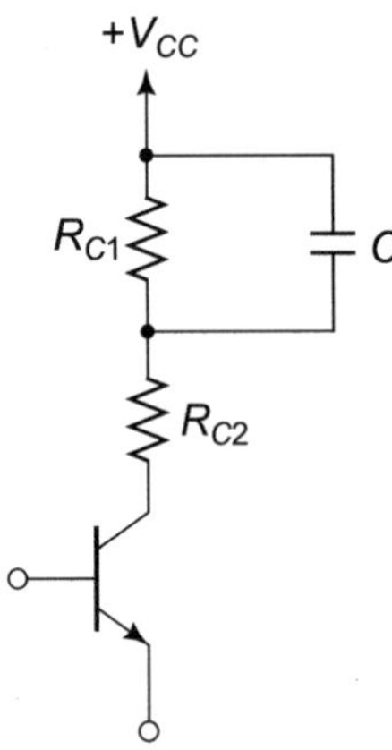

Fig. 23.16 L-F compensation

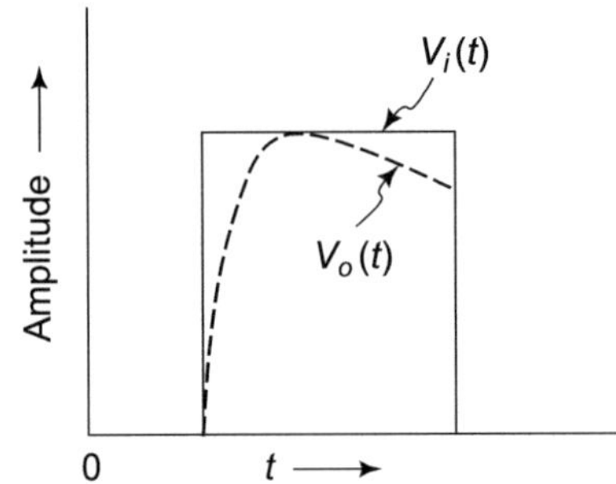

Fig. 23.17 Pulse response of an RC coupled amplifier

defined as the time required for the waveform to rise from 10 to 90% of the final value, can be related to ω_2, while the amount of sag can be related to ω_1.

Problems

P.1. Suppose in Fig. 23.1, R_1 is split into R_{1A} and R_{1B} and a resistor R_{1C} is connected between the joint of R_{1A} and R_{1B} and ground. Derive the necessary equations for the biasing condition of the transistor.

P.2. At frequencies at which $r \gg 1/(\omega C)$ and $r_0 \gg (1/\omega C_0)$, find an expression for the frequency response of a transistor, assuming a load R_L and a source of resistance R_S.

P.3. Consider a two-stage cascaded amplifier with source of resistance R_S and load R_L. Find an expression for the overall gain, if the stages interact with each other.

P.4. In Fig. 23.15, a capacitor C is connected from the joint of L and R_C and ground. Find an expression for the high-frequency gain.

P.5. In Fig. 23.16, C is neither open-nor short-circuit. Derive an expression for the low-frequency gain.

For further information on amplifiers, see [1]

Reference

1. A.S. Sedra, K.C. Smith, *Microelectronic Circuits* (Sanders College Publishing, Fortworth, 1992)

Appearances Can Be Deceptive: The Case of a BJT Biasing Circuit

24

It is shown that bias stability is the best with the four resistor circuit. A two-resistor BJT biasing circuit, which appears to be an attractive alternative to the familiar four resistor circuit, is shown to have serious limitations. It is also shown that even when augmented by one or two resistors, these limitations are only partially overcome and that the bias stability that can be achieved thereby is poorer than that of the four resistor circuit.

Keywords
Bias · Bias stability · 2, 3 and 4 resistor biasing

Introduction

Any student of Electronics should be familiar with the four resistors BJT biasing circuit shown in Fig. 24.1, to be called N_1, hereafter, and any standard textbook would give the derivation of the following expression for the collector current I_C (see, e.g. [1–3]):

$$I_C = \frac{[V_{CC}R_1/(R_1 + R_2)] - V_{BE} + I_{CBO}[1 + (1\beta)][(R_E + R_1 \parallel R_2)]}{R_E + \{[R_E + (R_1 \parallel R_2)]/\beta\}}, \tag{24.1}$$

where the symbols V_{BE}, I_{CBO} and β have their usual meanings. Stabilization of I_C against variations of these three parameters due to temperature change and/or replacement of transistor demands that

Source: S. C. Dutta Roy, "Appearances can be Deceptive: The Case of a BJT Biasing Circuit," *Students' Journal of the IETE*, vol. 37, pp. 3–6, January–June 1996.

S. C. Dutta Roy, *Circuits, Systems and Signal Processing*,
https://doi.org/10.1007/978-981-10-6919-2_24

$$R_1 \parallel R_2 \ll \beta R_E \tag{24.2}$$

and

$$V_{CC}R_1/(R_1+R_2) \gg V_{BE}+I_{CBO}[R_E+(R_1 \parallel R_2)], \tag{24.3}$$

where the usual assumption of $\beta \gg 1$ has been made. It would be recognized that Eq. 24.3 is a conservative condition because what we require is that the left-hand side should be much greater than both of the two terms on the right-hand side. It is also easy to see that the midband gain of the circuit is $-g_m R_c$ if R_E is bypassed for AC by a capacitor, as shown in Fig. 24.1.

Another circuit for BJT biasing, usually mentioned as an exercise for the student in textbooks, is that of Fig. 24.2, to be designated as N_2, for brevity. As compared to N_1, N_2 looks attractive because it uses only two, instead of four resistors. It is believed that bias stabilization occurs in N_2 due to feedback through R_F. We shall examine this circuit critically in this chapter and demonstrate that it has serious limitations, as compared to N_1. We also show that even with the addition of one or two resistors, these limitations can only be overcome partially, and that the bias stability achieved thereby compares poorly with that of N_1.

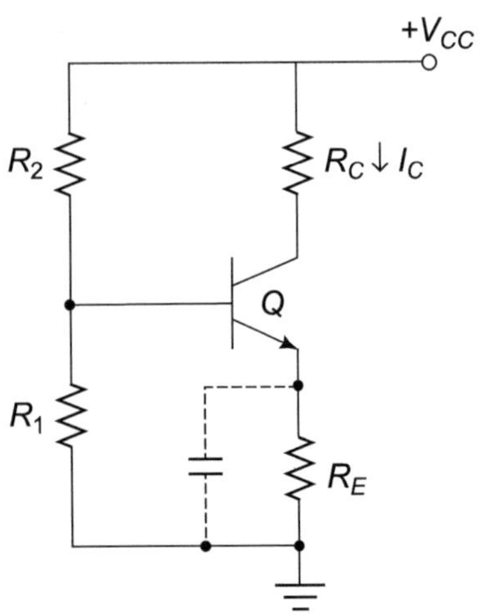

Fig. 24.1 N_1—the familiar four resistor BJT biasing circuit

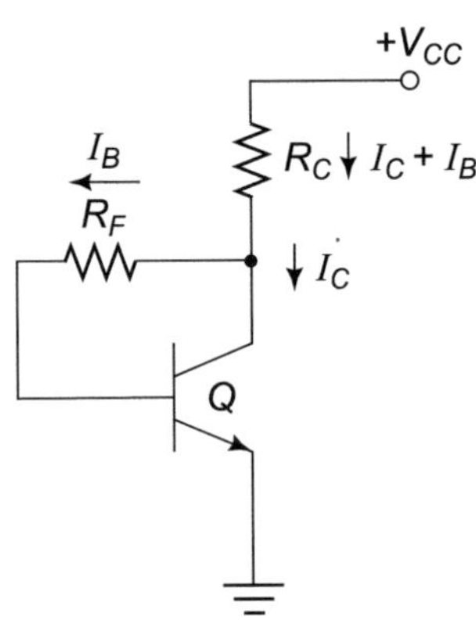

Fig. 24.2 N_2—an apparently attractive alternative to N_1

Analysis of N_2

Application of Kirchoff's voltage law to the loop formed by V_{CC}, R_C, R_F and the base–emitter junction gives

$$V_{CC} = (I_C+I_B)R_C + I_B R_F + V_{BE} \tag{24.4}$$

Also, the base and collector currents are related to the following equation

$$I_C = \beta I_B + (\beta+1)I_{CBO} \tag{24.5}$$

Combining Eqs. 24.4 and 24.5, we get

$$I_C = \frac{V_{CC}-V_{BE}+I_{CBO}[1+(1/\beta)](R_C+R_F)}{R_C+[(R_C+R_F)]/\beta} \tag{24.6}$$

Bias stability now demands that

$$R_F \ll \beta R_C \tag{24.7}$$

and

$$V_{CC} \gg V_{BE}+I_{CBO}(R_C+R_F) \tag{24.8}$$

under the usual assumption that $\beta \gg 1$ and the conservatism mentioned earlier. The midband gain of the circuit can be easily derived to be $-g_m$

$(R_F \parallel R_C)$ under the assumption of $g_m \gg (1/R_F)$, which is usually satisfied. The limitations of the circuit are best brought out through a numerical example, as worked out next.

An Example of Design

Let the transistor Q have the following parameters

$$V_{BE} = 0.6\,\text{V}, I_{CBO} = 10\,\text{nA} \quad \text{and} \quad \beta = 100 \tag{24.9}$$

at 25 °C, and let the Q-point be

$$V_{CE} = 4\,\text{V} \quad \text{and}\, I_C = 4\,\text{mA} \tag{24.10}$$

with V_{CC} = 12 V. Also, let the midband gain required be −160. The g_m of the transistor is

$$g_m = 4\ \text{mA/25 mV} = 0.16\ \mho \tag{24.11}$$

so that for a gain of 160, we need

$$R_C \parallel R_F = 1\text{K} \tag{24.12}$$

The specification on V_{CE} determines R_C as

$$R_C \cong 2\text{K} \tag{24.13}$$

From Eqs. 24.12 and 24.13, we get

$$R_F = 2\text{K} \tag{24.14}$$

Note that $1/R_F$ equals 0.001 $\mho$ which is indeed much smaller than $g_m = 0.16\ \mho$ thus validating the midband gain formula, but the problem arises elsewhere. With R_F = 2 K, the base current is

$$I_B = (V_{CE} - V_{BE})/R_F = 1.7\,\text{mA} \tag{24.15}$$

Since βI_B has a value of 170 mA, clearly, the transistor will be saturated!

What is the remedy? In [2], it is suggested that we split R_F into two parts and use a bypass capacitor, as shown in Fig. 24.3. This circuit, to be called N_3, has a gain $-g_m\ (R_C \parallel R_{F1})$ so that

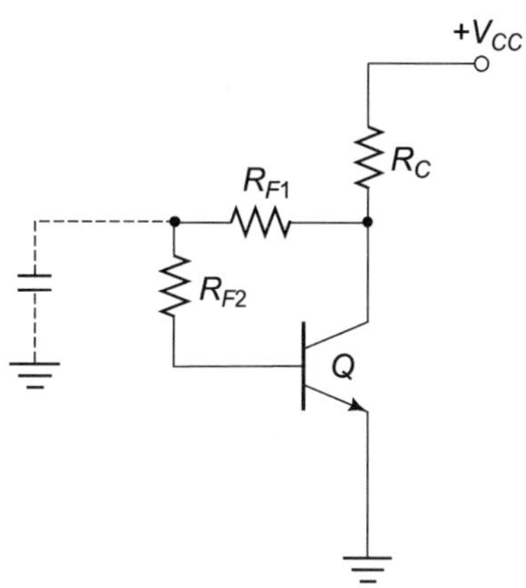

Fig. 24.3 N_3—modified form of N_2 to avoid saturation of Q

for a gain of −160, we need R_{F1} = 2 K. R_{F2} has to be chosen to satisfy the base current I_B = 0.04 mA. From the relation

$$I_B = (V_{CC} - V_{BE})/(R_{F2} + R_{F1}) \tag{24.16}$$

we calculate R_{F2} as 83 K. Thus, our R_F is 85 K, which, to our disappointment, does not satisfy Eq. 24.7, because $\beta R_C = 200\,\text{K}$. To satisfy Eq. 24.7, R_F should not exceed 20 K, taking the thumb rule of 1:10 for the sign '$\ll$' to be satisfied. Clearly, N_3 does not achieve bias stability!

What should we do now? Use one more resistor? Let us see.

This additional resistor can be either from the base to ground, as shown in Fig. 24.4, or from the emitter to ground, as shown in Fig. 24.5. These two circuits will be designated as N_4 and N_5, respectively. Consider N_5 first. It is not difficult to realize that since the same current passes through R_C and R_E, the expression for I_C will be the same as Eq. 24.6 except that R_C + R_E will take the place of R_C. Thus, for bias stability, we need $R_F \ll \beta\,(R_E + R_C)$ and since I_B is still given by Eq. 24.16, R_F does not change. Hence, this modification is of no use.

For N_4, given in Fig. 24.4, the currents in the various branches, as indicated, can be easily established. Kirchoff's voltage law can be used to write the following equation:

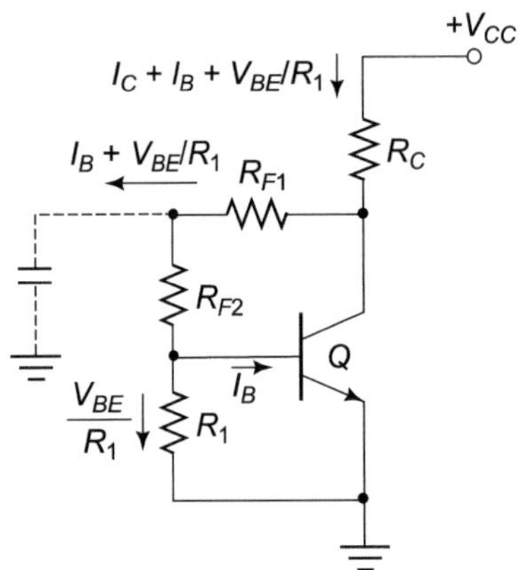

Fig. 24.4 N_4—a modification of N_3

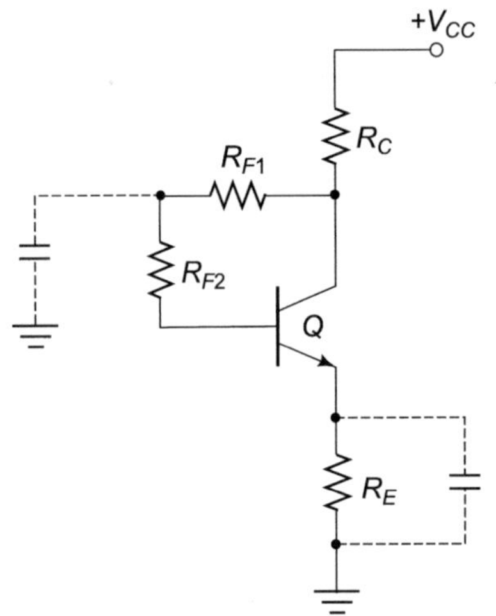

Fig. 24.5 N_5—another modification of N_3

$$V_{CC} = [I_C + I_B + (V_{BE}/R_1)]R_C + [I_B + (V_{BE}/R_1)]R_F + V_{BE} \tag{24.17}$$

Combining this with Eq. 24.5, and solving for I_C gives

With $\beta \gg 1$, bias stability requirements now become

$$R_F \ll \beta R_C \tag{24.19}$$

and

$$V_{CC}R_1/(R_1 + R_C + R_F) \gg V_{BE} + I_{CBO}[R_1 \parallel (R_C + R_F)] \tag{24.20}$$

Since R_C will have to be 2 K to satisfy the Q-point and R_{F1} will also have to be 2 K to satisfy the gain requirement, the question that arises is the following: is it possible to choose $R_{F2} < 83$ K?

Note that

$$[I_B + (V_{BE}/R_1)]R_F = V_{CE} - V_{BE} \tag{24.21}$$

Putting numerical values, this gives

$$(3.4/R_F) - (0.6/R_1) = 0.04 \times 10^{-3} \tag{24.22}$$

We want to have $R_F \leq 20$ K. If R_F is chosen, arbitrarily, as 17 K, Eq. 24.22 gives $R_1 = 3.75$ K. Now look at the other requirement, given by Eq. 24.20. Putting numerical values, the left-hand side is calculated as 1.978 V while the right-hand side is greater than 0.6 V. Hence, Eq. 24.20 is not satisfied. In fact, it can be shown that the highest value of the left-hand side of Eq. 24.20 under the constraint of Eq. 24.22 and $R_F \leq 20$ K occurs when $R_F = 20$ K, and that this value is only 2.075. Hence, we conclude that N_4, like N_5, is also not of much use in stabilizing the transistor Q-point.

$$I_C = \frac{[V_{CC}R_1/(R_1 + R_C + R_F)] - V_{BE} + I_{CBO}[1 + (1/\beta)][(R_1||R_C + R_F)]}{[R_1R_C/(R_1 + R_C + R_F)] + \{[R_1||(R_C + R_F)]/\beta\}} \tag{24.18}$$

Conclusion

The preceding example clearly demonstrates the limitations of N_2, and its modified versions—N_3, N_4 and N_5, in stabilizing the Q-point of a BJT. On the other hand, one can easily show that N_1 with $R_C = R_E = 1$ K and $R_1 = R_2 = 20$ K gives

$$R_1 \parallel R_2 = 10\text{ K}, \beta R_E = 100\text{ K},$$
$$V_{CC}R_1/(R_1 + R_2) = 6\text{ V}, \quad \text{and}$$
$$V_{BE} + I_{CBO}(R_E + R_1 \parallel R_2) \cong 0.6\text{V}$$

so that both Eqs. 24.2 and 24.3 are approximately satisfied. We conclude therefore that N_1 is the best choice for stabilizing the Q-point of a BJT.

Problems

You may have to couple these with the previous chapter

P.1. In Fig. 24.1 circuit, the capacitor is neither open nor short. Find an expression for the low-frequency gain, using, of course, the hybrid $-\pi$ equivalent circuit.

P.2. In Fig. 24.1 circuit, a resistor R_L is connected between the collector and ground. Comment on the bias stability of the circuit. Justify.

P.3. In Fig. 24.3 circuit, the capacitor is there and is neither a short circuit nor an open circuit. Deriver an expression for the low-frequency gain.

P.4. In Fig. 24.4 circuit, the capacitor is there and is neither a short circuit nor an open circuit. Derive an expression for the high-frequency gain.

P.5. What happens when the capacitor is shifted to have a position between the collector and ground? Carry out the necessary analysis for the low-frequency gain.

References

1. J. Millman, A. Grabel, *Microelectronics* (McGraw-Hill, New York, 1987)
2. S.G. Burns, P.R. Bond, *Principles of Electronic Circuits* (West Publishing Company, St. Paul, 1987)
3. A.S. Sedra, K.C. Smith, *Microelectronic Circuits* (Sanders College Publishing, Fortworth, 1992)

25 BJT Biasing Revisited

The familiar four resistor circuit for biasing a bipolar junction transistor (BJT) is generalized through simple reasoning, and transformed to yield a different topology. Three alternative four resistor circuits are derived as special cases of the transformed generalized circuit, which do not appear to have been widely known in the literature. A detailed and careful analysis reveals that the bias stability parameters of all alternative circuits are comparable to those of the conventional circuit. An illustrative example is included for demonstrating this fact.

Keywords
BJT · Biasing · Bias stability · Design

Introduction

Figure 25.1a shows the familiar BJT biasing circuit for linear class A amplification, and is to be called N_1 in the sequel. It uses four resistors, of which R_{E1} is usually by-passed for AC [1–3]. In this chapter, N_1 is generalized by a proper choice of a few additional resistors and then transformed to a different topology. The latter is shown to yield, as special cases, three alternative four resistor circuits, to be called N_2, N_3 and N_4, which do not appear to have been widely known in the literature in the context of biasing a BJT. From a detailed and careful analysis, it is shown that the bias stability parameters achieved in all the four circuits—N_1, N_2, N_3 and N_4—are comparable. An illustrative example of bias design is worked out to demonstrate this fact.

The Generalized Circuits and Special Cases

Let a resistor connected between nodes X and Y be denoted by $R(X, Y)$. In Fig. 25.1a, there are five nodes—A, B, C, E and G—and a most general biasing circuit would be the one in which every node is connected to every other node by a resistor. Several exceptions are to be made, however. The resistors $R(A, G)$, $R(A, E)$, $R(C, E)$ and $R(C, G)$ are not necessary for biasing and cause additional loss of power. If these four resistors are excluded, then the generalized biasing circuit looks like that shown in Fig. 25.1b.

If the two π-networks BCA and BEG are converted into T's, then the transformed circuit becomes that shown in Fig. 25.1c. Note that

Source: S. C. Dutta Roy, "BJT Biasing Revisited," *IETE Journal of Education*, vol. 46, pp. 27–33, January–March 2005.

S. C. Dutta Roy, *Circuits, Systems and Signal Processing*,
https://doi.org/10.1007/978-981-10-6919-2_25

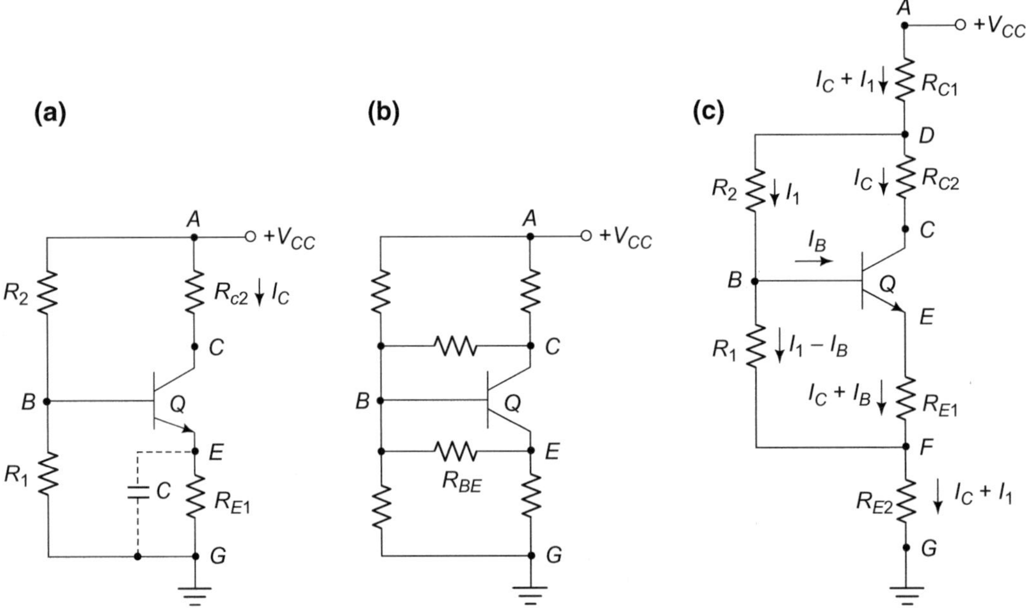

Fig. 25.1 **a** N_1—the familiar four resistor BJT biasing circuit; **b** a generalized BJT biasing circuit; **c** alternative generalized BJT biasing circuit obtained by transformation of the circuit of (**b**)

since π to T conversion does not involve any subtraction operation, all resistances in Fig. 25.1c are positive. Hence, both circuits can claim to be general canonic biasing circuits for the BJT. Four special cases of four resistor circuits can be derived from the two generalized circuits as follows.

(1) Let $R(B, C) = R(B, E) = \infty$ in Fig. 25.1b or $R(A, D) = R(F, G) = 0$ in Fig. 25.1c; then we get the conventional circuit N_1 of Fig. 25.1a.
(2) Let $R(E, F) = R(F, G) = 0$ in Fig. 25.1c; then we get the circuit of Fig. 25.2, henceforth referred to as N_2.
(3) Let $R(B, C) = R(B, G) = \infty$ in Fig. 25.1b or $R(A, D) = R(E, F) = 0$ in Fig. 25.1c; then we obtain the circuit of Fig. 25.3, which we shall refer to as N_3.
(4) Let $R(A, B) = R(B, E) = \infty$ in Fig. 25.1b or $R(D, C) = R(F, G) = 0$ in Fig. 25.1c; this gives the circuit of Fig. 25.4, henceforth to be referred to as N_4.

Note that (*i*) for the convenience of reference, we have designated the circuits such that N_i refers to Fig *i*, and that (*ii*) N_2 cannot be derived from Fig. 25.1b. Because of the latter, the circuit of Fig. 25.1c, therefore seems to have an edge over that of Fig. 25.1b in terms of topological generality. Also, note that another four resistor circuit can be obtained by setting $R(A, B) = R(B, G) = \infty$ or $R(D, C) = R(E, F) = 0$ in Fig. 25.1c. However, in the resulting circuit, R_{C1} and R_{E2} carry the same DC; hence for biasing purposes, they can be combined into one resistance and the circuit thereby behaves as a three resistor one. It has been found that such a circuit has a poorer bias stability than N_1, and will not, therefore, be considered further.

In N_2, it is of advantage to by-pass R_{C1} for AC, as shown, by connecting a large capacitor

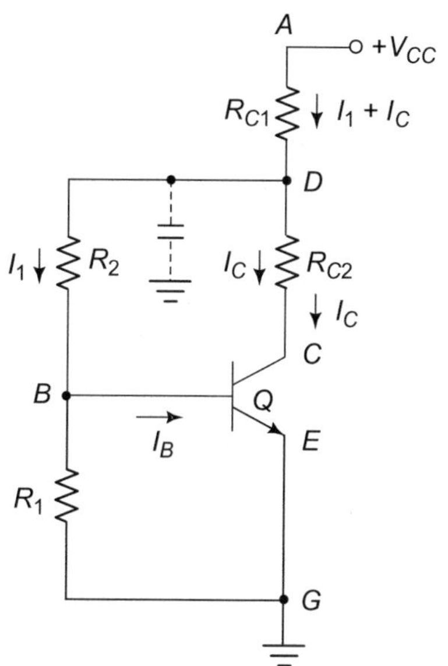

Fig. 25.2 N_2—an alternative BJT biasing circuit

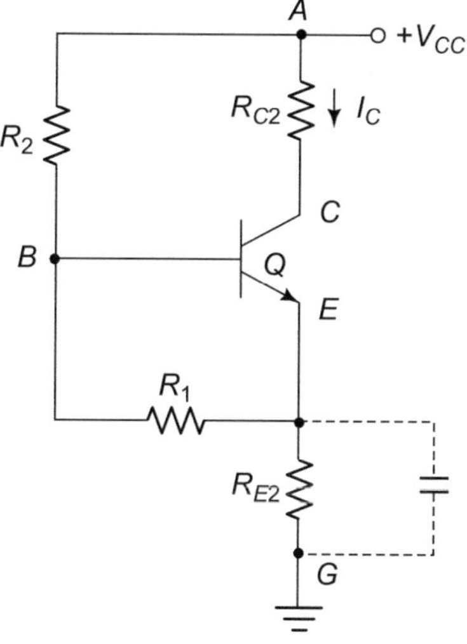

Fig. 25.3 N_3—another alternative BJT biasing circuit

between nodes D and G, so that the gain is determined by R_{C2} only. It may be noted that this form of the circuit finds use in multistage amplifiers for power supply decoupling, where R_{C1} is common to all the stages. For single-stage biasing, however, N_2 does not appear to have been used. As compared to N_1, N_2 trades an additional resistor at the collector side for that at the emitter side. In N_3 as well as N_4, the emitter resistance is by-passed for ac, as in N_1.

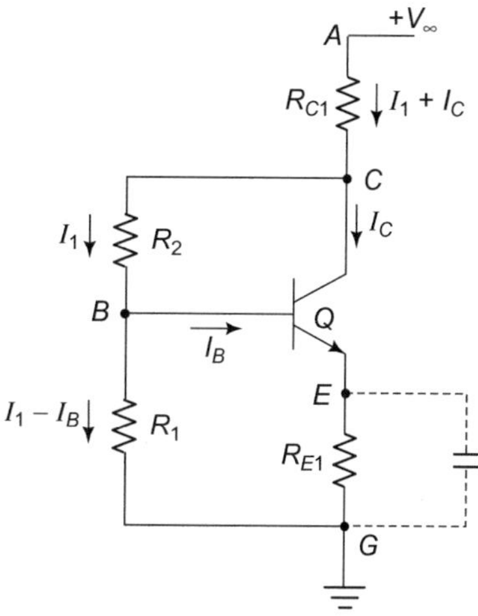

Fig. 25.4 N_4—yet another alternative BJT biasing circuit

Bias Stability Analysis

The generalized circuit of Fig. 25.1c will only be analyzed because it gives all the four special cases. For this purpose, all the resistors have been named and the currents in all of them have been identified. Let the transistor Q be characterized by the parameters V_{BE}, β and I_{CBO}, where the symbols have their usual meanings. The currents I_C and I_B are related by the following equation

$$I_C = \beta I_B + (\beta + 1) I_{CBO}. \tag{25.1}$$

By applying Kirchoff's voltage law, one obtains

$$V_{CC} - (I_C + I_1)(R_{C1} + R_{E2}) = I_1 R_2 + (I_1 - I_B) R_1, \tag{25.2}$$

and

$$(I_1 - I_B) R_1 = V_{BE} + (I_C + I_B) R_{E1} \tag{25.3}$$

From Eq. 25.3, I_1 can be obtained in terms of I_B and I_C. Substituting this value in Eq. 25.2, replacing I_B in terms of I_C and I_{CBO} from Eq. 25.1, and carrying out some algebraic manipulations, one obtains the following expression for I_C:

$$I_C = \frac{V_1 - V_{BE} + I_{CBO} r_1}{r_2 + (r_1/\beta)}, \tag{25.4}$$

where

$$V_1 = V_{CC} R_1/(R_1 + R_2 + R_{C1} + R_{E2}), \tag{25.5a}$$

$$r_1 = R_{E1} + R_1 \parallel (R_2 + R_{C1} + R_{E2}), \tag{25.5b}$$

$$r_2 = R_{E1} + [R_1(R_{C1} + R_{E2})/(R_1 + R_2 + R_{C1} + R_{E2})], \tag{25.5c}$$

and we have made the simplifying, but practical assumption that $\beta \gg 1$ so that the factor $[1 + (1/\beta)]$ can be approximated by unity.

Bias stability is achieved if I_C can be made insensitive to variations in V_{BE}, β and I_{CBO}. Note that I_C is independent of R_{C2}, but of course, R_{C2} has an important effect on V_{CE}. Also, note that R_{E2} always occurs in combination with R_{C1} in the form $R_{C1} + R_{E2}$. This is to be expected because, as is clear from Fig. 25.1c, the same current $I_C + I_1$ flows in them. As mentioned earlier, both resistors are not necessary; one can make either $R_{C1} = 0$ or $R_{E2} = 0$ without any loss of generality. This fact is also reflected in the circuits N_1–N_4.

Referring to Eq. 25.4, we observe that bias stability demands the following conditions to be met:

$$r_2/r_1 \gg 1/\beta, \tag{25.6}$$

$$V_1 \gg V_{BE}, \tag{25.7}$$

and

$$V_1 \gg I_{CBO} r_1 \tag{25.8}$$

As will be evident from the practical designs worked out later in the chapter, usually $V_{BE} \gg I_{CBO}\, r_1$ so that satisfying Eq. 25.7 automatically satisfies Eq. 25.8.

To obtain a quantitative measure of bias stability, consider the case in which the parameter set (V_{BE}, β, I_{CBO}) changes from (V_{BE1}, β_1, I_{CBO1}) to (V_{BE2}, β_2, I_{CBO2}) due to temperature variation and/or replacement of transistor, where the changes are not infinitesimal. One can find, from Eq. 25.4, the net change $\Delta I_C = I_{C2} - I_{C1}$ and divide by I_{C1} to determine the fractional (or percentage) variation of I_C. In most textbooks, however, this procedure is not followed because the resulting expression is considered to be 'very formidable and not too informative' [2, p. 412]. Instead, they consider the change of I_C due to each parameter separately, holding the other two constant. We shall also follow the same procedure to start with, and then show that considering all the changes simultaneously is not as difficult as it is made out to be.

To follow the conventional procedure, let δ, δ_v and δ_I denote the partial fractional changes in I_C due to changes in β, V_{BE} and $I_{CBO,}$ respectively. When all the three parameters vary simultaneously, and each δ is small (<0.1), the total fractional change in I_C, to be denoted by δ_T, is estimated as the sum of δ, δ_v, and δ_I.

From Eq. 25.4, one can easily derive expressions for the fractional deviations. The results are:

$$\delta_\beta = \frac{\Delta\beta/\beta_1}{1 + \beta_2(r_2/r_1)}, \tag{25.9}$$

$$\delta_v = \frac{-\Delta V_{BE}}{V_1 - V_{BE1} + I_{CBO} r_1}, \tag{25.10}$$

and

$$\delta_1 = \frac{\Delta I_{CBO}}{I_{CBO1} + [(V_1 - V_{BE1})/r_1]} \tag{25.11}$$

It is clear that the resistance r_1, given by Eq. 25.5b determines δ_v and δ_I, while δ is determined by the ratio r_2/r_1, which can be obtained from Eqs. 25.5b and 25.5c as

$$\frac{r_2}{r_1} = \frac{R_{E1}(R_1 + R_2 + R_{C1} + R_{E2}) + R_1(R_{C1} + R_{E2})}{R_{E1}(R_1 + R_2 + R_{C1} + R_{E2}) + R_1(R_{C1} + R_{E2}) + R_1 R_2} \tag{25.12}$$

The values of V_1, r_1 and r_2/r_1 for the four circuits are given in Table 25.1. In practical circuits, R_{C1}, R_{E1} and R_{E2} will be of the same orders of magnitude ($\cong$1 K) while R_1 and R_2 will be one order higher. It can, therefore, be observed that r_1 and r_2/r_1 are comparable for all the four circuits, which makes them comparable in terms of bias stability performance. In particular, if R_{C1} of N_2 is the same as R_{E2} of N_3, then N_2 and N_3 will have identical behaviour.

When all the three parameters vary simultaneously, as is usually the case in practice, one can easily show, using Eq. 25.4, that

$$\delta_0 = \frac{\Delta I_C}{I_{C1}} = \left(1 + \frac{-\Delta V_{BE} + I_{CBO} r_1}{V_1 - V_{BE} + I_{CBO} r_1}\right) \left(1 + \frac{\Delta\beta}{\beta_1}\right)\left(1 + \frac{\Delta\beta}{\beta_1 + (r_1/r_2)}\right)^{-1} - 1 \quad (25.13)$$

This expression does indeed look formidable, but is not difficult for computation once the numerical values are available. Also, to compare the competing circuits, all that changes are the values of r_1 and r_2/r_1.

In the next section, an illustrative example of design is worked out for absolute as well as comparative performances of the four circuits.

An Example

For a fair comparison of the four circuits, one should design each circuit for the same Q point and the same gain. First, consider N_1, N_2 and N_3, in all of which, the gain is approximately $-g_m R_{C2}$. Since the Q points are the same, one should have identical R_{C2} in each circuit. This ensures that the output resistance is also equal, under the usual assumption of r_0 of the BJT being much greater than R_{C2}. The input resistance in each circuit is approximately r in the usual situation of base biasing resistances having a negligible shunting effect. Let the various circuit and transistor parameters be as follows:

$$V_{CE} = 4\,\text{V}, I_C = 4\,\text{mA}, V_{CC} = 12\,\text{V}, V_{BE} = 0.6\,\text{V}, I_{CBO} = 10\,\text{nA} \text{ and } \beta = 100, \quad (25.14)$$

The last two quantities being measured at 25 °C. Also, let the gain required be −160, so that the required $R_{C2} = 160/g_m = 160/(40\ I_C) = 1$ K.

Design of N_1

Since $\beta = 100 \gg 1$, we require $R_{C2} + R_{E1} \cong (V_{CC} - V_{CE})/I_C = 2$ K. Thus $R_{E1} = 1$ K. From Eq. 25.6 and Table 25.1, it is required to have $R_{E1}/(R_{E1} + R_1 \parallel R_2) \gg 1/\beta$; with numerical values substituted, this translates to $R_1 \parallel R_2 \ll 99$ K. Let R_1 and R_2 be arbitrarily chosen as 20 K each. Then V_1 becomes 6 V so that Eq. 25.7 is satisfied. Also, r_1 is calculated as 11 K so that $I_{CBO}\ r_1 = 11 \times 10^{-5}$; thus Eq. 25.8 is also satisfied. The design is summarized in column 2 of Table 25.2.

Design of N_2

The gain requirement fixes R_{C2} as 1 K. From Eqs. 25.6–25.8 and Table 25.1, the requirements of bias stability become

$$R_2 \ll 99R_{C1} \text{ and } V_{CC}R_1/(R_1 + R_{C1} + R_2) \gg 0.6, 10^{-8}R_1 \parallel (R_{C1} + R_2). \quad (25.15)$$

Also, for this circuit,

Table 25.1 Values of r_1 and r_2/r_1 for N_1–N_4

Circuit	r_1	r_2/r_1
N_1	$R_{E1} + R_1 \parallel R_2$	$R_{E1}/(R_{E1} + R_1 \parallel R_2)$
N_2	$R_1 \parallel (R_{C1} + R_2)$	$R_{C1}/(R_{C1} + R_2)$
N_3	$R_1 \parallel (R_{E2} + R_2)$	$R_{E2}/(R_{E2} + R_2)$
N_4	$R_{E1} + R_1 \parallel (R_2 + R_{C1})$	$\frac{R_{E1}(R_1 + R_2 + R_{C1}) + R_1 R_{C1}}{R_{E1}(R_1 + R_2 + R_{C1}) + R_1 R_{C1} + R_1 R_2}$

Table 25.2 Bias stability of example designs

Circuit	N_1	N_2	N_3	N_4
Design parameters	$R_{C2} = R_{E1} = 1$ K $R_2 = R_1 = 20$ K	$R_{C2} = R_{C1} = 1$ K $R_1 = 10$ K $R_2 = 9$ K	$R_{C2} = R_{C1} = 1$ K $R_1 = 10$ K $R_2 = 9$ K	$R_{E1} = 0.896$ K $R_1 = 32.5$ K $R_2 = 20$ K $R_{C1} = 1.05$ K
V_1(volts)	6	6	Same	7.28
r_1(K)	11	5	Values	12.76
r_2(K)	1	0.5	As	1.43
δ_β	0.0342	0.0313	Those	0.0281
δ_v	0.0463	0.0463	Listed	0.0374
δ_I	0.0208	0.0095	For	0.0195
δ_T	0.1013	0.0871	Circuit	0.0850
δ_0	0.1034	0.0883	N_2	0.0863

$$V_{CE} = V_{CC} - (I_1 + I_C)R_{C1} - I_C R_{C2}, \quad (25.16)$$

and

$$I_1 = I_B + (V_{BE}/R_1) \cong (I_C/\beta) + (V_{BE}/R_1). \quad (25.17)$$

Combining Eqs. 25.16 and 25.17 and substituting numerical values (note $R_{C2} = 1$ K) gives the condition

$$(4/R_{C1}) - (0.6/R_1) = 4.04 \times 10^{-3} \quad (25.18)$$

Let $R_1 = 10$ K; then R_{C1} is obtained from Eq. 25.18 as 0.976 K $\cong$ 1 K. Now from Eq. 25.15, we should have $R_2 \ll 99$ K. With $R_2 = 9$ K, V_1, which is the left-hand side of Eq. 25.15, becomes 6 V, and $I_{CBO}\ r_1$, which is the second expression on the right-hand side of Eq. 25.15, becomes 5×10^{-8}; thus Eq. 25.15 is satisfied. The design is complete and is given in column 3 of Table 25.2.

Design of N_3

As mentioned earlier, the design of N_2 will also work for N_3 if R_{E2} is taken as 1 K.

Performances of N_1, N_2 and N_3

Let β change from $\beta_1 = 100$ to $\beta_2 = 150$ and let the temperature change from 25 to 125 °C. As is well known, V_{BE} decreases with temperature at the rate of 2.5 mV/°C, and I_{CBO} doubles for every 10 °C rise in temperature. Thus, V_{BE} changes by $\Delta V_{BE} = -250$ mV while the corresponding change in I_{CBO} is $\Delta I_{CBO} = (2^{10} - 1)\ I_{CBO} = 10.23$ μA.

For each of the three designs, the values of δ, δ_v and δ_I can now be calculated from Eqs. 25.9–25.11 and Table 25.1. These values are given Table 25.2 along with other necessary information. Note that no partial fractional deviation is more than 0.1 so that in each circuit, the total fractional deviation can be obtained by summing the three partial ones. The circuits N_2 and N_3 cause a smaller change in I_C than N_1, although no special care was taken to show N_2 and N_3 in a brighter light than N_1. However, no generalization can be made from this specific design; it is possible that with a redesign of N_1, balance can be tilted in its favour. All that can be said, and it has been said earlier, is that bias stabilities that can be achieved by the three circuits are comparable.

Design and Performance of N_4

N_4 is different from N_1, N_2 and N_3 in that it had DC as well as AC feedback through R_2. The ac equivalent circuit is shown in Fig. 25.5, from which the gain can be calculated as

$$\frac{V_o}{V_i} = \frac{(1/R_2) - g_m}{(1/R_{C1}) + (1/R_2)} \tag{25.19}$$

By Miller's theorem, the input impedance would be $R_1 \| r_\pi \| R_2/(1 + |\text{gain}|)$. While the shunting effect of R_1 can be ignored, that of $R_2/(1 + |\text{gain}|)$ cannot; in fact, the latter will, in practice, be one order smaller than r ! The output impedance is approximately $R_{C1} \parallel R_2/(1 + |\text{gain}|^{-1}) \cong R_{C1}$ if $R_2 \gg R_{C1}$ as is usually the case.

With a gain requirement of -160, and R_2 arbitrarily chosen as 20 K, Eq. 25.19 gives R_{C1} = 1.05 K. Referring to Fig. 25.4, we see that $I_1 = (V_{CE} - V_{BE})/R_2$ = 0.17 mA. Hence, voltage drop across R_{E1} is $V_{CC} - V_{CE} - (I_C + I_1)R_{C1}$ = 3.62 V; since the current through R_{E1} is 4.04 mA, we get R_{E1} = 0.896 K. Finally, $R_1 = (V_{BE}$ + voltage drop across $R_{E1})/(I_1 - I_B)$ = 32.5 K.

With the above design, r_1 and r_2/r_1 are calculated from Table 25.1 as 12.76 K and 0.112; thus Eq. 25.6 is satisfied. Also, V_1 here becomes 7.28 V so that Eq. 25.7 is also satisfied. Finally, as in the other designs, the value of $I_{CBO}\ r_1$ is negligible compared to 0.6 so that Eq. 25.8 is satisfied.

The complete design along with values of δ, δ_v, δ_I and δ_T are given in Table 25.2.

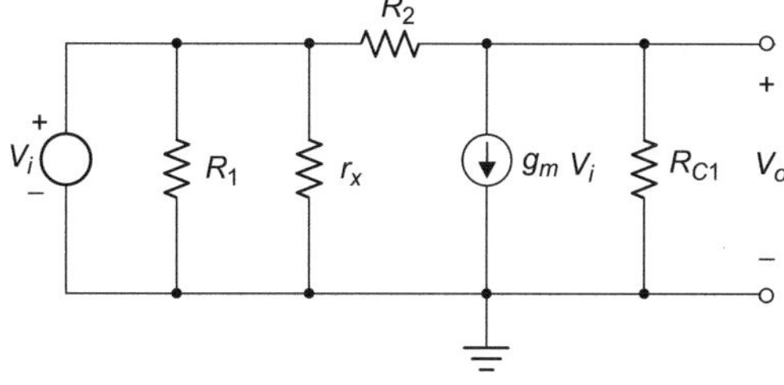

Fig. 25.5 AC equivalent circuit of N_4

Using the Total Change Formula

For the example under consideration, the total change formula Eq. 25.13 was used for each design, and the values of δ_0 are found to be 0.1034, 0.0883, 0.0883 and 0.0863 for N_1, N_2, N_3 and N_4, respectively. Clearly, $\delta_0 > \delta_T$. Thereby showing that the estimate of total fractional change on the basis of partial changes is an optimistic one.

Conclusion

The BJT biasing circuit has been generalized and transformed to yield three alternative four resistor circuits, whose bias stability performance is comparable to that of the commonly used four resistor circuit. It has been shown that the commonly used performance measure δ_T obtained by summing the partial fractional changes is an optimistic one. It has also been shown that the calculation of the total fractional change δ_0 poses no problem even though the formula looks formidable. Another general guideline in designing a bias circuit that has been revealed in our designs is that once $V_1 \gg V_{BE}$ has been established, $V_1 \gg I_{CBO}\ r_1$ need not be checked because V_{BE} is usually a few orders greater than $I_{CBO}\ r_1$.

The circuits N_1, N_2 and N_3 have similar performance in terms of gain, input and output impedances, while for the same gain and output impedance, N_4 has one order lower input impedance.

Problems

P.1. Replace the dotted capacitor C by a firm connection. Choose C such that its impedance is comparable to R_{E1}. What happens to the biasing? Analyze.

P.2. What if R_{BE} is absent in Fig. 25.1b?

P.3. What if $R_{BE} = 0$ in Fig. 25.1b?

P.4. Let $R_{EZ} = 0$ in Fig. 25.1c. What is the effect on biasing?

P.5. What if C in Fig. 25.2 is not too large to become a short circuit at AC?

References

1. J. Millman, A. Grabel, *Microelectronics* (McGraw-Hill, New York, 1987)
2. S.G. Burns, P.R. Bond, *Principles of Electronic Circuits* (West Publishing House, St Paul, 1987)
3. A.S. Sedra, K.C. Smith, *Microelectronic Circuits* (Oxford University Press, New York, 1998)

Analysis of a High-Frequency Transistor Stage

26

It is shown that, contrary to popular belief, classical two-port network theory is adequate for an exact analysis of a general high-frequency transistor stage, including emitter feedback, almost by inspection.

Keywords
Two-port analysis · High-frequency stage

Introduction

Consider the high-frequency amplifier circuit shown in Fig. 26.1, which includes an un-bypassed emitter resistance R_E. The capacitor C_μ is traditionally singled out as the troublesome element, but for which the analysis would have been much simpler. In most textbooks on electronics, therefore, the circuit is unilateralized through application of Miller's theorem, and simplified by assuming a resistive load, and ignoring the reflected Miller admittance on the load side [1]. These assumptions, as one readily appreciates, are not always valid; further, as pointed out by Yeung [2], the Miller equivalent circuit is not valid for output impedance calculations. There exist several other methods for carrying out the analysis: the classical node or mesh analysis, analysis using feedback concepts, driving point impedance technique [3], and the recently proposed open and short circuit technique [2]. Of these, the last one appears attractive, and is based on the calculation of two simpler gain functions and a driving point impedance.

The purpose of this chapter is to show that classical two-port network theory is adequate for analyzing the circuit exactly, almost by inspection. The method has been tested in the undergraduate classes and has been well received.

Two Port Analysis

Let as indicated in Fig. 26.1,

$$R_x = R_s + r_z \quad \text{and } Z_\pi = 1/(g_\pi + sC_\pi) = 1/Y_\pi \tag{26.1}$$

where $g_\pi = 1/r_\pi$. We shall carry out the analysis in several steps. First, consider the two-port shown in Fig. 26.2a. By inspection, its y-matrix is

$$[y]_a = \begin{bmatrix} y_\pi & 0 \\ g_m & 0 \end{bmatrix} \tag{26.2}$$

Source: S. C. Dutta Roy, "Analysis of a High Frequency Transistor Stage," *Students' Journal of the IETE*, vol. 29, pp. 5–7, January 1988

S. C. Dutta Roy, *Circuits, Systems and Signal Processing*,
https://doi.org/10.1007/978-981-10-6919-2_26

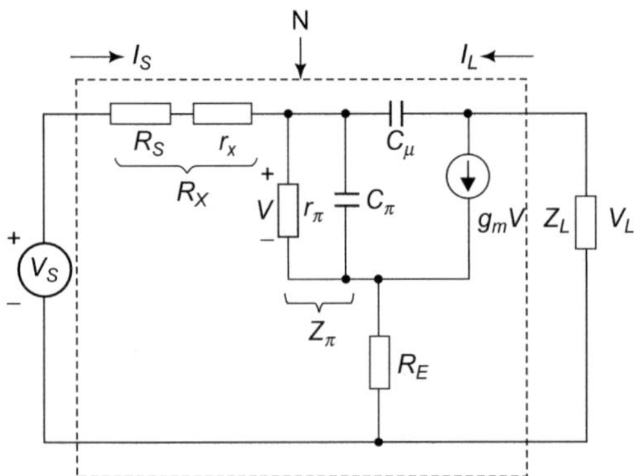

Fig. 26.1 Equivalent circuit incorporated along with the actual components

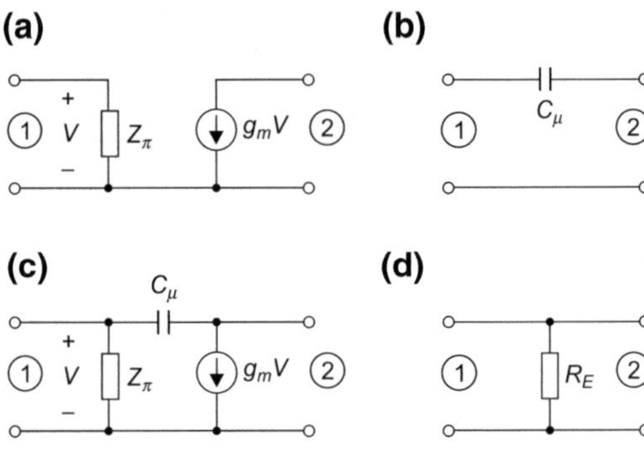

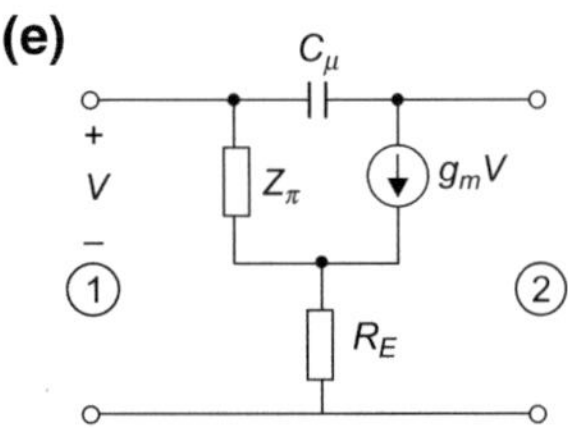

Fig. 26.2 Steps in derivation

Next, consider the two-port of Fig. 26.2b for which, again by inspection,

$$[y]_b = \begin{bmatrix} sC_\mu & -sC_\mu \\ -sC_\mu & sC_\mu \end{bmatrix} \tag{26.3}$$

Now, connect the two two-ports of Fig. 26.2a, b in parallel, as in Fig. 26.2c; the y-matrix of this two-port is the sum of Eqs. 26.2 and 26.3, i.e.,

$$[y]_c = \begin{bmatrix} y_\pi + sC_\mu & -sC_\mu \\ g_m - sC_\mu & sC_\mu \end{bmatrix} \tag{26.4}$$

The determinant of Eq. 26.4 is

$$|y|_c = sC_\mu(y_\pi + g_m) \tag{26.5}$$

Hence the z-matrix of the two-port of Fig. 26.2c is

$$[z]_c = \frac{1}{|y|_c}\begin{bmatrix} y_{22c} & -y_{12c} \\ -y_{21c} & y_{11c} \end{bmatrix} = \begin{bmatrix} \frac{1}{y_\pi + g_m} & \frac{1}{y_\pi + g_m} \\ \frac{-g_m + sC_\mu}{sC_\mu(y_\pi + g_m)} & \frac{y_\pi + sC_\mu}{sC_\mu(y_\pi + g_m)} \end{bmatrix} \tag{26.6}$$

Next consider the two-port of Fig. 26.2d. Its z-matrix is given by

$$[z]_d = \begin{bmatrix} R_E & R_E \\ R_E & R_E \end{bmatrix} \tag{26.7}$$

If we connect the two-ports of Fig. 26.2c, d in series, the two-port of Fig. 26.2e results, whose z-matrix is obtained by adding 26.6 and 26.7, i.e.,

$$[z]_e = \begin{bmatrix} R_E + \frac{1}{y_\pi + g_m} & R_E + \frac{1}{y_\pi + g_m} \\ R_E + \frac{sC_\mu - g_m}{sC_\mu(y_\pi + g_m)} & R_E + \frac{y_\pi + sC_\mu}{sC_\mu(y_\pi + g_m)} \end{bmatrix} \tag{26.8}$$

Adding R_x in series at the input port in Fig. 26.2e gives the two-port N, indicated by dashed outline in Fig. 26.1. The z-matrix of N is therefore the same as given in Eq. 26.8 except for an increase of z_{11} by R_x. Hence

$$[z]_N = \begin{bmatrix} R_x + R_E + \frac{1}{y_\pi + g_m} & R_E + \frac{1}{y_\pi + g_m} \\ R_E + \frac{sC_\mu - g_m}{sC_\mu(y_\pi + g_m)} & R_E + \frac{y_\pi + sC_\mu}{sC_\mu(y_\pi + g_m)} \end{bmatrix} \tag{26.9}$$

Now postulate the currents I_S and I_L as shown in Fig. 26.1. Then

$$V_s = I_s z_{11N} + I_L z_{12N} \tag{26.10}$$

$$V_L = I_s z_{21N} + I_L z_{22N} \quad (26.11)$$

$$V_L = -I_L Z_L \quad (26.12)$$

Solving for I_L from Eqs. 26.11 and 26.12, we get

$$I_L = -I_s z_{21N}/(z_{22N} + Z_L) \quad (26.13)$$

Substituting this in Eqs. 26.11 and 26.12 gives the voltage transfer function $H(s) = V_L/V_s$ and the input impedance $Z_{in} = V_s/I_s$ as

$$H(s) = Z_L z_{21N}/(Z_{11N} Z_L + |z|_N) \quad (26.14)$$

$$Z_{in} = (z_{11N} Z_L + |z|N)/(z_{22N} + Z_L) \quad (26.15)$$

The output impedance Z_{out} faced by the load is precisely $1/y_{22N}$, i.e.

$$Z_{out} = |z|_N / z_{11N} \quad (26.16)$$

Combining Eqs. 26.14–26.16 with Eq. 26.9 gives the desired expressions.

Conclusion

It has been demonstrated that simple two-port techniques are adequate for analyzing, exactly and almost by inspection, a general high-frequency transistor stage having an un-bypassed emitter resistor and a general load. It should be clear that the effect of r_μ could be taken account of by replacing sC_μ in Eq. 26.9 by $g_\mu + sC_\mu$ and that the effect of a parallel r_o, C_o combination across the current generator $g_m V$ could be taken account of by putting $y_{22}\ a = g_o + sC_o$, instead of zero, in Eq. 26.2 and continuing the analysis.

Problems

P.1. Rederive the equations by assuming $g_m \rightarrow \infty$ in Fig. 26.1.

P.2. What happens if $C_\mu = 0$ in Fig. 26.1?

P.3. What happens if $C_\mu \rightarrow \infty$ in Fig. 26.1?

P.4. Now, it's the turn of R_E. What happens when $R_E = 0$ in Fig. 26.1?

P.5. What happens when $R_E \rightarrow \infty$ in Fig. 26.1?

References

1. J. Millman, *Microelectronics* (McGraw-Hill, New York, 1979)
2. K.S. Yeung, An open and short circuit technique for analyzing electronic circuits. IEEE Trans. Educ. **E-30**, 55–56 (1987)
3. R.D. Kelly, Electronic circuit analysis and design by driving point impedance techniques. IEEE Trans. Educ. **E-13**, 154–167 (1970)

Transistor Wien Bridge Oscillator 27

Three possible circuits of transistor Wien bridge oscillator, derived from analogy with the corresponding vacuum tube circuit, are described. Approximate formulas for the frequency of oscillation and the voltage gain required for maintenance of oscillations are deduced. A practical circuit using two OC71 transistors is given. The frequency of oscillation is found to agree fairly well with that calculated from theory. The relative merits of the different forms have also been discussed.

Keywords
Transistor · Oscillator · Wien bridge

Introduction

The RC network shown in Fig. 27.1 is a degenerated form of the Wien bridge and will, henceforward, be referred to as the Wien network. Driven by an ideal voltage generator and working into an infinite impedance load, it has a transfer function given by

$$\frac{v_{\text{out}}}{v_{\text{in}}} = \frac{1}{1 + \frac{C_1R_1 + C_2R_2}{C_1R_2} + j\left(\omega C_2R_1 - \frac{1}{\omega C_1R_2}\right)} \tag{27.1}$$

where $\omega = 2\,\pi f$ denotes the angular frequency of the driving source. The phase shift produced by the network is zero at a frequency ω_o, where

$$\omega_o = \left(\frac{1}{R_1R_2C_1C_2}\right)^{1/2} \tag{27.2}$$

Figure 27.2 shows the circuit of a vacuum tube oscillator using the network of Fig. 27.1. It consists of a two-stage amplifier with positive feedback provided through the Wien network. Under open loop conditions, the amplifier has a flat gain and a phase shift of 360° over the frequency range of interest. Thus the circuit will oscillate at a frequency given by Eq. 27.2 provided that the open loop gain A_o of the amplifier satisfies the inequality

$$A_o \geq 1 + \frac{C_2}{C_1} + \frac{R_1}{R_2}$$

A common emitter (CE) transistor amplifier is analogous to a common cathode vacuum tube amplifier in that both give voltage amplification with a phase reversal. A transistor circuit,

Source: S. C. Dutta Roy, "Transistor Wien Bridge Oscillator," *Journal of the Institution of Telecommunication Engineers*, vol. 8, pp. 186–196, July 1962

S. C. Dutta Roy, *Circuits, Systems and Signal Processing*,
https://doi.org/10.1007/978-981-10-6919-2_27

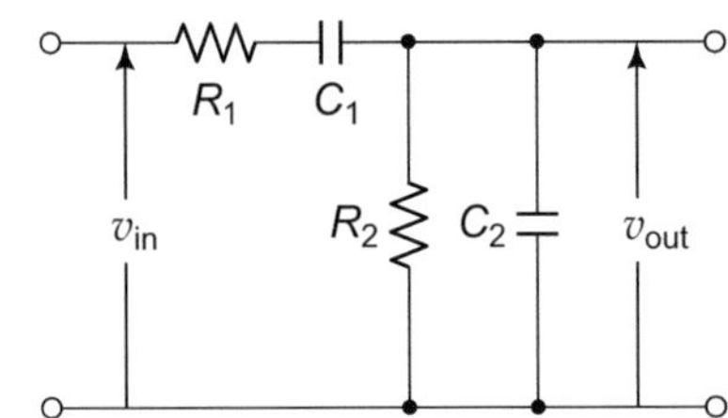

Fig. 27.1 Degenerated Wien bridge network

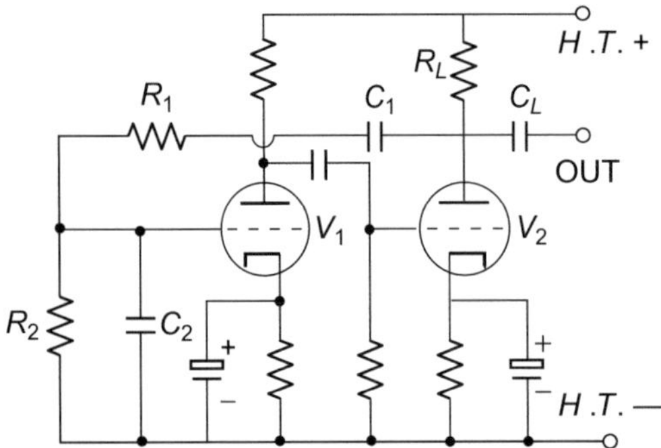

Fig. 27.2 Circuit of a vacuum tube Wien bridge oscillator

analogous to that of Fig. 27.2, will, however, have the following drawbacks.

(1) A transistor amplifier in the CE mode has a high output impedance so that the behaviour at the output terminals is essentially that of a current generator. The series arm of the Wien network will thus be supplemented by the output impedance of the equivalent voltage generator. In contrast, in Fig. 27.2, V_2 acts as a voltage generator with an internal impedance $r_pR_L/(r_p + R_L)$ (r_p = plate resistance of V_2) which is usually small compared to R_1.

(2) The input impedance of a CE transistor amplifier is low and causes a considerable loading of the shunt arm of the Wien network. In contrast, the input impedance of a vacuum tube is very high, ideally infinite, so that the Wien network in Fig. 27.2 works into an open circuit load.

(3) Since the shunt arm is heavily loaded and the impedance of the series arm is increased, a large voltage gain will be required for maintenance of oscillations. In view of the low input impedance of a CE amplifier, the gain of the first stage will be small; the second stage will then have to provide the necessary voltage gain.

(4) In an oscillator circuit, it is desirable that the frequency of oscillation should be controlled by varying the passive elements only. In the above circuit, since both the series and the shunt arms arc supplemented by transistor impedances, the frequency will be largely controlled by the transistor impedance parameters, which are again functions of the transistor operating point.

These difficulties may be obviated by using a current dual of the Wien network as the feedback network [1, 2]. In this chapter, it is shown that circuits analogous to that of Fig. 27.2 can be designed to overcome some or all the difficulties listed above. In all, three circuits have been discussed, using 4, 3 and 2 transistors. In the analysis of these circuits, a transistor amplifier has been assumed to be capable of representation by the equivalent circuit shown in Fig. 27.3 in which the effect of the collector capacitance has been assumed to be negligible. The expressions for the voltage gain (A) and the output impedance (Z_i) are:

$$A = \frac{v_2}{v_1} = \frac{-h_{21}Z_L}{h_{11} + \Delta^k Z_L} \tag{27.3}$$

$$Z_i = \frac{v_1}{i_1} = \frac{h_{11} + \Delta^k Z_L}{1 + h_{22}Z_L} \tag{27.4}$$

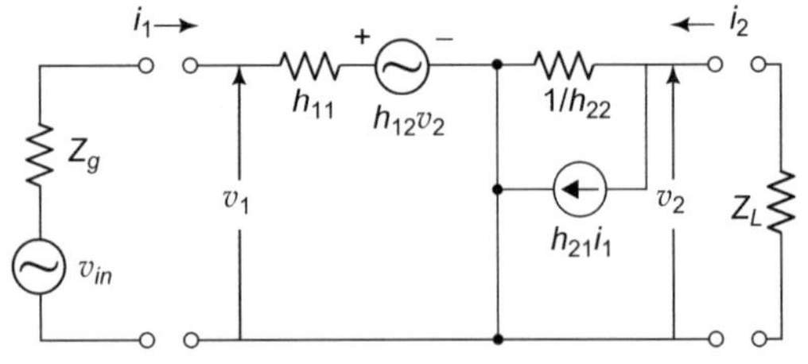

Fig. 27.3 Low frequency equivalent circuit of a transistor amplifier

Table 27.1 Low frequency h-parameters of OC71 at Vc = −2 V, Ic = 3 mA

Parameter	Configuration		
	CB	CE	CC
h11 (Ω)	17	800	800
h12	8×10^{-1}	5.4×10^{-4}	1
h21	−0.979	47	−47
h22 (℧)	1.6×10^{-6}	80×10^{-6}	80×10^{-6}
Δ^k	8.1×10^{-1}	0.0386	47.06

where

$$\Delta^k = h_{11}h_{22} - h_{12}h_{21}$$

A practical circuit, using two OC71 transistors, is given. Each transistor is maintained at the following operating point: collector to emitter voltage, $V_c = -2$ V; collector current, $I_c = 3$ mA. The h-parameters at the above operating point are given in Table 27.1. The frequency of oscillation is found to agree fairly well with that calculated from theory.

Circuit 1

A common collector (CC) transistor amplifier is analogous to a cathode follower circuit. It has a high input and a low output impedance while the gain is slightly less than unity. Thus the high impedance output of a CE stage can he transformed into a low impedance one by cascading a CC stage to it. The problem of loading of the shunt arm of the Wien network can be solved in a similar way. The transistor analogue of the Wien bridge oscillator of Fig. 27.2 will then look like that shown in Fig. 27.4. Only the A.C. equivalent circuit has been drawn on the assumption that the effect of the biasing resistors is negligible and that the coupling and decoupling condensers behave as short circuits at the frequency of oscillation.

The two-stage amplifier T_1T_2 gives an output voltage which is in phase with the input voltage. The first stage has a load approximately equal to the input impedance of the second and as such has a low gain. In view of the high input impedance of the CC stage T_3, T_2 has a load approximately equal to R_{L2} and can be designed to have a high gain. As the output impedance of T_3 is low, the Wien network is effectively supplied by a voltage generator. The network has a load equal to the input impedance of the CC stage T_4 and as such, if R_2 is not too high, the loading may be considered to be negligible. T_4 has a low output impedance and is not loaded when connected to the input of T_1. The CC stages produce no phase shift. Thus the circuit will oscillate at a frequency at which the phase shift through the network is zero.

An approximate solution for this circuit can be obtained as follows. Let us assume that the transistors T_1–T_4 are identical. If the effective load impedance of T_2 is not very large, then its input impedance is approximately h_{11e}, where the subscript e is used to mean a common emitter parameter. The effective load impedance of T_1 is

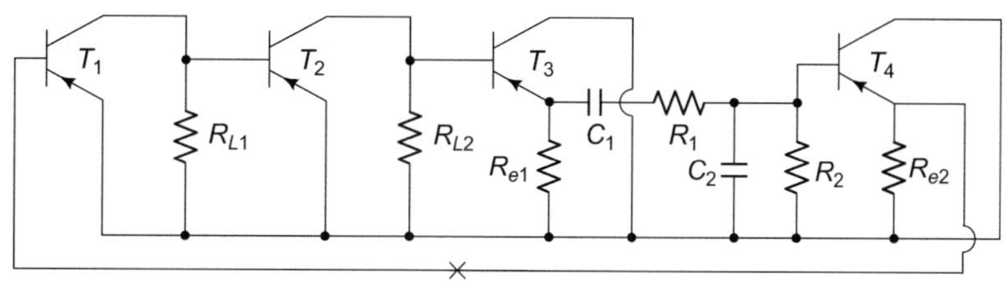

Fig. 27.4 Circuit 1 using four transistors (2 CE and 2 CC)

$$(R_L)_1 \simeq \frac{R_{L1} h_{11e}}{R_{L1} + h_{11e}}$$

The voltage gain of this stage is, by formula Eq. 27.3,

$$A_1 = \frac{-h_{21e}(R_L)_1}{h_{11e} + (\Delta^h)_e (R_L)_1}$$

Now, $(R_L)_1 < h_{11e}$ and from Table 27.1, $(\Delta^k)_e \ll 1$. Thus

$$A_1 \simeq -\frac{h_{21e}}{h_{11e}}(R_L)_1 = \frac{-h_{21e} R_{L1}}{R_{L1} + h_{11e}} \qquad (27.5)$$

The effective load impedance of T_2 is

$$(R_L)_2 = \frac{R_{L1}(Z_i)_3}{R_{L2} + (Z_i)_3}$$

where $(Z_i)_3$ is the input impedance of T_3. From Eq. 27.4,

$$(Z_i)_s \simeq \frac{h_{11e} + (\Delta^h)_e R_{e1}}{1 + h_{22e} R_{e1}}$$

The approximation involved in the above equation is that the loading of R_{e1} by the Wien network and the following stage is negligible. Substituting values from Table 27.1 and assuming R_{e1} = 1 kΩ, we get $(Z_i)_3 \simeq 48$ kΩ. Now, R_{L2} will be of the order of 3 kΩ or less so that we can put $(Z_i)_3 \gg R_{L2}$ and get $(R_L)_2 \simeq R_{L2}$. Thus the gain of the second stage is

$$A_2 \simeq \frac{-h_{21e} R_{L2}}{h_{11e} + (\Delta^k)_e R_{L2}} \qquad (27.6)$$

The gain of the third stage is

$$A_3 \simeq \frac{-h_{21e} R_{e2}}{h_{11e} + (\Delta^k)_e R_{e1}}$$

Assuming R_{e1} = 1 kΩ and putting the values of the parameters from Table 27.1, we get $A_3 \simeq 1$. Thus if the voltage at the input of T_1 is v_1, then that at the input of the Wien network is $A_1 A_2 v_1$.

Since R_{e2} is shunted by the input impedance of T_1 which is small ($\simeq h_{11e}$) the input impedance of T_4, $(Z_i)_4$, will not, in general, be negligible compared with the impedance of the shunt arm of the Wien network. $(Z_i)_4$, however, can be calculated as follows. The load impedance of T_4 is

$$(R_L)_4 \simeq \frac{R_{e2} h_{11e}}{R_{e2} + h_{11e}}$$

Thus from Eq. 27.4,

$$(Z_i)_4 = \frac{h_{11e} + (\Delta^k)_e (R_L)_4}{1 + h_{22e}(R_L)_4}$$

Now, $(R_L)_4 < h_{11e}$ so that from Table 27.1, $h_{22e}(R_L)_4 \ll 1$. Therefore,

$$(Z_i)_4 \simeq h_{11e} + (\Delta^k)\frac{R_{e2} h_{11e}}{R_{e2} + h_{11e}} \qquad (27.7)$$

Let

$$R_2' = \frac{R_2 (Z_i)_4}{R_2 + (Z_i)_4} \qquad (27.8)$$

Then from Eq. 27.1, the voltage at the input of T_4 is

$$\frac{A_1 A_2 v_1}{1 + \frac{C_1 R_1 + C_2 R_2'}{C_1 R_2'} + j\left(\omega C_2 R_1 - \frac{1}{\omega C_1 R_2'}\right)} \qquad (27.9)$$

The gain of T_4 is, by formula Eq. 27.3,

$$A_4 = \frac{-h_{21e}(R_L)_4}{h_{11e} + (\Delta^k)_e (R_L)_4}$$

Assuming R_{e2} = 1 kΩ and substituting for the parameters from Table 27.1, we get $A_4 \simeq 1$. Thus the output voltage of T_4 is given by Eq. 27.9; but, this is equal to v_1 so that

$$A_1A_2 = 1 + \frac{C_1R_1 + C_2R_2'}{C_1R_2'} + j\left(\omega C_2R_1 - \frac{1}{\omega C_1R_2'}\right) \quad (27.10)$$

Equating the imaginary parts on either side of Eq. 27.10 gives the frequency of oscillation as

$$\omega_o = \left(\frac{1}{C_1C_2R_1R_2'}\right)^{1/2} \quad (27.11)$$

Combining Eqs. 27.7 and 27.8 with Eq. 27.11, we have

$$\omega_o = \omega_n\left[1 + \left\{\frac{R_2(R_{e2}+h_{11e})}{h_{11e}(h_{11e}+R_{e2})+(\Delta^k)_eR_{e2}h_{11e}}\right\}\right]^{1/2} \quad (27.12)$$

where ω_n denotes the angular frequency at which the phase shift through the isolated network is zero. The quantity within the second bracket in Eq. 27.12 may be looked upon as a correction factor, α_1. If $R_2 < 3$ kΩ and $R_{e2} = 1$ kΩ, then for the OC71 transistor, $\alpha_1 < 0 \cdot 14$ and we can write

$$\omega_o \simeq \omega_n\left(1 + \frac{1}{2}\alpha_1\right)$$

The condition for maintenance of oscillations is obtained by equating the real parts on either side of Eq. 27.10. Combining this with Eqs. 27.5 and 27.6 gives

$$\frac{h_{21e}^2R_{L1}R_{L2}}{(R_{L1}+h_{11e})\{h_{11e}+(\Delta^k)_eR_{L2}\}} = 1 + \frac{C_2}{C_1} + \frac{R_1}{R_2}(1+\alpha_1) \quad (27.13)$$

Normally, the left-hand side of eq. 27.13 will be far in excess of the right-hand side, so that the output waveform will be distorted. A good waveform can be obtained by inserting a suitable resistance R_f at the point marked X in Fig. 27.4. It is, however, better to reduce the gain by negative feedback. Local negative feedback may be applied through unbypassed emitter resistance. Total negative feedback may be applied by connecting a suitable resistance from the output of T_2, T_3 or T_4 to the first emitter, the biasing resistance connected to it being partly or wholly unbypassed. If a suitable non-linear element is used for the feedback resistance, amplitude stabilization of the oscillator output will result.

The use of negative feedback raises the input impedance of T_1 and thus reduces the loading of R_{e2}. As a result, the input impedance of T_4 increases and the loading of the shunt arm of the Wien network decreases. If sufficient negative feedback can be used, then $\alpha_1 \to 0$ and $\omega_0 \to \omega_n$.

The output of the oscillator is to be taken from the third emitter, as the output impedance of T_3 is small ($\simeq 64\ \Omega$ for OC71 transistor if $R_{L2} = 2$ 2 kΩ and $R_{e1} = 1$ kΩ). The output impedance of T_4 is also small, but a load connected at this point will reduce $(Z_i)_4$ and as such R_2'. Thus a variation in the load impedance will result in a change in the frequency of oscillation.

Circuit 2

A common base transistor amplifier can give voltage amplification with zero phase shift. Thus the two-stage CE amplifier in Fig. 27.4 can be replaced by a single stage CB amplifier as shown in Fig. 27.5. As in the previous case, only the A. C. equivalent circuit has been drawn on the same assumptions as made before.

Let the voltage at the input of T_1 be v_1. As in the previous case, the input impedance of T_2 will be high compared to R_L so that the gain of T_1 is

$$A_1 \simeq \frac{-h_{21b}R_L}{h_{11b}+(\Delta^k)_bR_L}$$

Substituting values from Table 27.1, we note that $h_{21b} \simeq -1$ and that if $R_L < 3$ kΩ then $(\Delta^k)_bR_L < 2 \cdot 43$. Thus, to a first approximation, we can neglect $(\Delta^k)_bR_L$ compared to h_{11b} and get

$$A_1 \simeq \frac{R_L}{h_{11b}} \quad (27.14)$$

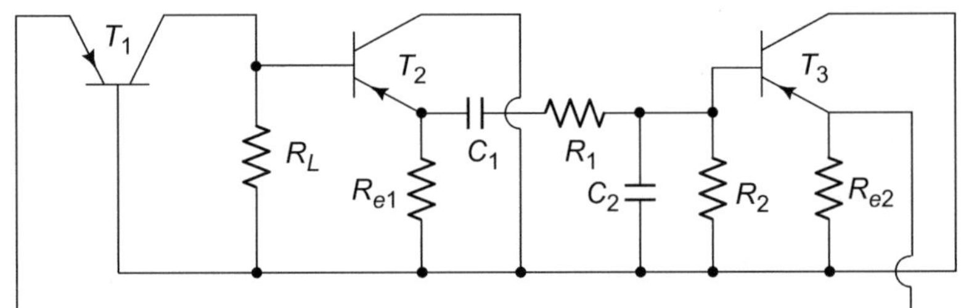

Fig. 27.5 Circuit 2 using three transistors (1 CB and 2 CC)

The gain of T_2 is approximately unity so that the input to the Wien network is $A_1 v_1$. The input impedance of T_3, $(Z_i)_3$, will be small compared with that of T_2 because of the heavy loading of R_{e2} by the input circuit of T_1. The load impedance of T_3 is $\simeq h_{11b}$ and from Table 27.1, $h_{22e}h_{11b} \ll 1$; thus

$$(Z_i)_3 \simeq h_{11e} + (\Delta^k)_e h_{11b}$$

Let

$$R'_2 = \frac{R_2\{h_{11e} + (\Delta^k)_e h_{11b}\}}{R_2 + h_{11e} + (\Delta^k)_e h_{11b}} \tag{27.15}$$

Then the output of the Wien network will be given by

$$v_3 = \frac{A_1 v_1}{1 + \frac{C_1R_1 + C_2R'_2}{C_1R'_2} + j\left(\omega C_2 R_1 - \frac{1}{\omega C_1 R'_2}\right)}$$

The gain of T_3 is

$$A_3 \simeq \frac{-h_{21e}h_{11b}}{h_{11e} + (\Delta^k)_e h_{11b}} \tag{27.16}$$

The output of T_3 is given by $v_4 = A_3 v_3$. But $v_4 = v_1$; thus

$$A_1A_3 = 1 + \frac{C_2}{C_1} + \frac{R_1}{R'_2} + j\left(\omega C_2 R_1 - \frac{1}{\omega C_1 R'_2}\right) \tag{27.17}$$

Equating the imaginary parts on either side of Eq. 27.17 gives the frequency of oscillation as

$$w_o = w_n(1 + a_2)^{1/2}$$

where

$$\alpha_2 = \frac{R_2}{h_{11e} + (\Delta^k)_e h_{11b}}$$

The correction factor α_2 in this case is quite large. For the OC71 transistor, if R_2 = 3.2 kΩ, then α_2 = 2. Equating the real parts on either side of Eq. 27.17 gives the condition for maintenance of oscillations. Combining this with Eqs. 27.14 and 27.16, we get

$$\frac{-h_{21e}R_L}{h_{11e} + (\Delta^k)_e h_{11b}} = 1 + \frac{C_2}{C_1} + \frac{R_1}{R_2}(1 + \alpha_2) \tag{27.18}$$

As in the previous case, the left-hand side of Eq. 27.18 will usually be in excess of the right-hand side. A suitable resistance may be used between the emitters of T_3 and T_1 to get a good waveform. Alternatively, negative feedback may be applied by connecting a suitable resistance from the output of T_1, T_2 or T_3 to the base of T_1, the biasing resistances at this point being partly or wholly unbypassed. As in the previous case, negative feedback reduces the correction factor, α_2.

As the output impedance of T_3 and the input impedance of T_1 are both small, the output voltage may be taken from the third emitter, if the load impedance is not too small. Alternatively, the output may be taken from the second emitter as in the previous case.

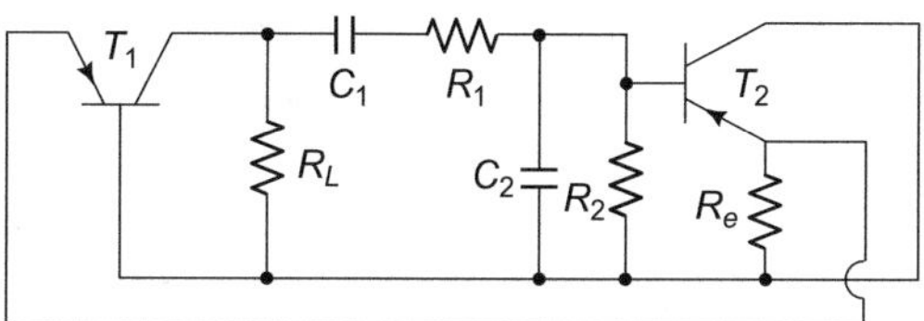

Fig. 27.6 Circuit 3 using two transistors (1 CB and 1 CC)

Circuit 3

A further simplification of the Wien bridge oscillator circuit is possible if we omit the transistor T_2 in Fig. 27.5. The resulting circuit is shown in Fig. 27.6. In view of the high output impedance of T_1, the series arm of the Wien network will also be supplemented by an extra impedance in this circuit.

As in circuit 2, the effective load impedance of T_2 is approximately h_{11b} and since $h_{22c}h_{11b} \ll 1$, its input impedance is

$$(Z_i)_2 \simeq h_{11c} + (\Delta^k)_c h_{11b}$$

Let

$$R_2' = \frac{R_2(Z_i)_2}{R_2 + (Z_i)_2} = \frac{R_2}{1 + R_2/\{h_{11c} + (\Delta^k)_c h_{11b}\}} \tag{27.19}$$

$$Z_1 = R_1 + \frac{1}{j\omega C_1} \tag{27.20}$$

and

$$Z_2 = \frac{R_2'}{j\omega C_2 R_2' + 1} \tag{27.21}$$

The effective load impedance of T_1 is

$$Z_L = \frac{R_L(Z_1 + Z_2)}{R_L + Z_1 + Z_2} \tag{27.22}$$

so that the voltage gain of T_1 is

$$A_1 = \frac{-h_{21b}Z_L}{h_{11b} + (\Delta^k)_b Z_L}$$

Now $Z_L < R_L$ and R_L is of the order of 3 kΩ. Thus $(\Delta^k)_b Z_L \ll h_{11b}$ and since $h_{21b} \simeq -1$,

$$A_1 \simeq \frac{Z_L}{h_{11b}} \tag{27.23}$$

The voltage across R_1 is $A_1 v_1$, v_1 being the voltage at the input of T_1. The output voltage of the Wien network is, therefore,

$$v_2 = \frac{A_1 v_1}{1 + \frac{C_1 R_1 + C_2 R_2'}{C_1 R_2'} + j\left(\omega C_2 R_1 - \frac{1}{\omega C_1 R_2'}\right)} \tag{27.24}$$

The voltage gain of T_2 is

$$A_2 \simeq \frac{-h_{21c}h_{11b}}{h_{11c} + (\Delta^k)_c h_{11b}} \tag{27.25}$$

The output of T_2 is $A_2 v_2 = v_1$. Combining this with Eqs. 27.22–27.25, we get

$$\frac{-h_{21c}R_L}{h_{11c} + (\Delta^k)_c h_{11b}} = D\left(1 + \frac{R_L}{Z_1 + Z_2}\right) \tag{27.26}$$

where D denotes the denominator of the right-hand side of Eq. 27.24. Now from Eqs. 27.20 and 27.21,

$$Z_1 + Z_2 = \frac{R_2' D}{j\omega C_2 R_2' + 1} \tag{27.27}$$

Combining 27.26 and 27.27 gives

$$\frac{-h_{21c}R_L}{h_{11c} + (\Delta^h)_c h_{11b}} = 1 + \frac{C_2}{C_1} + \frac{R_1 + R_L}{R_2'} + j\left\{\omega C_2(R_1 + R_L) - \frac{1}{\omega C_1 R_2'}\right\} \tag{27.28}$$

Equating the imaginary parts on either side of Eq. 27.28 and substituting for R_2' from Eq. 27.19 gives the frequency of oscillation as

$$\omega_o = \omega_n \left[\frac{1 + R_2/\{h_{11c} + (\Delta^k)_c h_{11b}\}}{1 + (R_L/R_1)} \right]^{1/2} \tag{27.29}$$

Equation 27.29 shows that ω_o can be made equal to ω_n by choosing

$$R_1 R_2 = R_L\{h_{11c} + (\Delta^k)_c h_{11b}\} \tag{27.30}$$

Equating the real parts on either side of Eq. 27.28 gives the condition for maintenance of oscillations as

$$\frac{-h_{21c} R_L}{h_{11c} + (\Delta^k)_c h_{11b}} = 1 + \frac{C_2}{C_1} + \frac{R_1}{R_2}\left(1 + \frac{R_L}{R_1}\right) \times \left\{1 + \frac{R_2}{h_{11e} + (\Delta^k)_e h_{11b}}\right\} \tag{27.31}$$

Here also the left-hand side of Eq. 27.31 will be in excess of the right-hand side, and the gain can be reduced by the same methods as employed in circuit 2. If degeneration is used, then the input impedance of T_1 will be raised and the output impedance lowered. The former reduces the loading of the shunt arm and the latter reduces the impedance adding to the series arm of the Wien network. If sufficient negative feedback can be applied, then ω_o can be made to approach ω_n very closely.

For simplicity's sake, let us suppose that the gain is reduced by inserting a resistance R_f between the emitters of T_2 and T_1. Then the load impedance of T_2 is

$$(Z_L)_2 \simeq \frac{(R_f + h_{11b})R_e}{R_e + R_f + h_{11b}}$$

Therefore

$$(Z_i)_2 \simeq h_{21e} + (\Delta^k)_e (Z_L)_2$$

because even if $(Z_L)_2 = 1\ \mathrm{k}\Omega$, $h_{22c}(Z_L)_2 = 80 \times 10^{-3} \ll 1$. The frequency of oscillation will be given by

$$\omega_o = \omega_n \left\{ \frac{1 + \frac{R_2}{h_{11c} + (\Delta^k)_c R_e (R_f + h_{11b})/(R_e + R_f + h_{11b})}}{1 + (R_L/R_1)} \right\} \tag{27.32}$$

The condition of oscillation is modified to the following:

$$\frac{-h_{21c} R_L}{h_{11b}\left\{(\Delta^k)_c + \frac{h_{11c}(R_e + R_f + h_{11b})}{R_e(R_f + h_{11b})}\right\}} = 1 + \frac{C_2}{C_1} + \frac{R_1}{R_2}\left(1 + \frac{R_L}{R_1}\right) \times \left\{1 + \frac{R_2}{h_{11c} + \frac{(\Delta^k)_c R_e (R_f + h_{11b})}{(R_e + R_f + h_{11b})}}\right\}$$

This can be solved to find the appropriate value of R_f. It is, however, more convenient to put a variable resistance for R_f and to adjust it experimentally.

The output voltage is taken from the emitter of T_2 from the same considerations as stated in the previous case.

Practical Circuit

A practical version of the two-transistor circuit is shown in Fig. 27.7a. Each transistor is maintained at the operating point at which the parameters of Table 27.1 apply. This was done for comparing the actual frequency with that calculated from theory. By choosing a smaller value of I_c, the circuit could be designed to work on a 6 V. battery. A 9 V battery could also be used for establishing the required operating point, but the biasing resistors required are so small that besides drawing a large power from the battery, their effect on the A.C. operation becomes quite appreciable.

A slightly lower value of resistance was used at the collector of T_2 than that at the collector of T_1 to establish a slight difference of potential between the two emitters.

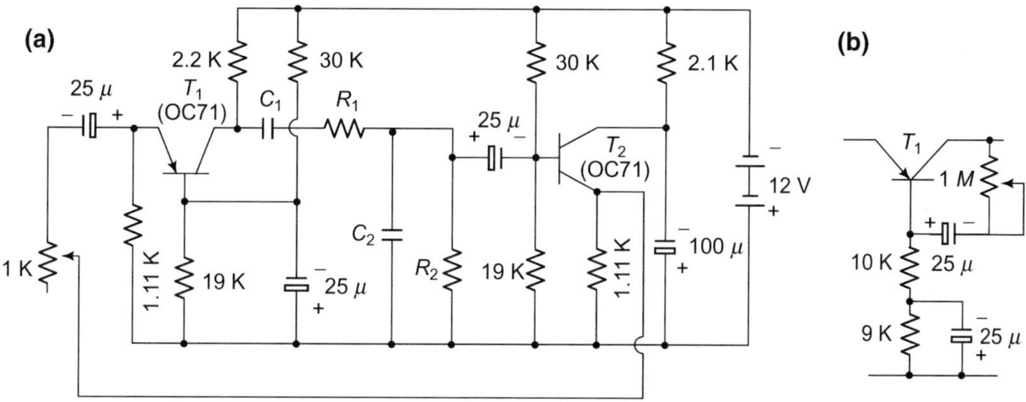

Fig. 27.7 **a** Practical oscillator circuit using two transistors; **b** arrangement for negative feedback

Table 27.2 Comparing the actual frequency with that calculated from theory

R_1 (kΩ)	C_1 (μF)	R_2 (kΩ)	C_2 (μF)	R_4 (Ω)	f_a (c/s)	f_c (c/s)
40	0.106	89	0.105	140	115	113
0.97	0.106	10.4	0.105	395	215	212
0.97	0.106	1.48	0.105	415	755	730
1.4	0.022	1.48	0.0208	375	3365	3400
0.97	0.011	1.48	0.0095	410	7223	7580
0.97	0.0065	1.48	0.00642	385	11,312	12,000

In Fig. 27.7a, the gain is shown to be reduced by inserting a variable resistance R_f in the positive feedback line. Thus for this circuit, formula 27.32 will be applicable. The arrangement for reducing the gain by negative feedback is shown in Fig. 27.7b.

The values of the Wien network components (R_1, C_1, R_2 and C_2), the feedback resistance (R_f). the actual frequency of oscillation (f_a) and the frequency calculated from Eq. 27.32 (f_c) are shown in Table 27.2. In calculating f_c for the first two cases, the effects of the biasing resistances were also taken into account. It will be seen that f_a agrees fairly well with f_c in the frequency range shown.

Discussions

From economic considerations, circuit 3 should be preferred as it uses the least number of transistors and other components.

The dependence of frequency of oscillation on the transistor operating point is most pronounced in circuit 2, because the correction factor α_2 is usually greater than unity. The correction factor α_1 in circuit 1 is generally less than unity while that in circuit 3 can be made a minimum by choosing R_1, R_2 and R_L such that Eq. 27.30 is satisfied. Condition Eq. 27.30 cannot, however, be maintained in the lower audio range because of the large values of condensers required.

The change in the frequency of oscillation due to a given change of load impedance will be the highest in circuit 3 and the least in circuit 1.

The lower limit of frequency in either of the three circuits considered will be set by the maximum value of the coupling and bypass capacitors that can be used while the high frequency limit will be set primarily by the collector capacitance.

Note that OC71 is obsolete. So do not search for one in the market. Instead wire of a circuit with a commonly available tansistor.

When this paper was written, [3] was our Bible for transistor circuits. Also see [4] for an early form of transistor oscillator.

Problems

P.1. Analyze the circuit of Fig. 27.3 with hybrid-π parameter. h-parameters are not used anymore. Do you know the reason?
P.2. Same for the circuit of Fig. 27.4.
P.3. Same for the circuit of Fig. 27.5.
P.4. Same for the circuit of Fig. 27.6.
P.5. What if there is no negative feedback?

References

1. D.E. Hooper, A.E. Jackets, Current derived resistance capacitance oscillators using junction transistor. Electron. Eng. **28**, 333 (1956)
2. R. Hutchins, Selective RC amplifier using transistors. Electron. Eng. **33**, 84 (1961)
3. R.F. Shea, *Principles of Transistor Circuits* (Wiley, 1953), p. 336
4. P.G. Sulzer, Low distortion transistor audio oscillator. Electronics **26**, 171 (1953)

28 Analysing Sinusoidal Oscillator Circuits: A Different Approach

Conventionally, in analysing sinusoidal oscillator circuits, one uses the Berkhausen's criterion, viz. $A\beta = 1$ in a positive feedback amplifier whose gain without feedback is A and whose feedback factor is β. However, the identification of A and β poses problems because of mutual loading of the amplifier and the feedback network. A different approach is presented here which does not require such identification. The method is based on assuming a voltage at an arbitrary node and coming back to it through the feedback loop.

Keywords
Sinusoidal oscillator · Different approach

Introduction

In most textbooks on analog electronic circuits, sinusoidal oscillator circuits are analysed by using the Berkhausen's criterion, viz $A\beta = 1$ in a positive feedback amplifier, where A is the gain of the amplifier without feedback and β is the feedback factor. However, except where the amplifier is nearly ideal, e.g. in op-amp circuits, the amplifier and the feedback networks load each other and the identification of A and β poses a problem. In this chapter, we propose a different approach which does not require such identification. In fact, we do not use feedback concepts at all. Instead, we assume a voltage at an arbitrary node and come back to the same node through the feedback loop. This results in the so-called characteristic equation of the oscillator. By putting $s = j\omega$ and equating the real and imaginary parts of the equation to zero, we get the condition for, and the frequency of oscillation.

An Op-Amp Oscillator

Consider the Wien bridge RC oscillator, shown in Fig. 28.1, using an op-amp as the gain element. Here, A refers to the gain between the nodes N_1 and N_2, and clearly,

$$A = 1 + (R_2/R_1) \tag{28.1}$$

which is independent of the feedback network because the input impedance of the op-amp tends to infinity and the output impedance tends to zero.

Source: S. C. Dutta Roy, "Analyzing Sinusoidal Oscillator Circuits: A Different Approach," *IETE Journal of Education*, vol. 45, pp. 9–12, January–March 2004.

S. C. Dutta Roy, *Circuits, Systems and Signal Processing*,
https://doi.org/10.1007/978-981-10-6919-2_28

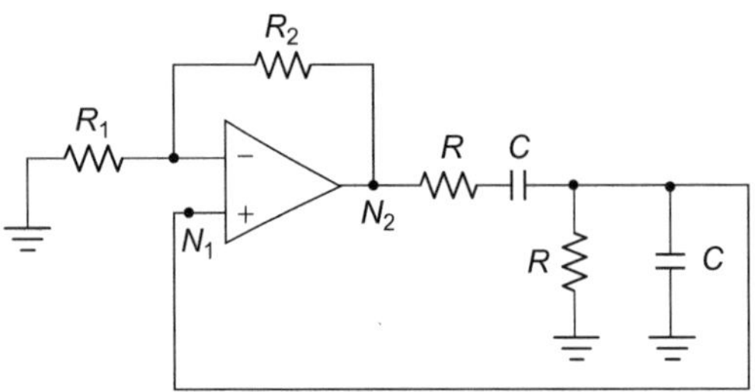

Fig. 28.1 An op-amp Wien bridge oscillator

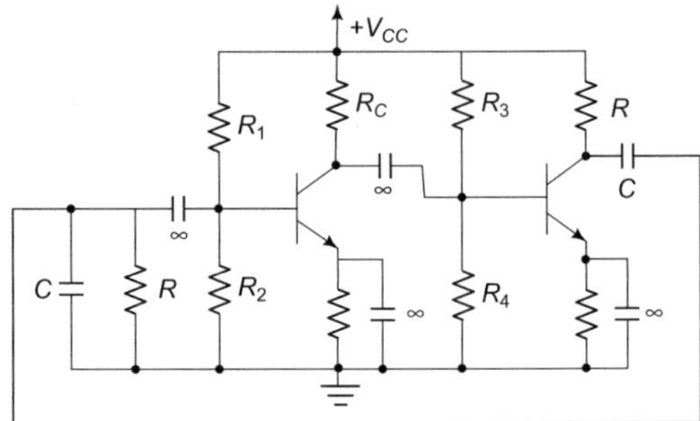

Fig. 28.2 Transistor Wien bridge oscillator

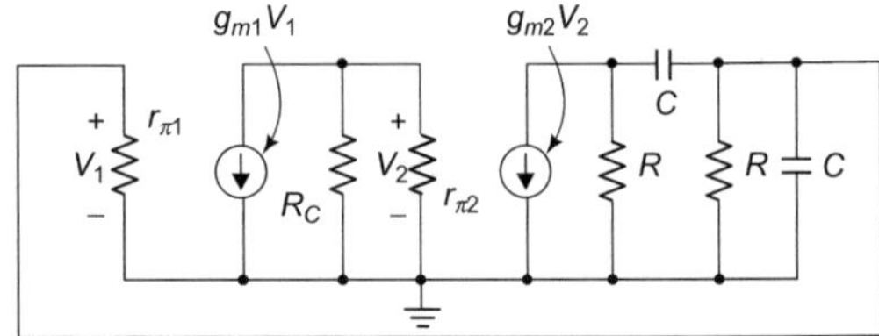

Fig. 28.3 AC equivalent circuit of Fig. 28.2

Because of the same reason, the transfer function of the β network is, by inspection,

$$\beta = [R/((sCR+1))]/\{R+[(1/sC)] + R/[(sCR+1)]\} = sCR/\left(s^2C^2R^2+3sCR+1\right). \tag{28.2}$$

Putting the values of A and β from Eqs. 28.1 and 28.2 in $A\beta = 1$, and simplifying, we get the characteristic equation as

$$s^2C^2R^2 + [2-((R_2/R_1)]sCR + 1 = 0. \tag{28.3}$$

Now putting $s = j\omega$ in Eq. 28.3 and equating its real and imaginary parts, we get the condition of oscillation as

$$R_2 = 2R_1 \tag{28.4}$$

and the frequency of oscillation as

$$\omega_0 = 1/(RC). \tag{28.5}$$

Transistor Version of the Wien Bridge Oscillator

Now consider the transistorized version of the Wien bridge oscillator shown in Fig. 28.2. Assume that the shunting effects of R_1, R_2, R_3 and R_4 are negligible that the coupling and bypass capacitances behave as short circuits and that the transistor internal capacitances behave as open circuits at the frequency of oscillation. Then, the AC equivalent circuit becomes that shown in Fig. 28.3, where the symbols have their usual meanings. To analyse this circuit, we start at V_1, and return to V_1 through the feedback loop. Note that

$$V_2 = -g_{m1}V_1R_c', \tag{28.6}$$

where

$$R_c' = R_c||r_{\pi 2}. \tag{28.7}$$

The current generator g_{m2} V_2 in parallel with R can be converted to a voltage source $-g_{m2}$ V_2R in series with R. Then, one can find V_1 as

$$V_1 = -g_{m2}V_2R\frac{R'/(1+sCR')}{R+[1/(sC)]+[R'/((1+sCR')]}, \tag{28.8}$$

where

$$R' = R||r_{\pi 1}. \tag{28.9}$$

Combining Eq. 28.8 with Eq. 28.6, cancelling V_1 from both sides and simplifying, we get the following characteristic equation:

$$s^2C^2RR' + sC(R + 2R' - g_{m1}g_{m2}RR'R'_c) + 1 = 0 \quad (28.10)$$

Putting $s = j\omega$ in this equation and equating the real and imaginary parts, we get the condition of oscillation as

$$g_{m1}g_{m2}RR'R'_c = R + 2R' \quad (28.11)$$

and the frequency of oscillation as

$$\omega_0 = 1/\left(C\sqrt{RR'}\right). \quad (28.12)$$

Another Example

As another example, consider the transistor Colpitt's oscillator shown in Fig. 28.4. At the frequencies at which the Colpitt's circuit is used, the transistor internal capacitances also may have to be considered and we shall do so. Again, we assume the coupling and bypass capacitances to behave as shorts and ignore the shunting effects of R_1 and R_2. Then, the ac equivalent circuit becomes that shown in Fig. 28.5, which is redrawn in Fig. 28.6 in a form more suitable for analysis. The current generator g_mV_1 and its shunting elements R_c and C can be replaced by a Thevenin equivalent, and the resulting circuit is shown in Fig. 28.7, where

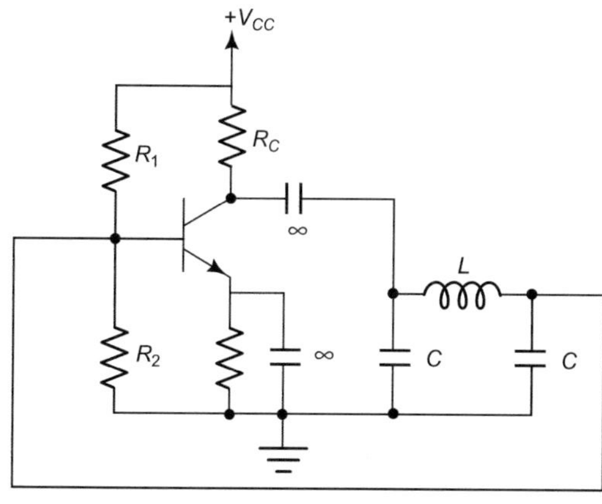

Fig. 28.4 Transistor Colpitt's oscillator

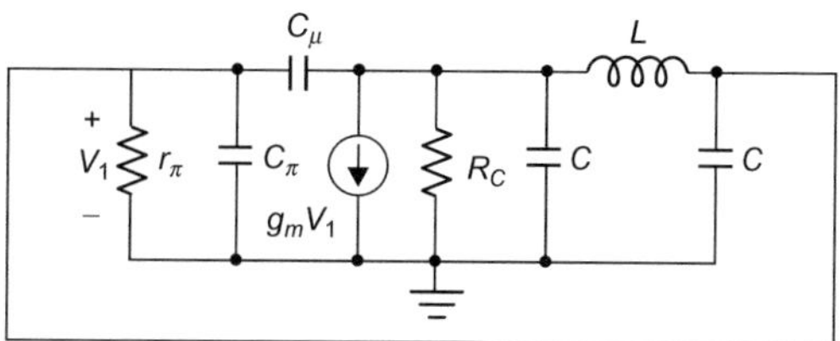

Fig. 28.5 AC equivalent circuit of Fig. 28.4

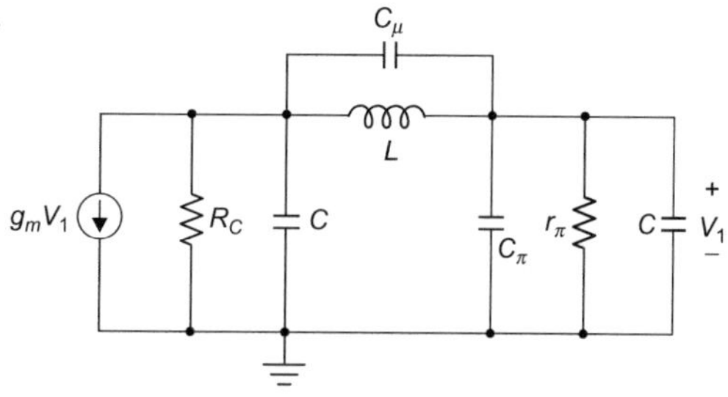

Fig. 28.6 Redrawn form of Fig. 28.5

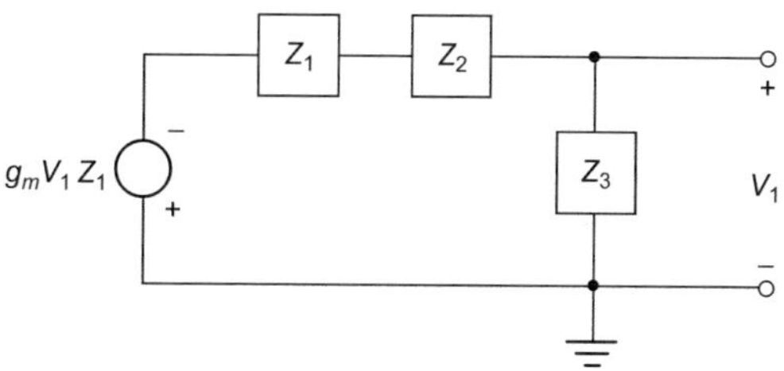

Fig. 28.7 Simplified version of Fig. 28.6

$$Z_1 = R_c/(sR_cC + l), Z_2 = sL/\left(s^2LC_\mu + 1\right) \text{ and } Z_3 = r_\pi/[sr_\pi(C + C_\pi) + 1]. \quad (28.13)$$

From Fig. 28.7, we get

$$V_1 = -g_mV_1Z_1Z_3/(Z_1 + Z_2 + Z_3). \quad (28.14)$$

Cancelling V_1 from both sides, combining with Eq. 28.13, and simplifying, one gets the following characteristic equation:

$$s^3L\left[C_m\left(2C+C_p\right)+C\left(C+C_p\right)\right] + s^2L\left[C_m\left(G_c+g_p+g_m\right)+G_c\left(C+C_p\right)+g_pC\right] + s\left(2C+C_p+LG_Cg_p\right)+\left(G_c+g_p+g_m\right)=0. \quad (28.15)$$

Putting $s = j\omega$ in Eq. 28.15 and equating the real and imaginary parts, we get the frequency oscillation as given by

$$\omega_0^2 = \frac{2C + C_\pi + LG_c g_\pi}{L[C_\mu((2C + C_\pi) + C(C + C_\pi)]} = \frac{G_c + g_\pi + g_m}{L[C_\mu(G_c + g_\pi + g_m) + G_c(C + C_\pi) + g_\pi C]}, \quad (28.16)$$

where the second part gives the condition of oscillation.

Concluding Comments

Rather than undertaking the involved task of identifying A and β in a sinusoidal oscillator circuit, we show that it is easier and less prone to mistake to start at a convenient node voltage and return to the same through the feedback loop. The characteristic equation is thus obtained; putting $s = j\omega$ in it and equating the real and imaginary parts on both sides give the frequency of, as well as the condition for oscillation.

Problems

P.1. What happens when series RC is interchanged with parallel RC in Fig. 28.1? Derive the necessary equations, and justify your conclusions.

P.2. Suppose in Fig. 28.1, the series R is absent, what will happen? Oscillations? Justify your answer with necessary equations.

P.3. What happens in Fig. 28.2 if the capacitor marked ∞ is not infinite? Again, justify your answer with equations.

P.4. In Fig. 28.3, if r_{π_1} and r_{π_2} are infinitely large, what will happen? Justify.

P.5. If in Fig. 28.1, the two C's are replaced by two L's and C is replaced by a single L, what would happen? Justify you answer with necessary derivations.

Triangular to Sine-Wave Converter

29

This chapter describes how a triangular wave is converted into a sine wave by using a piecewise linear transfer characteristic. A detailed analysis of the basic circuit is given, and its actual implementation in an available IC chip is briefly discussed.

Keyword
Conversion of waves

Introduction

Given a symmetrical triangular wave as shown in Fig. 29.1, is it possible to convert it into a sine wave by an electronic circuit? The answer turns out to be in the affirmative. Such a converter is, in fact, available as an analog IC chip, whose transfer or input–output characteristic consists of nine symmetrical, piecewise linear segments, as shown in Fig. 29.2. The central segment has a slope of unity, while the slope of the succeeding segments is in decreasing order as we go to the right or to the left. The last two segments, viz. those for $V_i > V_4'$ and $V_i < -V_4'$, have a slope of zero. If V_i is a symmetrical triangular wave with a peak value of $V_p = V_4'$, then the output shall be an approximation to a sine wave with a peak value of V_4' as shown in Fig. 29.3. If V_p exceeds V_4', it is obvious that the resulting sine wave shall have a clipped top and bottom (Fig. 29.3). On the other hand, if $V_p < V_4'$, then we get a reduced amplitude sine wave of poorer quality as compared to the case when $V_p = V_4'$.

The basic electronic circuit utilized to achieve the transfer characteristics (Fig. 29.2) is shown in Fig. 29.4, where $V_1 < V_2 < V_3 < V_4$, these being reference voltages derived from the power supply through an appropriate resistive voltage divider network. The circuit is functionally symmetrical about the centre line. The upper half of the circuit realizes the characteristic for $V_i > 0$, while the lower half takes care of the part $V_i < 0$. Because of symmetry, it suffices to consider only the part for $V_i > 0$.

To keep life simple, assume that all the diodes are ideal, i.e. they act as short circuits. We shall see later that in the actual chip, this is approximately ensured by a *pnp–npn* transistor combination. Suppose $0 < V_i < V_1$; then none of the diodes conduct and $V_o = V_i$. This is the situation for the central part of the characteristic in Fig. 29.2. When V_i is increased such that $V_1 \leq V_i < V_2'$, where V_2' is the input needed to make $V_o = V_2'$, diode D_1 conducts and the equivalent

Source: S. C. Dutta Roy, "Triangular to Sine Wave Converter," *Students' Journal of the IETE*, vol. 31, pp. 90–94, April 1990.

S. C. Dutta Roy, *Circuits, Systems and Signal Processing*,
https://doi.org/10.1007/978-981-10-6919-2_29

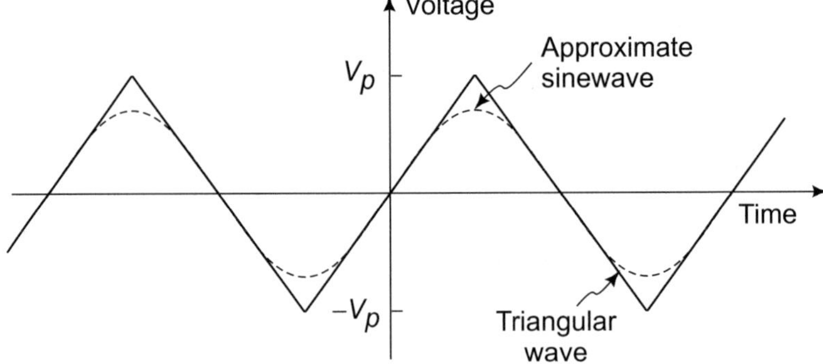

Fig. 29.1 A symmetrical triangular wave and the approximate sine wave that can be obtained by shaping it

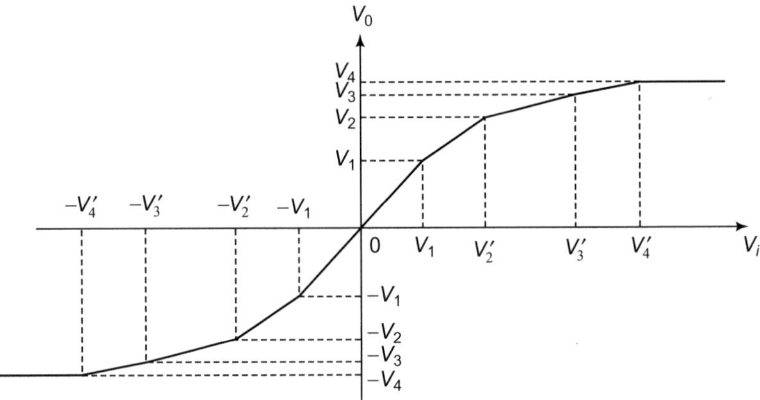

Fig. 29.2 Transfer characteristics of the triangular to sine-wave converter

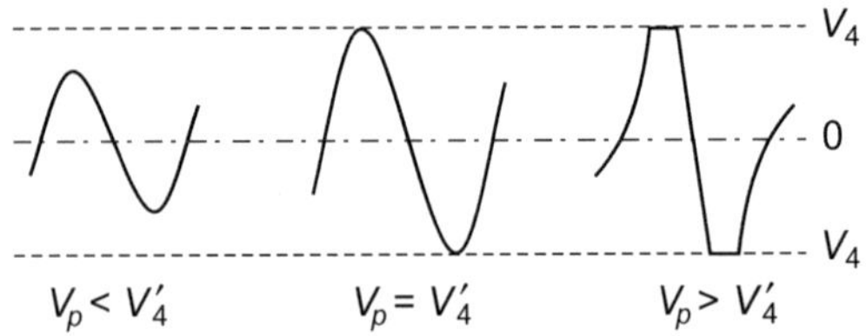

Fig. 29.3 Shape of the output waveform for various values of V_p in relation to V_4'

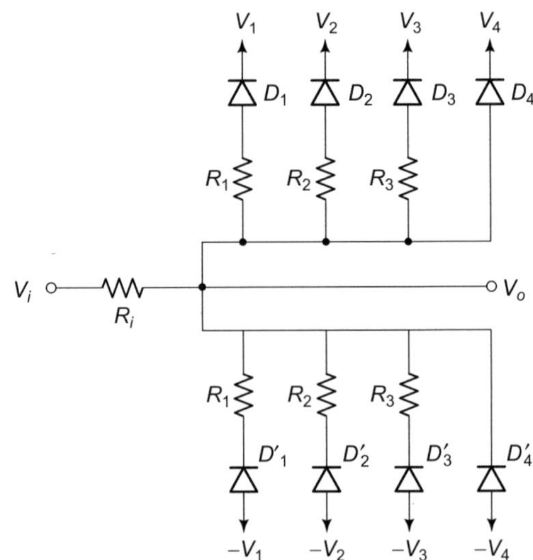

Fig. 29.4 The basic circuit of the triangular to sine-wave converter

circuit is shown in Fig. 29.5a. To find V_o, apply KCL at the node which gives

$$(V_i - V_o)G_i = (V_o - V_1)G_1, \tag{29.1}$$

where $G_x = 1/R_x$, $x = i$, 1, 2, 3. Solving for V_o gives

$$V_o = (V_1 G_1 + V_i G_i)/(G_1 + G_i) \tag{29.2}$$

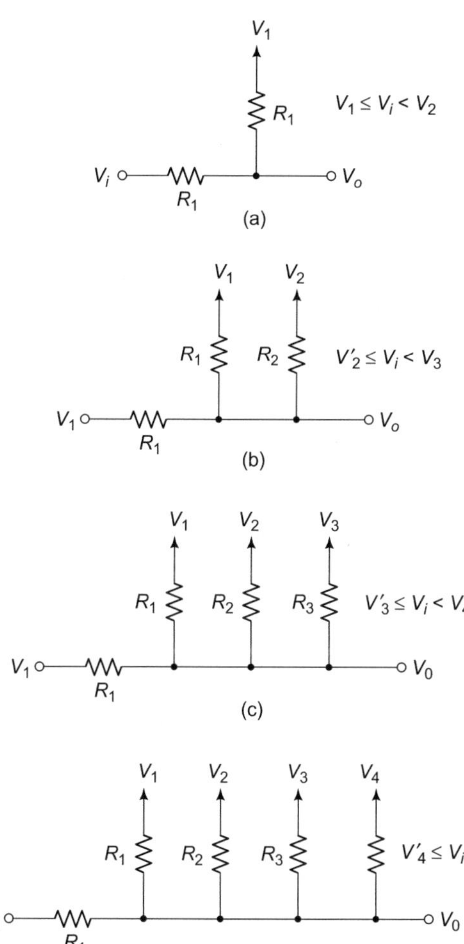

Fig. 29.5 Equivalent circuits for various input voltage ranges

This describes the second segment of the characteristic in Fig. 29.2. Whose slope is $G_i/(G_1+G_i) = R_1/(R_1+R_i) < 1$. To determine V_2', put $V_o = V_2$ and $V_i = V_2'$ in Eq. 29.2; this gives, on simplification.

$$V_2' = V_2(l + R_i/R_1) - V_1(R_i/R_1) \qquad (29.3)$$

Note that $V_2' > V_2$, which is of course expected.

When $V_2' \leq V_i < V_3'$, where V_3' is the input needed to make $V_o = V_3$, both of the diodes D_1 and D_2 conduct and the equivalent circuit is shown in Fig. 29.5b. Again applying KCL and simplifying, we get

$$V_o = (V_iG_i + V_1G_1 + V_2G_2)/(G_i + G_1 + G_2) \qquad (29.4)$$

This describes the third segment in Fig. 29.2, whose slope is $G_i/(G_i+G_1+G_2) = (R_1||R_2)/[(R_1||R_3)+R_i]$. To determine V_3', put $V_i = V_3'$ and $V_o = V_3$ in Eq. 29.4 and solve for V_3'. The result is

$$V_3' = V_3[1 + (R_i/R_1) + (R_i/R_2)] - V_1R_i/R_1 - V_2R_i/R_2 \qquad (29.5)$$

When V_i is further increased so that $V_3' \leq V_i < V_4'$, where V_4' is the input needed to make $V_o = V_4'$, diodes D_1, D_2 and D_3 conduct and the equivalent circuit is shown in Fig. 29.5c. From this, one can solve for V_o as

$$V_o = (V_iG_1 + V_1G_1 + V_2G_2 + V_3G_3)/(G_i + G_1 + G_2 + G_3) \qquad (29.6)$$

This characterizes the fourth segment of Fig. 29.2, which has a slope of

$$G_i/(G_i + G_1 + G_2 + G_3) = (R_1||R_2||R_3)/[(R_1||R_2||R_3) + R_i].$$

By putting $V_1 = V_4'$ and $V_o = V_4$ in Eq. 29.6, one can obtain V_4' as

$$V_4' = V_4[1 + (R_i/R_1) + (R_1/R_2) + (R_1/R_3)] - V_1R_i/R_1 - V_2R_i/R_2 - V_3R_1/R_3 \qquad (29.7)$$

Finally, when $V_1 \geq V_4'$, all four diodes D_1', D_2, D_3 and D_4 conduct, the equivalent circuit is shown in Fig. 29.5d and V_o settles at V_4. This corresponds to the last segment of the characteristic in Fig. 29.2.

For negative input voltages, a similar analysis can be performed with the part of the circuit below the centre line in Fig. 29.4 and it can be shown that the characteristic shown in Fig. 29.2 in the third quadrant is realized thereby.

In the Intersil 8038 chip implementation of the circuit shown in Fig. 29.4, the resistance values used are $R_i = 1$ K, $R_1 = 10$ K, $R_2 = 2.7$ K and $R_3 = 0.8$ K. The voltages V_1, V_2, V_3 and V_4 are derived from the +10 V, −10 V supplies through the resistive network shown in Fig. 29.6. It is readily calculated that $V_4 = 2.469$ V, $V_3 = 2.180$ V, $V_2 = 1.637$ V and $V_1 = 1.159$ V. The implementation of the diode is done in a clever way such that (i) the Thevenin impedance of each reference voltage is transformed to an insignificant value, and (ii) the voltage drop across the conducting diode is virtually reduced to zero. The actual circuit for the R_1, D_1, V_1 and R_1, D_1', $-V_1$ legs of Fig. 29.4 is shown in Fig. 29.7. The diode D_1 is realized by the complementary *pnp* (Q_2)–*npn* (Q_1) emitter follower pair. If Q_1 and Q_2 are matched, then their base-emitter drops will be equal and opposite. Thus, the voltage at the emitter of Q_2 will be $V_1 - V_{BE}, Q_1 - V_{BE}, Q_2 = V_1$. Also, because of the 33 K resistor in the emitter lead of Q_1, it will present an impedance of 33 K multiplied by its beta ($\beta_{Q1} \sim 100$) to the source V_1. This impedance will therefore be of the order 3300 K and should not affect the potential divider shown in Fig. 29.6 at all! On the other hand, the effective Thevenin impedance of the V_1 source, viz. $[(1.6 + 0.33 + 0.375 + 0.2 + 5.2)\|\ (0.33 + 0.375 + 0.2 + 5.2)]\ K \cong 3.4$ K will be transformed to 3.4 $K/_{Q1}$ at the emitter of Q_1, and to 3.4 $(K/_{Q1})/_{Q2}$ at the emitter of Q_2. Assuming $\beta_{Q1} = \beta_{Q2} = 100$, the resulting impedance reduces to 0.34 Ω only. Similarly, the impedances presented to D_2, D_3 and D_4 can be calculated as 0.34, 0.33 and 0.32 Ω, respectively. The scheme shown in Fig. 29.4 is therefore realized to a high degree of accuracy.

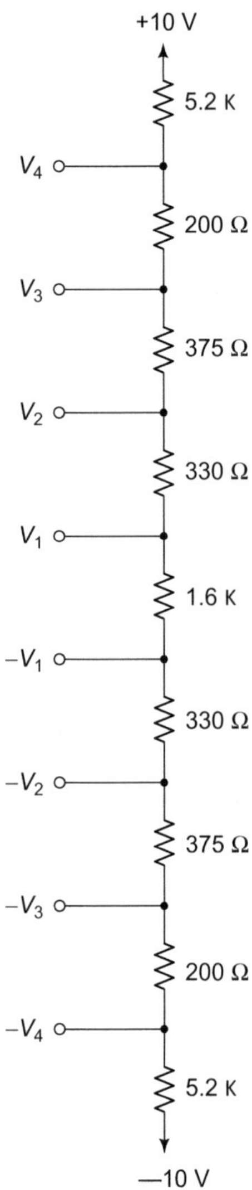

Fig. 29.6 The resistive voltage divider network for generating the reference voltages

Since all parameters are known, we can now calculate the input voltages at the various break points shown in Fig. 29.2 from Eqs. 29.3, 29.5 and 29.7. These are, respectively, $V_2' = 1.685$ V, $V_3' = 2.483$ V and $V_4' = 3.269$ V. The circuit can therefore be made optimum use of if the triangular wave peak voltage is 3.269 V; then a sine wave with a peak voltage of 2.469 V is obtained.

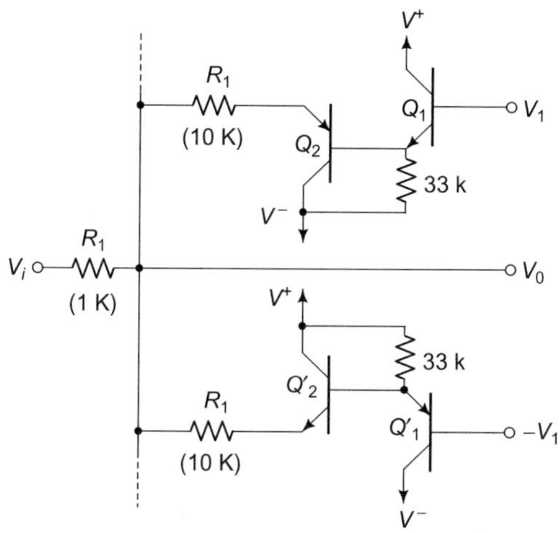

Fig. 29.7 Realization of diodes D_1 and D_1'

Why should one bother about generating a sine wave by conversion of a triangular wave? Instead, why should not one use an LC or RC sinusoidal oscillator? The reason is that it is very difficult to obtain a wide range variable frequency sinusoidal oscillator. In contrast, one can easily generate a 100: 1 frequency sweep with a voltage-controlled triangular wave oscillator. The resulting wave can then be shaped to a sine wave by a triangular to sine-wave converter as described in this chapter.

Problems

P.1. Convert a sine waveform to a triangular waveform.

P.2. Same, but a square waveform.

P.3. Same, but the desired waveform is shown in the figure below.

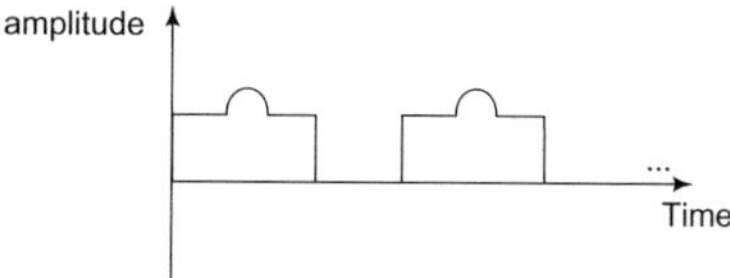

Fig P.3.

P.4. Suppose the diodes in Fig. 29.4 are all nonideal, but identical. What will happen?

P.5. Given a square wave going positive as well as negative. How would you generate a chain of positive impulses?

Bibliography

1. S. Soclof, *Applications of Analog Integrated Circuits.* (Prentice Hall, 1985)
2. A.B. Grebene, *Analog Integrated Circuit Design.* (Van Nostrand Reinhold, 1972)

Dynamic Output Resistance of the Wilson Current Mirror

30

A simple derivation is given for the dynamic output resistance of the Wilson current mirror, which forms a basic building block in many analog integrated circuits.

Keywords
Current mirror · Wilson circuit · Dynamic output resistance

Introduction

The Wilson current mirror, shown in Fig. 30.1b, is a basic building block in many analog integrated circuits. As compared to the simple current mirror shown in Fig. 30.1a, it has the advantage of achieving base current cancellation, so that $I_0 = I_1$, even if the base currents of the transistors (all assumed identical) are not negligible as compared to their respective collector currents. Further, its dynamic output resistance is greater than that of the simple current mirror by a factor of $\beta/2$. This has been mentioned by Wilson [1] but not proved. Grebene [2] follows Wilson, but refers to a hybrid-π equivalent circuit analysis made by Davidse [3]. Gray and Meyer [4] also do not prove this result. Soclof [5] attempted a simple proof but his result is higher by a factor of 2 due to a mistake in the assumed current distributions. In view of the importance of the Wilson current mirror and in view of the fact that Soclof's books [5, 6] are the most comprehensive texts available on the subject, we present here a simple and correct analysis leading to the result claimed by Wilson [1] and others.

Derivation

We adopt here the same approach as that of Soclof [5] and represent Q_3 by an ideal transistor Q_3' in parallel with its dynamic collector-to-emitter conductance $g_0 = 1/r_0$, as shown in Fig. 30.2. Let the output voltage change by a small amount ΔV_0 and let the consequent change in the output current be ΔI_0. If ΔI_0 can be determined in terms of ΔV_0 and transistor parameters, then the dynamic output conductance can be calculated as

$$g_0' = 1/r_0' = \Delta I_0/\Delta V_0.$$

Due to the incremental change ΔV_0, let the collector current of Q_2 change by ΔI; since Q_1 and Q_2 are matched and have the same base-to-emitter voltages, the collector current of Q_1 will also change by the same amount ΔI. Assuming that the current I_1 remains a constant, i.e. $\Delta I_1 = 0$, KCL dictates that the base current of Q_3 must change

Source: S. C. Dutta Roy, "Dynamic Output Resistance of the Wilson Current Mirror," *Students' Journal of the IETE*, vol. 31(4) 1990 and 32(1), pp. 165–168, 1991

S. C. Dutta Roy, *Circuits, Systems and Signal Processing*,
https://doi.org/10.1007/978-981-10-6919-2_30

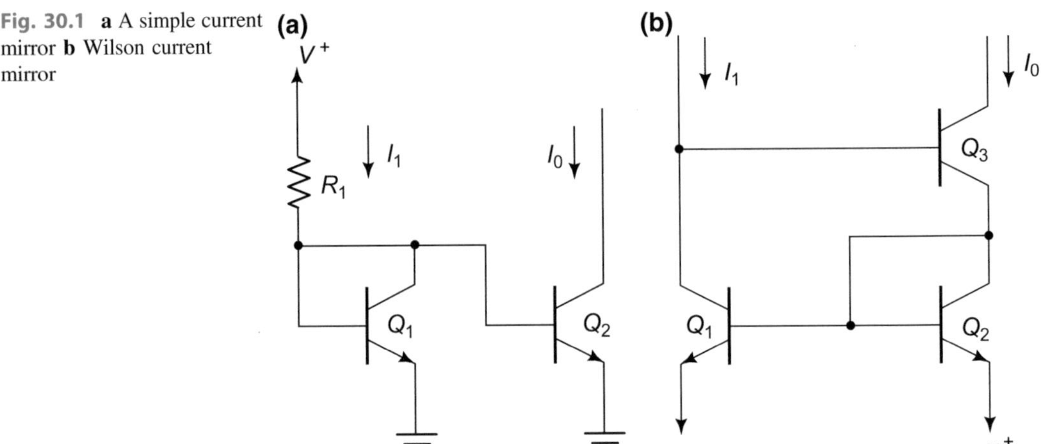

Fig. 30.1 **a** A simple current mirror **b** Wilson current mirror

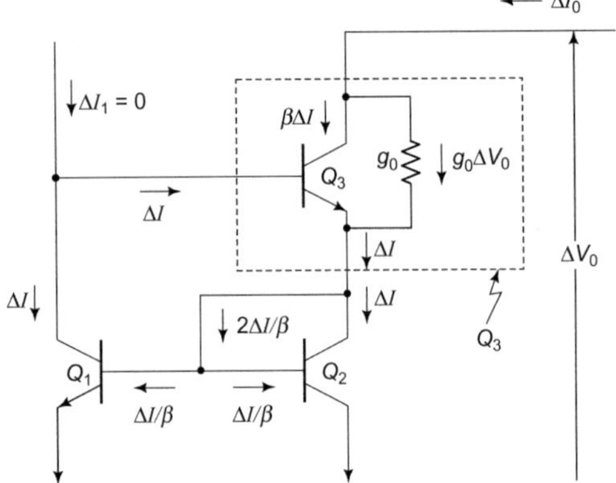

Fig. 30.2 Incremental equivalent circuit of the Wilson current mirror

by $-\Delta I$. This causes Q_3' collector current to change by $-\beta\ \Delta I$. The increment of current through g_0 will be $g_0\ \Delta V_0$, because the diode-connected transistor Q_2 offers a negligible dynamic resistance compared to that of Q_3 ($= r_0$). Now consider the dotted rectangular box in Fig. 30.2 (representing Q_3); it has two currents entering, viz. $-\Delta\ I$ and $\Delta\ I_0$, and one current leaving, viz, $\Delta I'$, so that by KCL,

$$\Delta I_0 = \Delta I = \Delta I' \tag{30.1}$$

Assuming all transistors to have the same β, we see that $\Delta I'$ supplies ΔI, and also the incremental base currents of Q_1 as well as Q_2, each of which is $\Delta I/\beta$. Thus

$$\Delta I' = \Delta I + 2\,\Delta I/\beta \tag{30.2}$$

Combining Eqs. 30.1 and 30.2, we get

$$\Delta I = \Delta I_0/[2(1 + 1/\beta)] \tag{30.3}$$

Now applying KCL at the collector of Q_3', we have

$$\Delta I_0 = g_0\Delta V_0 - \beta\Delta I \tag{30.4}$$

Combining Eqs. 30.3 and 30.4 and simplifying, we get

$$\Delta I_0/\Delta V_0 = g_0' = g_0/\{1 + \beta/[2(1 + 1/\beta)]\} \tag{30.5}$$

Assuming $\beta \gg 2$, as is the case in practice, this reduces to

$$g_0' = 2g_0/\beta \tag{30.6}$$

so that the equivalent dynamic output resistance becomes

$$r_0' = \beta r_0/2 \tag{30.7}$$

This agrees with the claim of Wilson [1] and others.

It may be mentioned here that Soclof's equivalent circuit assumed $\Delta I = \Delta I_0$, which, as is obvious, violates KCL and makes $\Delta I' = 0$!

Problems

P.1. Search out the literature for other current mirrors. Make a list of them and enumerate their merits and demerits as compared to the Wilson current mirror. Hint: Consult the references given above, and if you cannot find them, ask me.

P.2. Justify, with derivations, what happens when Q_1 and Q_2 emitters are connected to current generators.

P.3. Why is $\Delta I_1 \neq 0$ in Fig. 30.2. What happens when $\Delta I_1 \neq 0$?

P.4. Why is g_0 connected from the collector to emitter in Fig. 30.2?

P.5. What happens when g_0 end is shifted from emitter to base?

References

1. G.R. Wilson, A monolithic junction FET-NPN operational amplifier. IEEE J. Solid State Circ.SC-**3**, 341–348 (December 1968)
2. A.B. Grebene, *Bipolar and MOS Analog Integrated Circuit Design*. (John Wiley, 1984)
3. J. Davidse, *Integration of Analogue Electronic Circuits*. (Academic Press, 1979)
4. P.R. Gray, R.G. Meyer, *Analysis and Design of Analog Integrated Circuits*. (John Wiley, 1984)
5. S. Soclof, *Analog Integrated Circuits*. Prentice (Hall, 1985)
6. S. Soclof, *Applications of Analog Integrated Circuit*. (Prentice Hall, 1985)

Part IV
Digital Signal Processing

The field of Digital Signal Processing (DSP) has fascinated me for over three decades, since I first met it in the early 1970s. This is reflected in the eight chapters of Part IV. The first two articles, on which Chaps. 31 and 32 are based, were written while I was teaching at the University of Leeds, during 1972–1973. DSP was at its infant state at that time, and students had difficulty in understanding the basic concepts. That is why, I innovated the title as 'The ABCD's of Digital Signal Processing'. In these articles, I described DSP from common-sense arguments. These became popular with the students I taught, almost instantly. I hope beginning students of DSP will like them, even now. The article on second-order digital filters, described in Chap. 33, was again inspired by inquisitive questions from students in the class, and I took pains to give simple derivations of band-pass and band-stop filters, touching on the limits of selectivity attainable by them. Chapters 34 through 37 were also inspired by student's queries in the class. Chapter 34 topic was in fact solved by four M.Tech. students of IIT Delhi, with clues from me, and it was satisfying to note that they could come up with simple derivations of this important element of DSP, viz. all pass filters. I have always encouraged my teacher colleagues at IIT Delhi and elsewhere to involve students in their research by throwing challenges on unsolved problems to students in the class. This article was a result of such a challenge.

The FIR lattice structure is usually first described and then analysed to find the performance parameters. I took upon myself the task of viewing this as a synthesis problem and succeeded. This forms the content of Chap. 35. A special problem arose during the course of this development, and I solved it in the article on which Chap. 36 is based. FIR lattice, as described in the textbooks, uses twice the minimum number of multipliers, as required in a canonic realization. Long back, Johnson had answered the question with affirmative, but surprisingly, his paper, although published in IEEE Transactions on Audio and Electroacoustics, went completely ignored by later workers, even the famous ones. I took upon myself the job of giving due credit to this unsung hero in the article on which Chap. 36 is based.

In FFT signal flow graphs, there are some redundant multipliers. Also, there are identical multipliers which can be combined. Taking all these into account, the minimum possible number of multipliers is found out in Chap. 38.

That completes Part IV and the main contents of the book.

The ABCDs of Digital Signal Processing—PART 1

31

In this chapter, the basic concepts of digital signal processing will be introduced, leading to a mathematical description of a digital signal processor in terms of, first, a difference equation and, second, a z-domain transfer function. In the process, the effects of sampling and quantization will be briefly touched upon. Implementation of a processor by special purpose hardware and discrete Fourier transform technique will be discussed. The fast Fourier transform (FFT) will be introduced and several of its applications will be presented, along with the pitfalls and incorrect usage of the technique.

Keyword
Fundamentals of DSP · Difference equation
Z-transform · Sampling · Quantization
DFT · FFT and its pitfalls

Source: S. C. Dutta Roy, "The ABCD's of Digital Signal Processing (Part 1)," *Students' Journal of the IETE*, vol. 21, pp. 3–12, January 1980.

Introduction

The first textbook on digital signal processing, by Rader and Gold (1969), came out in 1969, and included the following sentence in its preface: 'The field of digital signal processing is too new to allow us to predict subsequent developments'. Today, more than four decades later, one cannot certainly claim the field to be new, particularly in view of the phenomenal progress made in the techniques of digital signal processing, leading to dramatic improvements in system efficiency, and its many applications in very diverse fields like biomedical engineering, geophysics, acoustics, radar and sonar, radio astronomy, etc. These advances, in turn, have been stimulated by fantastic advances in integrated circuit technology and computer hardware, in terms of volume, cost and speed.

One way of gauging the progress of a field is to look at the available literature related to it. By as early as the 1970s, there were more than ten textbooks [1–10], four collections of significant papers [11–14], a large number of special issues of professional research journals [15] and numerous journal conference articles reporting on new techniques, or improvements on known ones, or novel applications of digital signal processing [16].

On a subject as vast as the literature explosion suggests, it is not easy to decide as to what

S. C. Dutta Roy, *Circuits, Systems and Signal Processing*,
https://doi.org/10.1007/978-981-10-6919-2_31

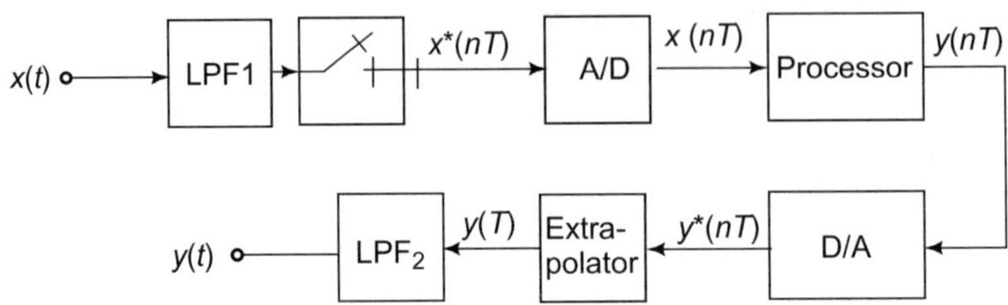

Fig. 31.1 Block diagram of a digital signal processor whose input and output are continuous signals

should be included in the 'ABCD's' because as the subject advances, so do the 'ABCD's'. The choice and organization of topics in this chapter have been greatly influenced by discussions with some members of potential readers and, of course, by personal preferences. Starting with the basic concepts involved in sampling and quantization, a mathematical description will be given of the digital signal processor in terms of a difference equation and then in terms of a z-domain transfer function. Implementation of the processor by special purpose hardware as well as by the discrete Fourier transform (DFT) technique will be discussed. In the latter, two basic forms of the fast Fourier transform (FFT) algorithm will be introduced. The presentation will conclude with some applications of the FFT, along with its pitfalls and potential incorrect usage.

Throughout the presentation, efforts will be made to keep the mathematics as simple as possible, and rigorous proofs and derivations will be avoided as much as practicable. System design aspects, starting from a given specification, will not be dealt with at all. This topic, along with others, will form the subject matter of a future chapter.

The Basic Digital Signal Processor

While digital signals and systems can be designed without reference to continuous systems, it is intuitively appealing, and often easier to understand, to build the theory of digital signal processing starting from continuous signals and systems which most engineers are more familiar with. Let us then consider a continuous signal $x(t)$, which is to be processed so as to facilitate the extraction of a desired information which, we will assume, is again a continuous signal. In other words, we wish to transform the given signal $x(t)$ into another signal $y(t)$ which is, in some sense, more desirable than the original. For example, $x(t)$ may, typically, be the desired signal contaminated by an undesired interference and our aim in processing may be to get rid of the latter. As another example, we may wish to enhance or estimate some component or parameter of a signal. Whatever our aim may be for the processing, if the processor is to be digital, $x(t)$ must be converted first to a discrete form $x^*(nT)$. A conceptually simple way of doing this is to have a switch in series which closes for a very short time τ after every T seconds, where $\tau \ll T$, as shown in Fig. 31.1. Ignore the box marked LPF, for the time being; also do not worry as to what should the value of T be. The resulting signal $x^*(nT)$ would appear as a series of narrow pulses at times $t = 0, T, 2T, \ldots$, the height of each pulse being equal to the value of the continuous signal at that instant of time. This sampled signal $x^*(nT)$ is then suitably coded or quantized in the analog-to-digital converter (A/D). The output of the A/D converter, $x(nT)$, is a series of coded numbers, coming out every T seconds (ignoring the small but non-zero A/D conversion time, as compared to T). The signal $x(nT)$ is now in a form which can be processed by digital hardware, indicated by the box labelled PROCESSOR in Fig. 31.1. The processor thus accepts a time series, i.e. series of numbers appearing at equal intervals of T seconds at its input, and performs some operations on them to produce another time series $y(nT)$. These operations, in the most general linear digital signal processor, can be described by the following linear[1] difference equation:

$$y(nT) = \sum_{i=0}^{M} a_i x(nT - iT) + \sum_{j=1}^{N} b_j y(nT - jT), \quad (31.1)$$

where a_i's and b_j's are suitable constants. $x(nT)$ and $y(nT)$ are, respectively, the input and output of the processor at the instant nT, while $x(nT - iT)$ and $y(nT - jT)$ represent, respectively, the input at some past instant $nT - iT$ and the output at another past instant $nT - jT$. At the instant nT, therefore, the computation of the output $y(nT)$ requires the past M inputs and N outputs; the processor therefore should have a memory in which these past input and output numbers can be stored. The constants a_i and b_j are also to be stored, of course. What the processor does after receiving the present input $x(nT)$, in fact, is then to recall the constants, past inputs and past outputs and perform the computation specified in Eq. 31.1, to deliver the output $y(nT)$. The computation does require some small but non-zero time, but this must be smaller than T in order that the processor may be ready to receive the next input with a clean *state*. We shall, in the following discussions, ignore this computation time and continue to call the output due to $x(nT)$ as $y(nT)$.

Before we proceed further, it is wise to recall the following sources of nonidealness in our signal processing system in Fig. 31.1. (i) τ, the duration for which the sampling switch remains closed, (ii) A/D conversion time, and (iii) processor computation time. The first two ideally should be zero, and the third should be at least less than T, if not significantly so.

By choosing the a_i's, b_j's, M and N in Eq. 31.1, one can achieve a variety of processing. If the emphasis is on shaping the spectrum of $x(t)$ in a desired fashion, we call the processing as filtering and the digital signal processor is then more commonly known as a digital filter.

The output of the processor in Fig. 31.1, $y(nT)$, is another sequence of numbers, which can be fed to a decoder, the digital-to-analog converter, to produce pulses of short duration, whose amplitudes are in proportion with the value of $y(nT)$. Continuous output can be obtained by passing this pulse sequence through an extrapolator or data reconstructor, which can be a simple zero-order hold, described by

$$y_1(nT + t) = y_1(nT), 0 \le t \le T, \quad (31.2)$$

i.e. the value between two sampling instants is held at the value of the immediately preceding sample.[2] The output of the extrapolator, $y_1(t)$, will contain some high-frequency ripples, which may be removed, if desired, by passing the signal through a low-pass filter, LPF_2. This filter is, however, a simple, inexpensive one and typically, a single *R–C* section serves the purpose.

Before we conclude this section, a discussion of the block LPF_1, is in order. The choice of T or the sampling frequency $f_s = 1/T$ should be in conformity with the sampling theorem, i.e. f_s should be greater than $2f_h$, where f_h is the highest frequency content of the signal $x(t)$ to be processed. If this is not the case, distortion occurs and $x^*(nT)$ will not be a true discrete representation of $x(t)$; another way of saying this is that it would not be possible to recover $x(t)$ from $x^*(nT)$. The sampling frequency should therefore be sufficiently high. However, for other reasons (e.g. coefficient quantization error, to be discussed later), the sampling frequency should not be too high. For a signal contaminated with high-frequency noise (impulse noise), the minimum required sampling frequency may be inordinately high. In such cases, it would be advantageous to filter the continuous signal $x(t)$, before sampling, by passing it through an inexpensive analog low-pass filter, LPF_1 as shown in Fig. 31.1. LPF_1 may also typically be a single *R–C* section

[1]We shall confine our attention to linear digital signal processing only.

[2]It can be shown that the zero-order hold has a lowpass frequency response with linear phase characteristic. Higher-order holds are not generally used because their implementation is difficult, their phase response is not linear and they introduce more delay to the signal.

The Sampling Process

As already mentioned, a sampler may be viewed as a switch which closes for a short duration τ, after every T seconds, where $\tau \ll T$, as shown in Fig. 31.2a. Implicit, of course, is also the assumption that the sampler can open and close instantaneously. A typical input $x(t)$ and output $x^*(nT)$ of the sampler are shown in Fig. 31.2b, c, respectively. A considerable simplicity in understanding and analysis is achieved if τ is assumed to be zero, i.e. if the sampler is assumed to be an ideal impulse sampler whose output is a series of impulses appearing at the sampling instants, as shown in Fig. 31.2d, the strength of the impulse at $t = nT$ being equal to $x(t)|_t = nT$. Then one can write

$$x*(nT) = x(t)\sum_{n=-\infty}^{\infty}\delta(t-nT)\underline{\underline{\Delta}}\,x(t)S(t),\text{ say} \tag{31.3}$$

What Eq. 31.3 says in essence is that the sampled sequence $x^*(nT)$ is obtained by multiplying the continuous signal $x(t)$ by an impulse train occurring at $2T$, $-T$, 0, T, $2T$,...; we have named this impulse train as another function $S(t)$. Obviously, $S(t)$ is periodic with period T and can be expanded in Fourier series. When this is done, one finds

$$S(t) = \frac{1}{T}\sum_{k=-\infty}^{\infty} e^{jk\omega st}, \tag{31.4}$$

where $\omega_s = 2\pi/T = 2\pi f_s$. Thus,

$$x*(nT) = \frac{1}{T}\sum_{k=-\infty}^{\infty} x(t)e^{jk\omega_s t} \tag{31.5}$$

If one now takes the Fourier transform of both sides and uses the notations $X^*(\omega)$ and $X(\omega)$ to represent the spectra of $x(t)$ and $x^*(nT)$ respectively, then one obtains.

$$X^*(\omega) = \frac{1}{T}\sum_{k=-\infty} X(\omega + k\omega_s) \tag{31.6}$$

Thus the spectrum of the sampled signal is obtained by superimposing, on the original signal spectrum, the same spectrum shifted on the frequency scale by k_s, where k takes all integer values, positive and negative, except zero. This is pictorially shown in Fig. 31.3 for a hypothetical $X^*(\omega)$ which is bandlimited such that $X(\omega) = 0$ for $|\omega| > \omega_s/2$. Note that $X^*(\omega)$ is the same as $X(\omega)$ in the band $|\omega| > \omega_s/2$, called the *base band*, and it is possible to recover $X(\omega)$ from $X^*(\omega)$ by using a low-pass filter whose transmission characteristic is shown by dotted lines. An example of a situation where $X(\omega)$ is not bandlimited and has considerable amplitude

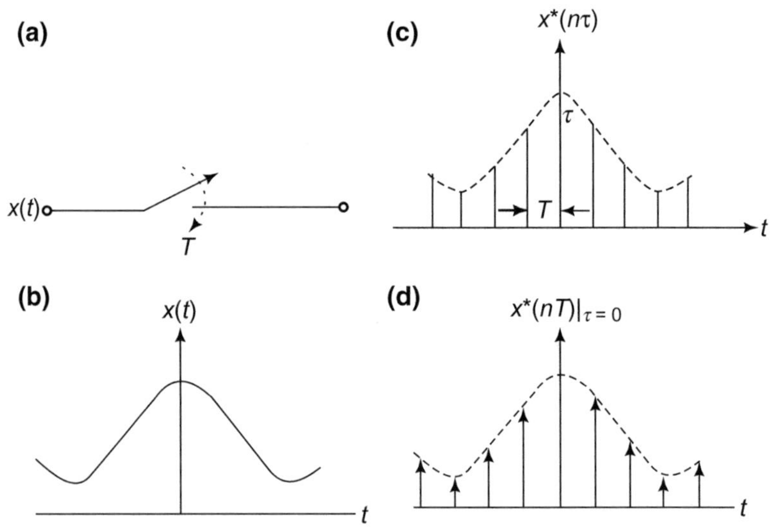

Fig. 31.2 The sampling process **a** sampler, **b** a continuous signal, **c** its sampled version and **d** idealized sampled version

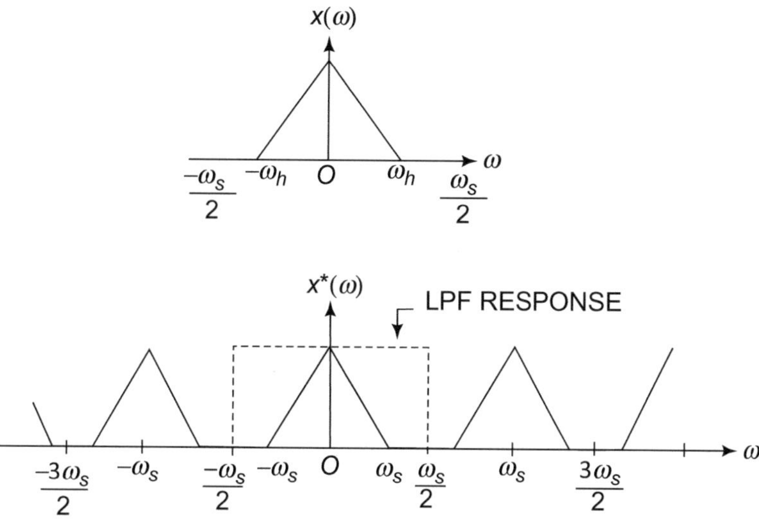

Fig. 31.3 Showing the spectrum of the sampled signal in relation to that of the continuous signal

beyond $\omega_s/2$, called the Nyquist frequency, is shown in Fig. 31.4. An alternative equivalent description of this situation is that the Nyquist frequency is lower than the highest frequency ω_h contained in $X(\omega)$. In this case obviously, $X(\omega)$ does not keep its identity in $X^*(\omega)$ and no filter can recover $X(\omega)$ from $X^*(\omega)$. Thus, to keep the information content of the signal intact in the sampled version, we must choose the sampling frequency ω_s to be at least twice the highest frequency, ω_h, contained in the signal. This, in essence, is the well-known sampling theorem.

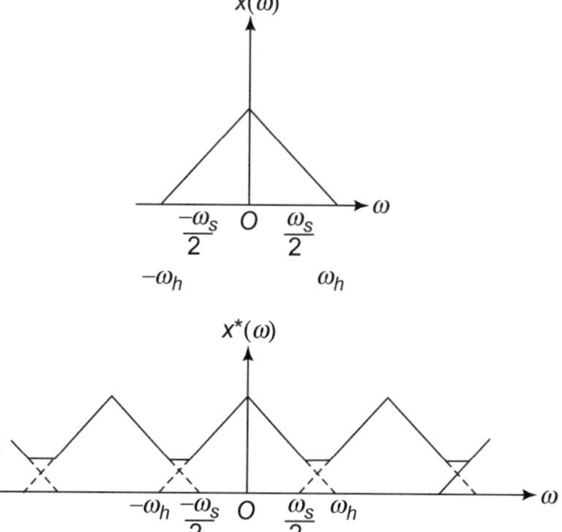

Fig. 31.4 Aliasing error

Practical signal are not, however, bandlimited and the signal spectrum recovered by passing $X^*(\omega)$ through an ideal low-pass filter will be different from $X(\omega)$ due to spillover from the adjacent bands. The error so introduced is called *aliasing* or *folding error*. Further, an ideal filter, with the brickwall characteristic, is not realizable in practice and this introduces an additional error. To keep these two errors within the tolerable limits, the sampling frequency is often required to be sufficiently high.

Another source of error is the fact that τ is not zero in practice; the nature and extent of its contribution to distortion has been discussed in Shapiro (1978) and will not be considered here.

Quantization Errors[3]

In this section, we would like to point out an inherent limitation on the accuracy of digital signal processors. This limitation arises due to

[3]This example is reprinted by kind permission of John Wiley & Sons Inc.

the fact that all digital systems operate with a finite number of bits, or a finite word length. Rather than going into a detailed theory, we prefer to illustrate the various errors, resulting from finite word length, through a simple example (Peled and Liu [7]). Suppose our processor is required to implement the following difference equation.

$$y(n) = x(n) - x(n-2) + 1.2727922y(n-1) - 0.81y(n-2),$$

$$y(n) = x(n) - x(n-2) + 1.2727922y(n-1) - 0.81y(n-2), \quad (31.7)$$

where for brevity, $x(nT)$ and $y(nT)$ have been represented as $x(n)$ and $y(n)$, respectively. A possible basic arrangement is shown in Fig. 31.5.

It consists of a memory for storing the coefficients; a set of data registers for storing the input and output samples; an arithmetic unit to perform the computation according to Eq. 31.7 and a control unit (not shown) for providing the timing signals.

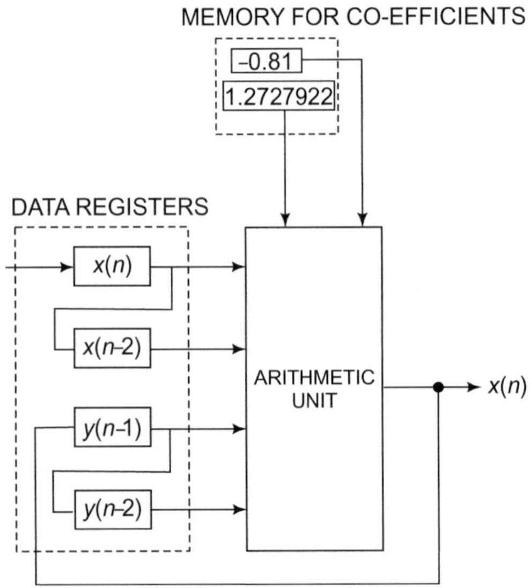

Fig. 31.5 Implementation of Eq. 31.7. Control unit is not shown

To begin with, the numbers we are dealing with must be represented in binary notation in order to be stored, manipulated and operated upon by digital hardware. Consider the coefficient 0.81, it can be written as

$$0.81 = \left(\tfrac{1}{2}\right)^1 + \left(\tfrac{1}{2}\right)^2 + \left(\tfrac{1}{2}\right)^5 + \left(\tfrac{1}{2}\right)^6 + \cdots$$

i.e. in base 2, 0.81 can be represented as 0.11001110101 … An infinite number of bits are needed to represent this coefficient exactly. Since all practical memory circuits have a finite number of bits for each word, the infinite binary string must be modified. If one uses a memory with a 6-bit word length, a simple way to store our number will be to keep only the 6 most significant bits, that is, 0.11001 as the approximate value for 0.81. However, 0.11001 in base 2 represents the number 0.78125, thus introducing an error of 0.02875 in this coefficient. Similarly, 1.2727922 has a 6-bit base 2 representation of 1.01000 or 1.25, resulting in an error of 0.0227922. Obviously, ± 1 or 0 can be represented exactly; finally, therefore, the equation that the processor actually implements is

$$y(n) = x(n) - x(n-2) + 1.25y(n-1) - 0.78125y(n-2) \quad (31.8)$$

The resulting error is called the *coefficient quantization error.*

Another source of error is the quantization of the input data in the A/D converter. Suppose $x(t)$ in Fig. 31.1 is sinusoidal and consider the following input segment… 0.2955, 0.5564, 0.8912, 0.9320… Suppose the A/D converter yields 8 bits and let the data registers in Fig. 31.5 be of the same capacity. Truncated to 8 bits, the above input data segment becomes, in binary form,

$$0.0100101, 0.1000111, 0.111010, 0.111011$$

corresponding to the values… 0.2890625, 0.5546875, 0.890625, 0.9265,… which, obviously, differ from the actual samples of the sinusoidal signal. The resulting error is called the *input quantization error.*

The third source of error is due to the limited accuracy with which arithmetic operation can be performed. In computing the term—0.78125*y* (*n* − 2) in Eq. 31.8, for example, the product of a 6-bit number (−0.78125) and an 8-bit number [*y* (*n* − 2)] will give 14 significant bits. This must be shortened to 8 bits so that the result will fit in the 8-bit data register. The error thus committed is known as the *round-off error.* Further and more importantly, the previously computed output samples are used via Eq. 31.8 to compute later output samples, and this has a cumulative effect.

What are the overall effects of these errors? Unless carefully analysed and accounted for, the results can be very disappointing. For example, coefficient quantization error may convert a stable processor into an unstable one. The arithmetic round-off errors can result in low-level limit cycles and overflow oscillations.

Z-Transform

Recall that the input–output relation of a digital signal processor is expressed by a linear difference equation of the form of Eq. 31.1. It is well known that the solution of such an equation is greatly simplified by using the *z*-transform (just as the solution of linear differential equations, which can be used to characterize linear continuous systems, is greatly simplified by using Laplace transforms). Further, a better understanding of the digital signal processor, particularly its frequency domain or spectral behaviour, is obtained from its *z*-transformed description.

Consider the sampled signal $x^*(nT)$ described in Eq. 31.3, reproduced below in a slightly different, but equivalent form

$$X^*(nT) = \sum_{n=-\infty}^{\infty} x(t)\delta(t - nT) \qquad (31.9)$$

Also, to keep things simple, let us consider a causal signal, i.e. let $x(t) = 0,\ t < 0$. Further, since a delta function exists only when its argument is zero, we could rewrite Eq. 31.9 in the form

$$X^*(nT) = \sum_{n=0}^{\infty} x(nT)\delta(t - nT) \qquad (31.10)$$

If we take Laplace transform (LT) on both sides and call the LT of $x^*(nT)$ as $x^*(s)$, we get

$$X^*(s) = \sum_{n=0}^{\infty} x(nT)\mathrm{e}^{-snT} \qquad (31.11)$$

To get rid of the transcendental function e^{st}, let us replace it by z, i.e. let

$$z = e^{sT} \qquad (31.12)$$

Further, let $X * (s) \triangleq X(z)$; then

$$X(z) = X^*(s) \underline{\underline{\Delta}} \sum_{n=0}^{\infty} x(nT)z^{-n} \qquad (31.13)$$

The variable z need not necessarily be thought of as e^{sT}; it could be interpreted as an ordinary variable whose exponent (ignoring the negative sign) represents the position of the particular pulse in the sequence $\{x(nT)\}$. When viewed in the latter light, $X(z)$ is a 'generating function' and may be treated without identification with a Laplace transform.

The infinite summation Eq. 31.13 defines the *z*-transform of the sequence $\{x(nT)\}$ or more concisely $\{x(n)\}$. (Note the use of $\{.\}$ to represent a sequence and the dropping of T for brevity). Thus, formally,

$$Z\{x(n)\} = X(z) \underline{\underline{\Delta}} \sum_{n=0}^{\infty} x(n)z^{-n} \qquad (31.14)$$

We shall not go into the details of existence, convergence and other mathematical properties of the z-transform here, but it is better to remember that the series Eq. 31.14 converges outside a circle in the z-plane whose radius equals the n-th root of maximum $x(n)$ in $\{x(n)\}$.

Given $X(z)$, one can recover $\{x(n)\}$, in general, by applying the inversion integral and Cauchy's residue theorem; however, for rational $X(z)$, as is usually the case, a long division is adequate. As an example, if

$$X(z) = \frac{z}{z-k} = \frac{1}{1-kz^{-1}} \tag{31.15}$$

then

$$X(z) = 3 + kz^{-l} + k^2z^{-2} + \cdots \tag{31.16}$$

and obviously $x(n) = k^n$.

Three important properties of z-transforms will now be stated without proof. First, *the z-transform is a linear operation,* i.e. if $Z\{x_i(n)\} = X_i(z)$, then

$$Z\left[\sum_{i=1}^{P}\{a_i x_i(n)\}\right] = \sum_{i=1}^{P} a_i x_i(z) \tag{31.17}$$

The second property concerns the z-transform of a shifted sequence, viz.

$$Z\{x(n-m)\} = z^{-m}X(z) \tag{31.18}$$

The third concerns the z-transform of a convolution of two sequences. Before we state this, however, let us understand what we mean by the convolution of two sequences. In the continuous domain, convolution $x(t)$ of two functions $x_1(t)$ and $x_2(t)$ is defined by

$$\begin{aligned} x(t) &= x_1(t) * x_2(t) \\ &= \int_0^t x_1(t)x_2(t-T)dT \\ &= \int_0^t x_1(t-T)x_2(T)dT \end{aligned} \tag{31.19}$$

In the discrete case, by analogy with Eq. (19), we define convolution of two sequences $\{x_1(n)\}$ and $\{x_2(n)\}$ as another sequence $\{x(n)\}$ such that

$$\begin{aligned} x(n) &= \sum_{r=0}^{n} x_1(r)x_2(n-r) \\ &= \sum_{r=0}^{n} x_1(n-r)x_2(r) \end{aligned} \tag{31.20}$$

The discrete convolution can be given a graphical interpretation, analogous to continuous convolution, but we would not discuss this here. Instead, we now state the third important property of z-transforms, viz. that if Eq. 31.20 is true, then so is,

$$X(z) = X_1(z)X_2(z) \tag{31.21}$$

This result is exactly analogous of the Laplace transform of the convolution of two continuous functions. The proofs of Eqs. 31.17, 31.18 and 31.21 follow easily from the definition of z-transform given in Eq. 31.14.

Transfer Function of a Digital Signal Processor

Consider the digital signal processor described in Eq. 31.1 once again and let $Z[\{x(n)\}; \{y(n)\}] = X(z); Y(z)$. If one takes the z-transform of both sides and uses the shifting property of z-transforms mentioned in the preceding section, it is easy to see that

$$Y(z) = \sum_{i=0}^{M} a_i X(z)z^{-i} + \sum_{j=0}^{N} b_j Y(z)z^{-j} \tag{31.22}$$

This can be put in the form

$$H(z) = \frac{Y(z)}{X(z)} = \frac{\sum_{i=0}^{M} a_i z^{-1}}{1 - \sum_{j=0}^{N} b_j z^{-1}} \tag{31.23}$$

The quantity $H(z)$, defined 'as the ratio of z-transform of output sequence to the z-transform of the input sequence', obviously is a characteristic of the processor only and is an adequate representation of it. It is, by analogy with continuous systems, called the z-domain transfer function or simply the transfer of the digital signal processor. Note that

$$Y(z) = H(z) \text{ when } X(z) = 1 \qquad (31.24)$$

What does $X(z) = 1$ signify? We can write

$$X(z) = 1 + 0.z^{-1} + 0.z^{-2} \qquad (31.25)$$

If one compares Eq. 31.25 with Eq. 31.14, it is obvious that $X(z) = 1$ corresponds to an input sequence.

$$x(n) = 1 \quad \text{for } n = 0 = 0 \quad \text{otherwise} \qquad (31.26)$$

i.e. $x(n) = \{1, 0, 0, \ldots\}$. This is called the unit pulse and we see that it plays the same role as the impulse function $\delta(t)$ in a continuous system. The inverse z-transform of $H(z)$, denoted by $\{h(n)\}$, is obviously the output sequence of the processor when the input is a unit pulse. $\{h(n)\}$ is called the impulse response of the digital signal processor.

From Eq. 31.23 and the z-transform property for convolution, it should also be apparent that for a general input $\{x(n)\}$, the output sequence $\{y(n)\}$ should be given by the convolution of the input sequence with the impulse response, i.e.

$$y(n) = \sum_{r=0}^{N} x(r)h(n-r) \qquad (31.27)$$

$$= \sum_{r=0}^{n} x(n-r)h(r)$$

Suppose we have a digital signal processor, characterized by an impulse response $\{h(n)\}$, where $h(n) = 0,\ n > M$. Such a processor is said to be of the *Finite Impulse Response* (*FIR*) type. For this, Eq. 31.27 becomes

$$y(n) = \sum_{r=0}^{M} h(r)x(n-r) \qquad (31.28)$$

Note that this is of the same form as Eq. 13.1 with b_j's equal to zero, and $a_r = h(r)$. Under this condition, the transfer function $H(z)$ given in Eq. 31.23 becomes a polynomial in z^{-1}. A processor which is not FIR is of the *Infinite Impulse Response* (*IIR*) type. For this, at least one b_j in Eq. 31.1 is non-zero and $H(z)$ in Eq. 31.23 is a rational function in z^{-1}.

There are two other terms which are very commonly used in digital signal processing terminology; these are *non-recursive* and *recursive.* A very common mistake that has been perpetuated in the literature is to identify FIR with non-recursive and IIR with recursive. As pointed out by Gold and Jordan [17], the terms recursive and non-recursive should be used only to describe the method of realization. A realization in which no past values of the output have to be called back to compute the present output is called non-recursive; if one or more past values of the output are required for computing the present value of the output, the realization is called recursive. Obviously, FIR processors are realized most conveniently in non-recursive form, while recursive form is to be preferred for IIR processors. But, as shown by Gold and Jordan (*loc. cit.*), FIR processors can be realized recursively, and IIR processors can be realized non-recursively.

The transfer function Eq. 31.23 can be expressed as

$$H(z) = A\frac{(z^{-1}-\alpha_1)(z^{-1}-\alpha_2)\ldots(z^{-1}-\alpha_M)}{(z^{-1}-\beta_1)(z^{-1}-\beta_2)\ldots(z^{-1}-\beta_N)}, \qquad (31.29)$$

where A is a real constant, and $\alpha's$ and $\beta's$ are either real or complex if they are complex they occur in conjugate pairs, $\alpha's$ are called *zeros* and $\beta's$ are called *poles* of the digital signal processor in the z^{-1} plane. A digital signal processor is

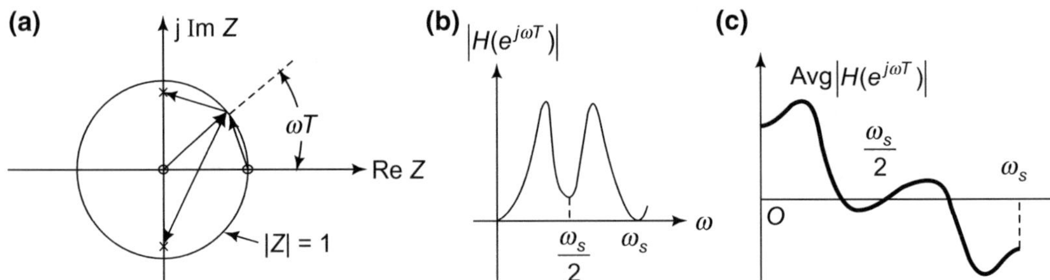

Fig. 31.6 Pole-zero sketch of Eqs. 31.30 or 31.31, **b** Magnitude response, **c** Phase response (To be continued)

stable if the poles in the z^{-1} plane *all* lie outside the unit circle, which is equivalent to having all poles inside the unit circle with z-plane; this comes from the correspondence of the $j\omega$-axis in the s-plane to the unit circle in the z^{-1} plane, see Eq. 31.12. The frequency response of the digital signal processor is obtained by putting $z = e^{j\omega T}$ in $H(z)$. As an example, let a digital signal processor be described by the difference equation

$$y(n) = x(n) - x(n-l) - 0.8y(n-2) \quad (31.30)$$

The transfer function of the system is

$$\begin{aligned} H(z) &= \frac{1 - z^{-1}}{1 + 0.81z^{-2}} \\ &= \frac{1 - z^{-1}}{(1 + j0.9z^{-1})(1 - j0.9z^{-1})} \\ &= \frac{z(z-1)}{(z + j0.9)(z - j0.9)}, \end{aligned} \quad (31.31)$$

where the last form has been used to facilitate a pole-zero sketch in the z-plane, as shown in Fig. 31.6a. Putting $z = e^{j\omega t}$ in Eq. 31.31 and evaluating the amplitude and phase, one can obtain the plots shown in Figs. 31.6b, c. It can also be done graphically by drawing the vectors shown in Fig. 31.6a for a particular frequency. Obviously, our signal processor described by Eq. 31.30 represents a band-pass filter in the baseband. It should be mentioned that the magnitude response is symmetrical while the phase response is antisymmetrical around the Nyquist frequency $\omega_s, /2$ and that both responses repeat after every ω_s radians.

Problems

P.1. What happens to the spectrum if the impulses in Fig. 31.2d are replaced by thin rectangular pulses?

P.2. If the base spectrum in Fig. 31.3 is a full sinusoid form—ω_h and $\omega_h > \frac{\omega_s}{2}$, what will happen to the sampled spectrum?

P.3. Why are all powers of z in z-transform negative? What is the meaning of positive powers? Comment on their realizability in real time and virtual time.

P.4. Can a z-transform with negative powers of z have a numerator of degree higher than that of the denominator? What will be its inverse transform?

P.5. Can you realize a difference equation with term like $x(n + 1)$, $x(n + 2)$ …?

References

1. A.V. Oppenheim, R.W. Schafer, *Digital Signal Processing*. (Prentice-Hall, 1975)
2. L.R. Rabiner, B. Gold, *Theory and Applications of Digital Signal Processing*. (Prentice Hall, 1975)

3. W.D. Stanley. *Digital Signal Processing.* (Reston, 1975)
4. M.H. Ackroyd, *Digital Filters.* (Butterworth, 1973)
5. E.O. Brigham, *The Fast Fourier Transform.* (Prentice-Hall, 1974)
6. K. Steiglitz, *An Introduction to Discrete Systems.* (Wiley, 1974)
7. A. Peled, B. Liu, *Digital Signal Processing.*(Wiley, 1976)
8. S.A. Tretter, *Introduction to Discrete Time Signal Processing.* (Wiley, 1976)
9. R.E. Bogner, A.G. Constantinides, *Introduction to Digital Filtering.* (Wiley Interscience, 1975)
10. D. Childers, A. Durling, *Digital Filtering and Signal Processing.* (West Pub. Co., 1975)
11. A.V. Oppenheim (ed.), *Papers on Digital Signal Processing.* (MIT Press, 1969)
12. L.R. Rabiner, C. Rader (eds.), *Digital Signal Processing.* (IEEE Press, 1972)
13. B. Liu, (ed.), *Digital Filters and the Fast Fourier Transform.* (Dowden Hutchinson Ross, 1975)
14. A.V. Oppenheim et al. (eds.), *Selected Papers in Digital Signal Processing II.* (IEEE Press, 1976)
15. See e.g. *IEEE Transactions on Audio and Electroacoustics;* June 1967, September 1968, June 1969, June 1970, December 1970, October 1972, June 1973, June 1975. *IEEE Transactions on Circuit Theory:* November 1971, July 1973. *IEEE Transactions on Circuits and Systems:* March 1975. *Proceedings of IEEE:* July 1972, October 1972, April 1975. *IEEE Transactions on Computers:* July 1972, May 1974. *IEEE Transactions on Communication Technology*: December 1971
16. Digital signal processing papers appear. in *Proceedings of IEEE, IEEE Transaction on Acoustics, Speech and Signal Processing* (formerly *Audio and Electroacoustics*), *IEEE Transaction on Circuits and Systems* (formerly *Circuit Theory*), *IEEE Transaction on Communications, IEEE Transaction on Computers, Bell System International Journal of Circuit Theory and Applications, Proceedings IEEE, IEEE, Journal on Electronic Circuits and Systems, Electronics Letters, Radio Electronic Engineer, Journal on Acoustical Society of America.,* and many others. Conferences which devote a significant portion of time to digital signal processing papers are *IEEE International Conference on ASSP, IEEE International Conference on CAS, Allerton, Asilomar, Midwest Symposium., European Conference on Circuit Theory and Design, NATO Special Conferences, Summer Schools on Circuit Theory* held at Prague, etc. etc
17. B. Gold, K.L. Jordan, *Digital Signal Processing.* (McGraw-Hill, 1968)

Bibliography

18. B. Gold, C. Rader, *Digital Processing of Signals.* (McGraw-Hill, 1969)
19. H.D. Helms, et al. (eds.), *Literature in Digital Signal Processing.* (IEEE Press, 1976)
20. L. Shapiro, Sampling Theory in Digital Processing. *Electron. Eng.* 45–50 (May 1978)
21. B. Gold, K. Jordan, A Note on Digital Filter Synthesis *Proceedings on IEEE* 56, October 1968. pp. 1717–1718
22. J.W. Cooley, J.W. Tukey, An algorithm for the Machine Calculation of Complex Fourier Series. *Math. Comput.* **19**, 297–301 (April 1965)
23. W.M. Gentlemen, G. Sande, Fast Fourier Transforms–for Fun and Profit. in *1966 Fall Joint Computer Conference on AFIPS Proceedings*, pp. 563–578

The ABCDs of Digital Signal Processing–PART 2

32

Here, we deal with the realizations of DSP's DFT, FFT, application of FFT to compute convolution and correlation, and application of FFT to find the spectrum of a continuous signal.

Keyword
DSP realization · DFT · FFT and its applications · Convolution · Correlation Picket fence effect

Realization of Digital Signal Processors

It should be clear from what has been discussed so far that a digital signal processor may be realized by use of the storage registers, arithmetic unit and the control unit of a general-purpose computer. Alternatively, special digital hardware may be designed to perform the required computations; this would result in a special-purpose processor (e.g. for radar or sonar signals) that would more or less be committed to a specific job. In either case, the digital signal processor can be represented by a variety of equivalent realization diagrams or structures. When implemented in a general-purpose computer, the structure may be thought of as the representation of a computational algorithm, from which a computer program is derived. When implemented by special-purpose hardware, it is often convenient to think of the structure as specifying a hardware configuration.

Corresponding to the basic operations required for implementation of a digital signal processor, the basic elements required to represent a difference equation pictorially are an adder, a delay and a constant multiplier, the commonly used symbols for which are shown in Fig. 32.1. Physically, Fig. 32.1a represents a means for adding together two sequences, Fig. 32.1b represents a means for multiplying a sequence by a constant and Fig. 32.1c represents a means for storing the previous value of a sequence. The representation used for a single sample delay arises from the fact that the z-transform of $x(n-1)$ is simply z^{-1} times the z-transform of $x(n)$.

As an example of the representation of a difference equation in terms of these elements, consider the second-order equation

$$y(n) = b_1 y(n-1) + b_2 y(n-2) + ax(n). \quad (32.1)$$

A realization structure for Eq. 32.2 is shown in Fig. 32.2. In terms of a computer program, Fig. 32.2 shows explicitly that storage must be provided for the variables $y(n-1)$ and $y(n-2)$ and also the constants b_1, b_2 and a. Further,

Source: S. C. Dutta Roy, "The ABCDs of Digital Signal Processing (Part 2)," *Students' Journal of the IETE*, vol. 21, pp. 60–70, April 1980.

Revised version of the text of a seminar series.

S. C. Dutta Roy, *Circuits, Systems and Signal Processing*,
https://doi.org/10.1007/978-981-10-6919-2_32

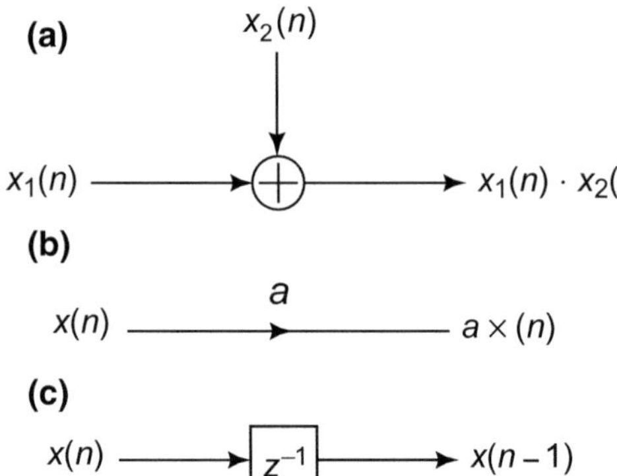

Fig. 32.1 Basic elements used in the realization diagram of a digital signal processor

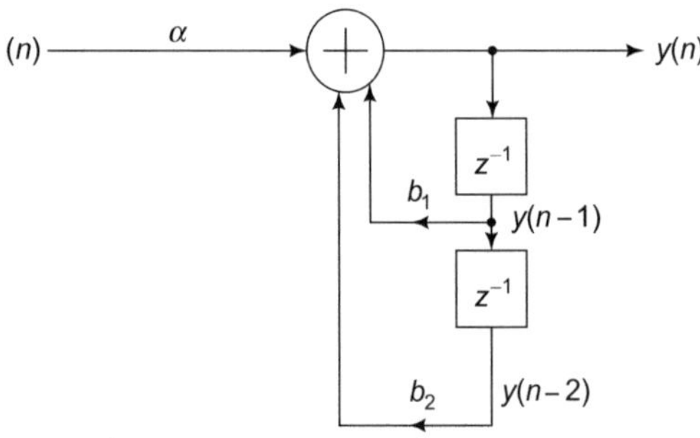

Fig. 32.2 A realization of Eq. 32.1

Fig. 32.2 shows that to compute an output sample $y(n)$, one must form the products $b_1y(n-1)$ and $b_2y(n-2)$, $ax(n)$, and add them together. In terms of special digital hardware, Fig. 32.2 indicates that we must provide storage for the variables and constants, as well as means for multiplication and addition. Thus, diagrams such as Fig. 32.2 serve to depict the complexity of a digital signal processor algorithm and the amount of hardware required to realize the processor.

As mentioned earlier, a variety of structures can be derived to implement a given difference equation. In case of Eq. 32.1, for example, the transfer function is

$$H(z) = a/(1 - b_1z^{-1} - b_2z^{-2}). \tag{32.2}$$

Let b_1 and b_2 be such that the poles of $H(z)$ are real. Then, one could write $H(z)$ in the following two equivalent forms:

$$H(z) = \frac{a_1}{1 - c_1z^{-1}} \cdot \frac{a_2}{1 + c_2z^{-1}} \tag{32.3}$$

$$= \frac{a^1}{1 - c_1z^{-1}} + \frac{a^{11}}{1 + c_2z^{-1}}, \tag{32.4}$$

where $c_1 - c_2 = b_1$, $c_1\ c_2 = b_2$, $a_1\ a_2 = a = a_1 + a_2$ and $a_1c_2 - a_2\ c_2 = 0$. Each first-order form

$$\frac{\text{constant}}{1 \pm c_1z^{-1}}$$

can be implemented by using a delay, two constant multiplications and an adder. To implement form Eq. 32.1, one needs to cascade two first-order realizations, as shown in Fig. 32.3a, while form Eq. 32.4 requires parallel connection of two first-order forms, as shown in Fig. 32.3b. Obviously, the three realizations of Figs. 32.1, 32.3a, b differ in computational algorithm and hardware requirements; what is more important, however, is that the quantization errors also, in general, differ from structure to structure. We shall not, however, explore this point further in this chapter, but shall be content with the knowledge that a digital signal processor can be implemented by various equivalent structures, and one should choose the one which is the optimum under the given set of constraints.

We shall not also discuss here methods of finding out the difference equation to suit a particular set of specifications; this we shall reserve for a future article. The rest of this chapter will deal with DFT technique in relation to digital signal processing.

The Discrete Fourier Transform

As we have already seen, one can implement a digital signal processor, described by Eq. 31.1, in a general-purpose computer or by a special-purpose hardware. Another way of implementing a digital signal processor is based on the fact that the output sequence $\{y(n)\}$ is the convolution of the input sequence $\{x(n)\}$ with the impulse response sequence $\{h(n)\}$ of the

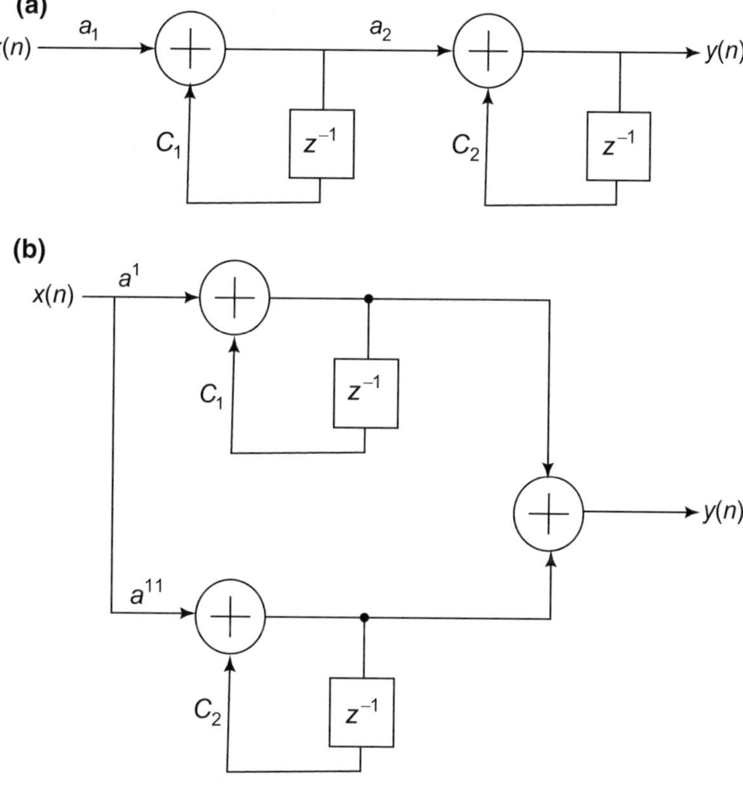

Fig. 32.3 **a** Cascade realization. **b** Parallel realization

processor, as given by Eq. 31.27. This equation is reproduced here for convenience as follows:

$$y(n) = \sum_{r=0}^{n} h(r)x(n-r) = \sum_{r=0}^{n} h(n-r)x(r). \tag{32.5}$$

Recall that in the continuous signal case, the convolution of the input signal $x(t)$ with the impulse response $h(t)$ gives the output $y(t)$ and that in the frequency domain, this amounts to a multiplication of the transforms (Laplace or Fourier) of $x(t)$ and $h(t)$ to give the transform of $y(t)$. $y(t)$ can then be obtained by the inverse transform operation. A similar operation can be performed with discrete time systems if we have a suitable transform. As we have already seen, the z-transform does provide such a vehicle; however, for numerical computation, a modified version of it, called the discrete Fourier transform (DFT) has been found most suitable. The signal processing operation then simply boils down to the following sequence of computations:

(i) Compute the DFT of $\{x(n)\}$.
(ii) Compute the DFT of $\{h(n)\}$.
(iii) Multiply the two.
(iv) Compute the inverse DFT (IDFT) of the product.

Let, for simplicity, the notation x_k be used for $x(k) \equiv x(kT)$, and consider a sequence $\{x_k\}$ of length N, i.e. $k = 0, 1, 2, \ldots N-1$. Then, the *DFT* of $\{x_k\}$ is defined by

$$A_r = \sum_{k=0}^{N-1} x_k e^{-j2\pi rk/N}, r = 0, 1, 2, \ldots N-1. \tag{32.6}$$

The DFT is thus also a sequence $\{A_r\}$ of length N. The x_k's may be complex numbers; the

A_r's are almost always complex. For notational convenience, let

$$W = e^{-j2p/N}, \quad (32.7)$$

so that

$$A_r = \sum_{k=0}^{N-1} x_k W^{rk}, r = 0, 1, \ldots N-1. \quad (32.8)$$

If one compares Eq. 32.6 with the continuous Fourier transform $A(\omega)$ of a signal $x(t)$, viz.

$$A(\omega) = \int_{-\infty}^{\infty} x(t) e^{-j2\pi ft} dt \quad (32.9)$$

then one way of interpreting the DFT is that it gives the N-point discrete spectrum of the N-point time series $\{x(kT)\}$ at the frequency points $r/(NT)$; $r = 0, 1, \ldots N-1$; the fundamental frequency, obviously, is $f_o = 1/(NT)$.

The inverse DFT (IDFT) of the complex sequence $\{A_r\}$, $r = 0, 1 \ldots N-1$, is given by

$$x_k = \frac{1}{N} \sum_{r=0}^{N-1} A_r W^{-rk}, k = 0, 1, \ldots N-1. \quad (32.10)$$

That this exists and is unique can be easily established by substituting Eq. 32.8 in Eq. 32.10 and carrying out some elementary manipulations.

Since e^j is periodic with a period 2π, it follows from Eq. 32.8 and Eq. 32.10 that

$$\begin{aligned} A_r &= A_{r+mN}, m = 0, \pm 1, \pm 2, \ldots \\ x_k &= x_{k+mN}. \end{aligned} \quad (32.11)$$

i.e. both DFT and IDFT yield sequences which are periodic, with periods $Nf_o = T^{-1} = f_s$ and NT respectively.

The Fast Fourier Transform

The fast Fourier transform (FFT) is a highly efficient method for computing the DFT of a time series. A direct computation from Eq. 32.8 would require N^2 complex multiplications; in contrast, application of FFT can reduce this number to $(N/2)\ \log_2 N$. For example, for $N = 512$, the ratio $(N/2)\ \log_2 N \div N^2$ becomes less than 1%. This drastic reduction in computation time through FFT has made the FFT an important tool in many signal processing applications.

The DFT, given by Eq. 32.8 and its inverse, given by Eq. 32.10 are of the same form so that any algorithm capable of computing one may be used for computing the other by simply exchanging the roles of x_k and A_r, and making appropriate scale factor and sign changes. There are two basic forms of FFT; the first, due to Cooley and Tukey [1], is known as *decimation in time,* while the other, obtained by reversing the roles of x_k and A_r, gives the form called *decimation in frequency,* and was proposed by Gentleman and Sande [2]. Clearly, they should be equivalent; it is, however, worth distinguishing between them and discussing them separately.

Let N be even and the sequence $\{x_k\}$ be decomposed as

$$\{x_k\} = \{u_k\} + \{v_k\}, \quad (32.12)$$

where

$$\begin{aligned} u_k &= x_{2k} \\ k &= 0, 1, 2, \ldots N/2-1 \\ v_k &= x_{2k+1}. \end{aligned} \quad (32.13)$$

Thus $\{u_k\}$ contains the even-numbered points and $\{v_k\}$ contains the odd-numbered points of $\{x_k\}$ and each has $N/2$ points. The DFTs of $\{u_k\}$ and $\{v_k\}$ are, therefore,

$$\begin{aligned} B_r &= \sum_{k=0}^{N/2-1} u_k e^{-j2\pi rk/(N/2)} \\ &= \sum_{k=0}^{N/2-1} u_k e^{-j4\pi rk/N}\ r = 0, 1, 2, \ldots N/2-1 \\ C_r &= \sum_{k=0}^{N/2-1} v_k e^{-j4\pi rk/N}. \end{aligned} \quad (32.14)$$

The DFT we want is

$$
\begin{aligned}
A_r &= \sum_{k=0}^{N-1} x_k e^{-j2\pi rk/N} \\
&= \sum_{k=0}^{N/2-1} x_{2k} e^{-j4r\pi k/N} + \sum_{k=0}^{N/2-1} x_{2k+1} e^{-j2\pi r(2k+1)/N}, \\
r &= 0, 1, 2, \ldots N-1 \\
&= B_r + e^{-j2\pi r/N} C_r, 0 \leq r < N/2.
\end{aligned} \tag{32.15}
$$

because B_r and C_r are defined for $r = 0$ to $(N/2) - 1$. Further, B_r and C_r are periodic with period $N/2$ so that

$$B_{r+N/2} = B_r \quad \text{and} \quad C_{r+N/2} = C_r. \tag{32.16}$$

Thus

$$
\begin{aligned}
A_{r+N/2} &= B_r + e^{-j2\pi(r+N/2)}/^N C_r \\
&= B_r - e^{-j2\pi r/N} C_r, 0 \leq r < N/2.
\end{aligned} \tag{32.17}
$$

Finally, using Eqs. 32.7, 32.15 and 32.17, we get

$$
\begin{aligned}
A_r &= B_r + W^r C_r \\
&0 \leq r < N/2 \\
A_{r+N/2} &= B_r - W^r C_r.
\end{aligned} \tag{32.18}
$$

A direct calculation of B_r and C_r, from Eq. 32.14 requires $(N/2)^2$ complex multiplications each. Another N such multiplications are required to compute $A_r s$ from Eq. 32.18, thus making a total of $2(N/2)^2 + N = N^2/2 + N$, which is less than N^2 if $N > 2$. This is illustrated in Fig. 32.4 by a signal flow diagram for $N = 8$, where we have used the fact that $W^{N/2} = -1$, so that $-W^r = W^{r+N/2}$.

The DFTs of $\{u_k\}$ and $\{v_k\}$, $k = 0, 1, \ldots (N/2) -1$, can now be computed through a similar decomposition if $N/2$ is even; thus the

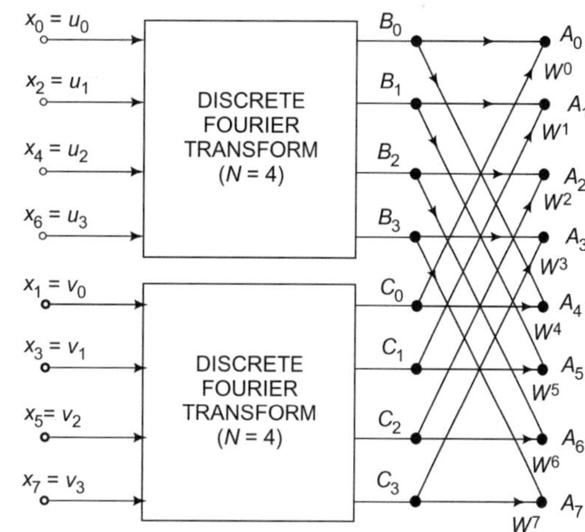

Fig. 32.4 Illustrating the first step in decimation in time form of FFT for $N = 8$

computation of $\{B_r\}$ and $\{C_r\}$ reduces to the task of finding the DFTs of four sequences, each of $N/4$ samples. These reductions can be continued as long as each sequence has an even number of samples. Thus if $N = 2^n$, one can make n such reductions by applying Eq. 32.13 and Eq. 32.18, first for N, then for $N/2$ and so on, and finally for a two-point function. The DFT of a one-point function is, of course, the sample itself. The successive reduction of an 8-point DFT, which began in Fig. 32.4, is continued in Figs. 32.5 and 32.6. In Fig. 32.6, the operation has been completely reduced to complex multiplications and additions. The number of summing nodes is (8) (3) = 24 and 24 complex additions are, therefore, required; the number of complex multiplications needed are also 24 = (no. of stages) (no. of multiplications in each stage) = (3) (8). Half of these multiplications are easily eliminated by noting that $W^7 = -W^3$, $W^6 = -W^2$, $W^5 = -W^1$ and $W^4 = -W^0$. Thus, in general, $N \log_2 N$ complex additions and, at most, (1/2) $N \log_2 N$ complex multiplications are required for the computation of an N-point DFT, when N is a power of 2.

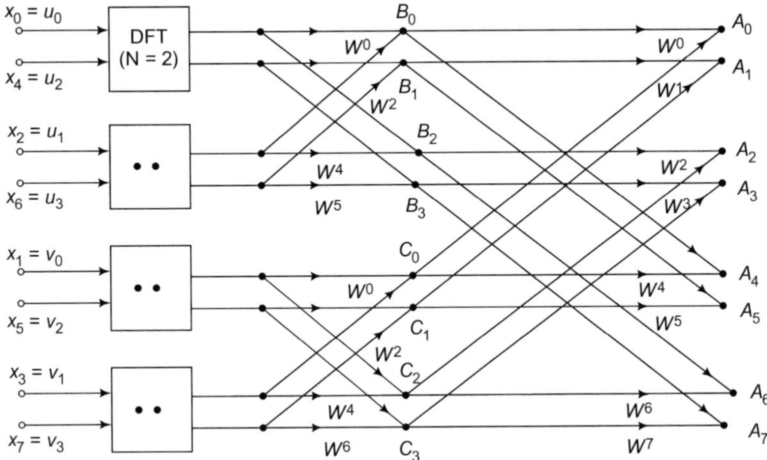

Fig. 32.5 Illustrating two steps of decimation in frequency form of FFT for $N = 8$

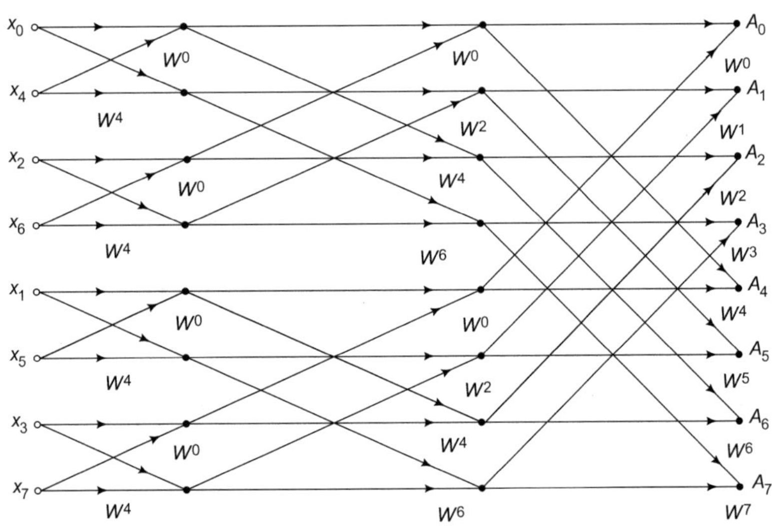

Fig. 32.6 Illustrating decimation in time form of FFT for $N = 8$

When N is not a power of 2, but has a factor of p, one can develop equations analogous to 32.13 through 32.18 by forming p different sequences, $\{u_k^{(i)}\} = \{x_{pk+i}\}$, $i = 0$ to $p - 1$, each having N/p samples. For example, if $N = 15$ having a factor 3, we can form three sequences as follows:

$$\begin{aligned} \{u_k^{(0)}\} &= \{x_0, x_3, x_6, x_9, x_{12}\} \\ \{u_k^{(1)}\} &= \{x_1, x_4, x_7, x_{10}, x_{13}\} \\ \{u_k^{(2)}\} &= \{x_2, x_5, x_8, x_{11}, x_{14}\}. \end{aligned} \quad (32.19)$$

Each of these sequences has a DFT $B_r^{(i)}$, and the DFT of $\{x_k\}$ can be computed from p simpler DFT's. Further simplification occurs if N has additional prime factors.

In the decimation in frequency form of FFT, the sequence $\{x_k\}$, $k = 0, 1, \ldots, N - 1$ and N even, is decomposed as

$$\begin{aligned} u_k &= x_k k = 0, 1, \ldots, N/2{-}1 \\ v_k &= x_{k+N/2} \end{aligned} \quad (32.20)$$

i.e. $\{u_k\}$ is composed of the first $N/2$ points and $\{v_k\}$ is composed of the last $N/2$ points of $\{x_k\}$. Then one can write

$$A_r = \sum_{k=0}^{N/2-1} [u_k e^{-j2\pi rk/N} + v_k e^{-2\pi r(k+N/2)/N}]$$
$$= \sum_{k=0}^{N/2-1} (u_k + e^{-j\pi r} v_k) e^{-2\pi rk/N},$$
$$r = 0, 1, \ldots N-1. \tag{32.21}$$

Consider the even-numbered and odd-numbered points of the DFT separately; let

$$\begin{aligned} R_r &= A_{2r} \quad 0 \le r < N/2 \\ S_r &= A_{2r+1}. \end{aligned} \tag{32.22}$$

It is this step that may be called the decimation in frequency. Note that for computing R_r, Eq. 32.21 becomes

$$R_r = A_{2r} = \sum_{k=0}^{N/2-1} (u_k + v_k) e^{-j2r\pi k/(N/2)}. \tag{32.23}$$

which we recognize as the $N/2$-point DFT of the sequence $\{u_k + v_k\}$. Similarly,

$$S_r = A_{2r+1} = \sum_{k=0}^{N/2-1} [u_k + v_k e^{-j\pi(2r+1)}] e^{-j2\pi(2r+1)k/N}$$
$$= \sum_{k=0}^{N/2-1} (u_k - v_k) e^{-j2\pi k/N} e^{-j2\pi rk/(N/2)}. \tag{32.24}$$

which we recognize as the $N/2$-point DFT of the sequence $\{(u_k - v_k)e^{-j2\pi k/n}\}$.

Thus, the DFT of an N–sample sequence $\{x_k\}$, N even, can be computed as the $N/2$ point DFT of a simple combination of the first $N/2$ and the last $N/2$ samples of $\{x_k\}$ for even-numbered points, and a similar DFT of a different combination of the same samples of $\{x_k\}$ for the odd-numbered points. This is illustrated in Fig. 32.7 for $N = 8$.

As was the case with decimation in time, we can replace each of the DFTs indicated in Fig. 32.2 by two 2-point DFTs, and each of the 2-point DFTs by two 1-point transforms, these last being equivalency operations. These steps are indicated in Figs. 32.8 and 32.9.

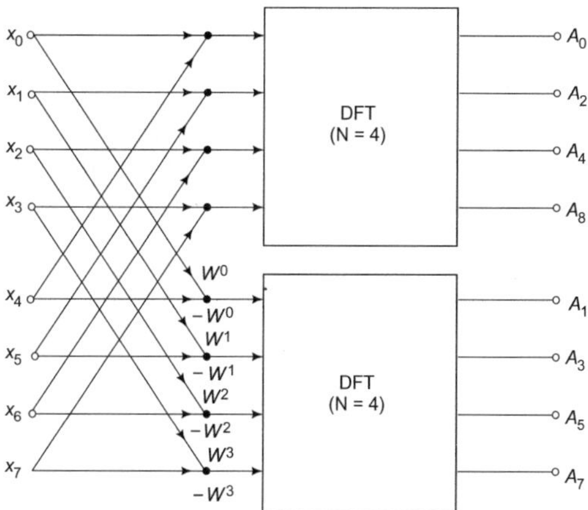

Fig. 32.7 Illustrating the first step in decimation in frequency form of FFT for $N = 8$

There are many variations and modifications of the two basic FFT schemes, which we would not discuss here.

Applications of FFT to Compute Convolution and Correlation

It may be recalled that our motivation for introducing the DFT and FFT was to convert the convolution relation Eq. 32.5, viz.

$$\begin{aligned} y_n &= \sum_{r=0}^{n} h_r x_{n-r} \\ &= \sum_{r=0}^{n} h_{n-r} x_r \end{aligned} \tag{32.25}$$

into a product form, through the DFT. To this end, assume that both the impulse response $\{h_n\}$, and the input $\{x_n\}$ are band limited to $1/2T$ Hz. Then the output $\{y_n\}$ is also frequency band limited. Also, if both $\{h_n\}$ and $\{x_n\}$ are defined for the range $0 \le n \le N-\mathrm{I}$, then $\{y_n\}$ is defined for the range $0 \le n \le 2N-1$. For example, if $\{h_n\} = \{h_0, \ h_1\}$ and $\{x_n\} = \{x_0, \ x_1\}$, then $\{y_n\} = \{h_0x_0, h_0x_1 + h_1x_0, h_1x_1\}$. Let the DFT's of $\{x_n\}$ and $\{h_n\}$ be $\{A_r\}$ and $\{H_r\}$ respectively.

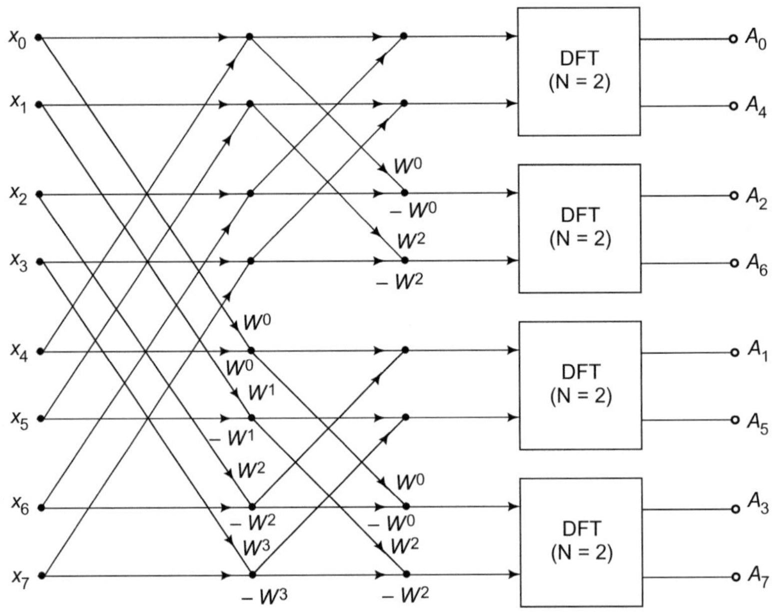

Fig. 32.8 Illustrating two steps of decimation in time form of FFT for $N = 8$

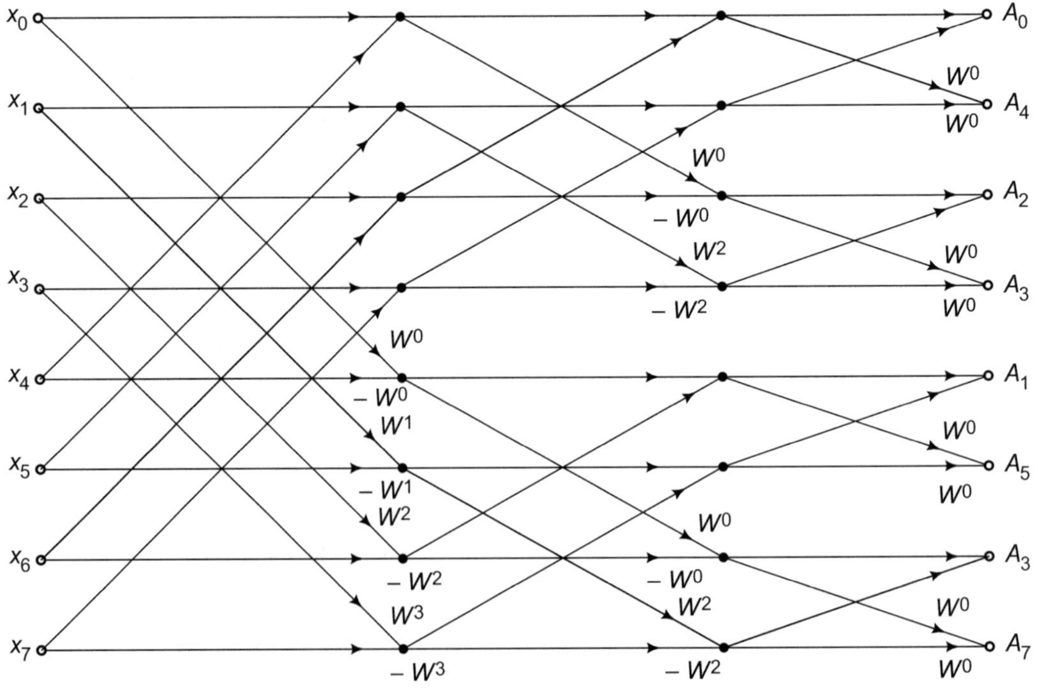

Fig. 32.9 Illustrating decimation in frequency form of FFT for N = 8

Then, the nth sample in the IDFT of the product $\{A_r H_r\}$ is

$$y'_n = 1/N \sum_{r=0}^{N-1} A_r H_r W^{-rn}, \quad n = 0, 1 \ldots N-1. \tag{32.26}$$

Substituting in Eq. 32.26,

$$\begin{aligned} A_r &= \sum_{k=0}^{N-1} x_k W^{rk}, \\ H_r &= \sum_{l=0}^{N-1} h_l W^{rl}. \end{aligned} \tag{32.27}$$

and carrying out some elementary manipulations, it is not difficult to show that Eq. 32.26 simplifies to

$$\begin{aligned} y'_n &= \sum_{k=0}^{N-1} x_k h_{n-k} \\ &= \sum_{k=0}^{n} x_k h_{n-k} + \sum_{k=n+1}^{N-1} x_k h_{N+n-k} \end{aligned} \tag{32.28}$$

$$= y_n + \text{perturbation term.} \tag{32.29}$$

The last form is obtained by comparison with Eq. 32.25, while the last term in Eq. 32.28 represents the 'cyclical' part of the convolution, arising out of the periodicity of DFT; and IDFT; h is the cyclical variable passing from h_o to h_{N-1} as k passes from n to $n + 1$. The convolution can be made cyclical in x instead of h by inter-changing x and h in Eq. 32.28.

The procedure outlined for implementing a digital signal processor, viz. taking the DFTs of $\{x_n\}$ and $\{h_n\}$, multiplying them, and taking IDFT of the product, does not, therefore, give the desired output sequence $\{y_n\}$ unless the perturbation term in 32.28 can be made zero. This term arises due to the fact that the DFT assumes both $\{x_n\}$ and $\{h_n\}$ to be periodic. Further, $\{y'_n\}$ is of length N instead of $2N - 1$. Note that if we extend both $\{x_n\}$ and $\{h_n\}$ to a length $2N$ by adding N zeros to each, i.e. if we change $\{x_n\}$ to $\{\hat{x}_n\} = \{x_0, x_1 \ldots x_{N-1}, 0, \ldots 0\}$ and similarly for $\{h_n\}$, then the perturbation term becomes zero. Further, the sequence $\{y_n\}$ will be $N + N - 1 = 2N - 1$ terms long, i.e. y_{2N-2} will be the last, non-zero term in $\{y_n\}$. As an example, let $N = 4$, i.e.

$$\begin{aligned} \{x_n\} &= \{x_0, x_1, x_2, x_3\} \\ \{h_n\} &= \{h_0, h_1, h_2, h_3\}. \end{aligned} \tag{32.30}$$

The true convolution of $\{x_n\}$ with $\{h_n\}$ gives

$$\begin{aligned} y_0 &= x_0 h_0 \\ y_1 &= x_0 h_1 + x_1 h_0 \\ y_2 &= x_0 h_2 + x_1 h_1 + x_2 h_0 \\ y_3 &= x_0 h_3 + x_1 h_2 + x_2 h_1 + x_3 h_0 \\ y_4 &= x_1 h_3 + x_2 h_2 + x_3 h_1 \\ y_5 &= x_2 h_3 + x_3 h_2 \\ y_6 &= x_3 h_3. \end{aligned} \tag{32.31}$$

On the other hand, the DFT procedure, leading to Eq. 32.28 gives

$$y'_n = \sum_{k=0}^{3} x_k h_{n-k}, \tag{32.32}$$

so that

$$\begin{aligned} y'_0 &= x_0 h_0 + x_1 h_{-1} + x_2 h_{-2} + x_3 h_{-3} \\ &= x_0 h_0 + (x_1 h_3 + x_2 h_2 + x_3 h_1) \\ y'_1 &= x_0 h_1 + x_1 h_0 + (x_2 h_3 + x_3 h_2) \\ y'_2 &= x_0 h_2 + x_1 h_1 + x_2 h_0 + (x_3 h_1) \\ y'3 &= x_0 h_3 + x_1 h_2 + x_2 h_1 + x_3 h_0. \end{aligned} \tag{32.33}$$

where the perturbation terms are bracketed. Also, $\{y'_n\}$ consists of only 4 terms. Now let

$$\{\hat{x}_n\} = \{x_0, x_1, x_2, x_3, 0, \ 0, \ 0, \ 0\}$$
$$\{\hat{h}_n\} = \{h_0, h_1, h_2, h_3, \ 0, \ 0, \ 0, \ 0\}. \quad (32.34)$$

Then the DFT procedure gives

$$y'_n = \sum_{k=0}^{7} x_k h_{n-k}, \quad (32.35)$$

so that

$$\begin{aligned} y'_0 &= x_0 h_0 \\ y'_1 &= x_0 h_1 + x_1 h_0 \\ y'_2 &= x_0 h_2 + x_1 h_1 + x_2 h_0 \\ y'_3 &= x_0 h_3 + x_1 h_2 + x_2 h_1 + x_3 h_0 \\ y'_4 &= x_1 h_3 + x_2 h_2 + x_3 h_1 \\ y'_5 &= x_2 h_3 + x_3 h_2 \\ y'_6 &= x_3 h_3. \end{aligned} \quad (32.36)$$

By comparing with Eq. 32.31, we see that

$$\{y'_n\} = \{y_n\}$$
$$n = 0, \ 1, \ 2, \ldots 7.$$

Thus, the modification does give correct results.

Before stating this simple remedy in formal terms, we would like to emphasize that blind use of FFT for computing the convolution of two sequences will lead to incorrect results, because the DFT introduces a periodic extension of both data and processor impulse response. This results in cyclic or periodic convolution, rather than the desired noncyclic or aperiodic convolution. If $\{x_n\}$ and $\{h_n\}$ contain N samples each, then the true convolution should result in 2 N – 1 samples for $\{y_n\}$. If DFT is used, then $\{A_r\}$ and $\{H_r\}$ each consist of N samples, so does $\{A_r H_r\}$ and hence its IDFT. Hence, $\{y_n'\}$ found by DFT is not the same as $\{y_n\}$ because of folding (or aliasing or cycling) occurring in the time domain. This can be corrected, as demonstrated by the example, by adding zeros to both $\{x_n\}$ and $\{h_n\}$ and thereby increase their lengths sufficiently so that no overlap occurs in the resultant convolution.

We now state formally the steps for computing convolution by DFT:

(i) Let $\{x_n\}$ be defined for

$$0 \le n \le M - 1$$

and $\{h_n\}$ be defined for

$$0 \le n \le P - 1$$

(ii) Select N such that

$$N \le P + M - 1$$
$$N = 2^k$$

(iii) Form the new sequences $\{\hat{x}_n\}$ and $\{\hat{h}_n\}$ such that

$$\hat{x}_n = \begin{cases} x_n, 0 \le n \le M - 1 \\ 0, M \le n \le N - 1 \end{cases}$$
$$\hat{h}_n = \begin{cases} h_n, 0 \le n \le P - 1 \\ 0, P \le n \le N - 1 \end{cases}$$

(iv) Compute the DFTs $\{\hat{A}_r\}$ and $\{\hat{H}_r\}$ of $\{\hat{x}_n\}$ and $\{\hat{h}_n\}$ by FFT.

(v) Compute

$$\{\hat{B}_r\} = \{\hat{A}_r \hat{H}_r\}$$

(vi) Find the IDFT of $\{\hat{B}_r\}$ by FFT; the result is $\{y_n\}$.

This technique is referred to as *select-saving*.

Next, we consider the application of FFT to compute the cross-correlation sequence $\{R_{xy}(k)\}$ of two given sequences $\{x_n\}$ and $\{y_n\}$, each of length N, where

$$R_{xy}(k) \triangleq 1/N \sum_{n=0}^{N-1} x_n y_{n-k} \quad (32.37)$$

and the auto-correlation sequence $\{R_{xx}(k)\}$ of a sequence $\{x_n\}$, where

$$R_{xx}(k) \triangleq 1/N \sum_{n=0}^{N-1} x_n x_{n-k}. \quad (32.38)$$

Note that the essential difference between convolution, as given by Eq. 32.25, and correlation, as given by Eqs. 32.37 and 32.38 is that one of the sequences is reversed in direction for one operation as compared with the other. Thus, if FFT is to be used to compute correlation, the same kind of precautions, as discussed for convolution, are to be exercised. The procedure here, is based on the fact that if

$$\text{DFT}\{x_n\} = \{A_r\}$$

and

$$\text{DFT}\{y_n\} = \{B_r\}$$

then

$$\text{DFT}\left\{\sum_{n=0}^{N-1} x_n y_{n-k}\right\} = \{A_r \overline{B_r}\}, \quad (32.39)$$

where bar denotes complex conjugate. Thus applied to Eqs. 32.37 and 32.38, one obtains

$$\begin{aligned}\{R_{xy}(k)\} &= \text{IDFT}\{A_r \overline{B}_r/N\}\\ &= \text{IDFT}\{S_{xy}(r)\}\end{aligned} \quad (32.40)$$

$$\begin{aligned}\{R_{xx}(k)\} &= \text{IDFT}\{{}^{1}A_r/^{2}/N\}\\ &= \text{IDFT}\{S_{xx}(r)\}\end{aligned} \quad (32.41)$$

where $\{S_{xy}(r)\}$ and $\{S_{xx}(r)\}$ are the cross-power spectrum sequence and auto power spectrum sequences respectively.

Application of FFT to Find the Spectrum of a Continuous Signal

The DFT, as we have seen, is specifically concerned with the analysis and processing of discrete periodic signals, and that it is a zero-order approximation of the continuous Fourier transform. It is, therefore, tempting to apply the DFT directly to provide, through FFT, a numerical spectral analysis of sampled versions of continuous signals. This would be a perfectly valid application, if the continuous, signal is periodic, band limited and sampled in accordance with the sampling theorem. Deviations from these cause errors, and most of the problems in using the DFT to approximate the CFT (C for continuous) are caused by a misunderstanding of what this approximation involves.

There are, essentially, three phenomena, which contribute to errors in relating the DFT to the CFT. The first, called *aliasing*, has already been discussed (Part 1). The solution to this problem is to ensure that the sampling rate is high enough to avoid any spectral overlap. This requires some prior knowledge of the nature of the spectrum, so that the appropriate sampling rate may be chosen. In absence of such prior knowledge, the signal must be prefiltered to ensure that no components higher than the folding frequency appear.

The second problem is that of *leakage*, arising due to the practical requirement of observing the signal over a finite interval. This is equivalent to multiplying the signal by a window function. The simplest window is a rectangular function as shown in Fig. 32.10a, and its effect on the spectrum of a sine signal, shown in Fig. 32.10b, is displayed in Fig. 32.10c. Note that there occurs a spreading or leakage of the spectral components away from the correct frequency; this results in an undesirable modification of the total spectrum.

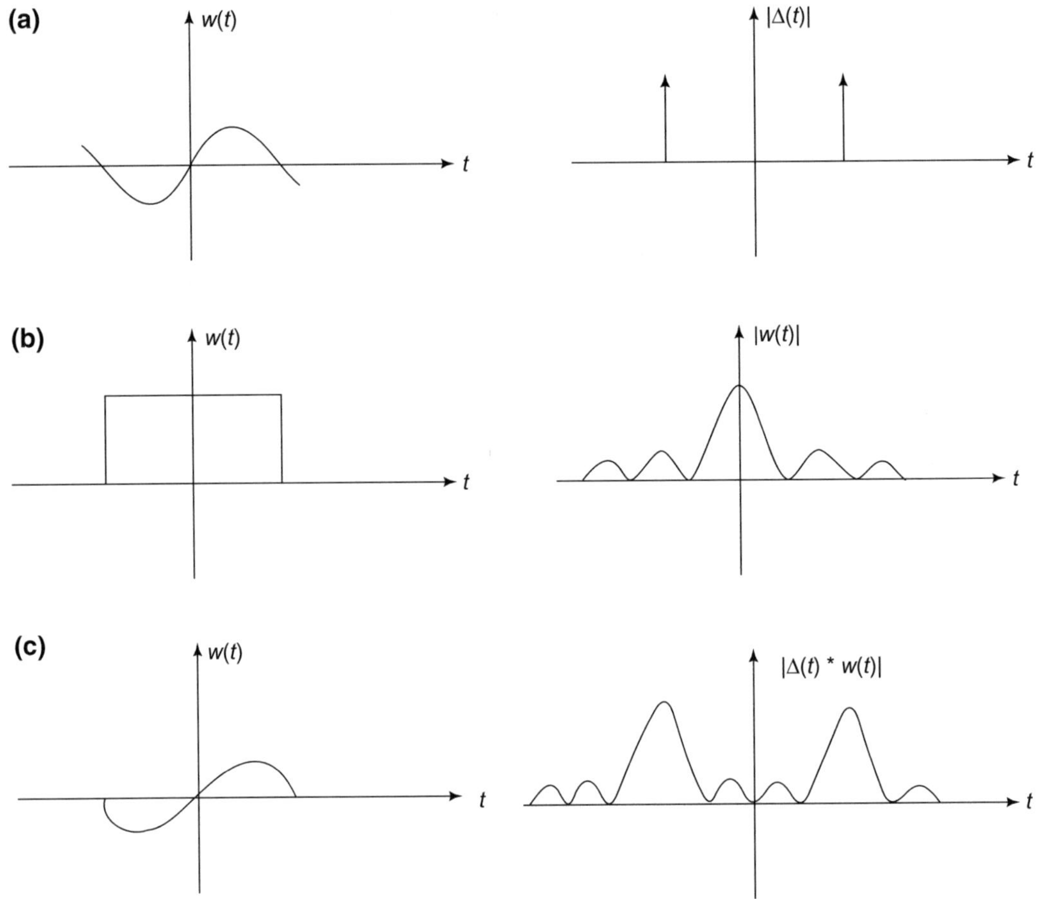

Fig. 32.10 Illustrating 'leakage' due to finite observation time

The leakage effect cannot always be isolated from the aliasing effect because leakage may also lead to aliasing if the highest frequency of the composite spectrum moves beyond the folding frequency. This possibility is particularly significant in the case of a rectangular window, because the tail of the window spectrum does not converge rapidly.

The solution to the leakage problem is to choose a window function that minimizes the spreading. One example is the so-called 'raised cosine' window in which a raised cosine wave is applied to the first and last 10 per cent of the data and a weight of unity is applied in between. Since only 20% of the terms in the time series is given a weight other than unity, the computation required to apply this window in the time domain is relatively small, as compared to other continuously varying weight windows, e.g. the Hamming window.

The third problem in relating the DFT to the CFT is the *picket fence effect*, resulting from the inability of the DFT to observe the spectrum as a continuous function, since the computation of the spectrum is limited to integer multiples of the fundamental frequency $f_o = 1/(NT)$. In a sense, the observation of the spectrum with the DFT is analogous to looking at it through a sort of 'picket fence' since we can observe the exact behaviour only at discrete points. It is possible that a major peak lies between two of the discrete transform lines, and this will go undetected without some additional processing.

One procedure for reducing the picket fence effect is to vary the number of points N in a time period by adding zeros at the end of the original record, while maintaining the original record intact. This process artificially changes the period, which, in turn, changes the locations of the spectral lines without altering the continuous form of the original spectrum. In this manner, spectral components originally hidden from view may be shifted to points where they may be observed.

Concluding Comments

The aim of this chapter was to introduce the basic concepts involved in digital signal processing, including an introduction to the FFT and its applications. We went through the sampling process carefully, and pointed out the various errors introduced by quantization. A brief discussion on structures was included to facilitate an understanding of the implementation of a digital signal processor in a general-purpose computer or by a special-purpose hardware. Two basic forms of FFT were introduced, and two of the most important applications of the FFT were discussed. It was pointed out that correct application of FFT requires a much more than casual understanding of the periodic extension introduced by the DFT process.

In conclusion, it is worth mentioning that digital signal processing is not an answer to all signal processing problems. Digital and analog techniques form 'two arrows in the quiver', which are always 'better than one'. '... And three are better still', the proverb continues; this third 'arrow' is provided by the charge transfer devices, which can perform analog as well as digital signal processing. As compared to the digital signal processors we have talked about, the charge transfer processing has the distinct advantage of not requiring an A/D conversion, and hence is less expensive, more versatile and more accurate.

Problems

P.1. Draw the equivalent of Fig. 32.6 for $N = 16$. You have to take an A3 paper with 90° turn around.

P.2. Draw the FFT diagram for N = 16 using decimation in frequency.

P.3. What is the minimum possible number of non-trivial multipliers in Fig. 32.8?

P.4. What is better? DIT or DIF?

P.5. What is a possible remedy for eliminating leakage altogether? Is it practicable?

References

1. J.W. Cooley, J.W. Tukey An algorithm for the machine calculation of complex fourier series. Math. Comput. **19**, 297–301 (April 1965)
2. W.M. Gentleman, G. Sande, Fast Fourier Transforms-for Fun and Profit. in *1966 Fall Joint Computer Conference of AFIPS Proceedings*, pp. 563–578

On Second-Order Digital Band-Pass and Band-Stop Filters

33

The chapter deals with the derivation, design, limitations and realization of second-order digital band-pass (BP) and band-stop (BS) filters with independent control of the centre frequency and the bandwidth in the BP case, and rejection frequency and the difference between the pass-band edges in the BS case.

Keywords

Digital filter · Band-stop · Band-pass
Second-order filters

Introduction

The second-order transfer function

$$H_1(z) = \frac{1-\alpha}{2}\frac{1-z^{-2}}{1-\beta(1+\alpha)z^{-1}+\alpha z^{-2}} \quad (33.1)$$

is that of a normalized digital band-pass filter (BPF) whose centre frequency ω_0 and 3-dB bandwidth B are given by

$$\omega_0 = \cos^{-1}\beta \quad \text{and} \quad B = \cos^{-1}\frac{2\alpha}{1+\alpha^2}. \quad (33.2)$$

Thus ω_0 and β are independently controllable by varying α and β, provided one can realize Eq. 33.1 by only two multipliers of the same values. Such a realization using the lattice structure has been given in [1]. The complement of Eq. 33.1, obtained by subtracting H_1 (z) from unity, is

$$H_2(z) = \frac{1+\alpha}{2}\frac{1-2\beta z^{-1}+z^{-2}}{1-\beta(1+\alpha)z^{-1}+\alpha z^{-2}}. \quad (33.3)$$

It represents a band-stop filter (BSF) whose rejection frequency ω_0 and the parameter $B = \omega_2 - \omega_1$, where ω_2 and ω_1 are the two 3 dB frequencies, are also given by Eq. 33.2. Hence, the BPF realization can also be used for realizing the BSF.

While one appreciates the elegance of the transfer function Eq. 33.1, the question of how it was conceived of has not been answered in textbooks. Another question that arises is the following: Is the α-controllability of the bandwidth valid for any arbitrary pass-band tolerance? Yet another relevant question concerns the realization of a canonic structure with multipliers α and β. Are structures other than that given in

Source: S. C. Dutta Roy, "On Second-Order Digital Band-Pass and Band-Stop Filters," *IETE Journal of Education*, vol. 49, pp. 59–63, May–August 2008.

S. C. Dutta Roy, *Circuits, Systems and Signal Processing*,
https://doi.org/10.1007/978-981-10-6919-2_33

Eq. 33.1 possible? This chapter presents answers to all these questions.

Derivation

For a BPF, we argue that the response at $\omega = 0$ and π should both be zero. With $z = e^{j\omega}$, $\omega = 0$ and π translate to $z = 1$ and -1, respectively. Hence, the numerator polynomial of the transfer function must be of the form $K(1 - z^{-2})$, K being a constant. Also, we know that real poles severely limit the selectivity of a BPF. Hence, we let the poles be at $re^{\pm j\theta}$, where r is close to but less than unity for high selectivity. The denominator polynomial of the BPF transfer function is, therefore,

$$\begin{aligned}&(1-re^{j\theta}z^{-1})(1-re^{-j\theta}z^{-1})\\&\quad=1-2r\cos\theta z^{-1}+r^2z^{-2}.\end{aligned} \tag{33.4}$$

The required transfer function is, therefore,

$$H_1(z)=\frac{K(1-z^{-2})}{1-2r\cos\theta z^{-1}+r^2z^{-2}}. \tag{33.5}$$

where K will be chosen to normalize the maximum magnitude to unity. The frequency response of Eq. 33.5 is given by

$$H_1(e^{j\omega})=\frac{2jK\sin\omega}{[(1-r^2)\cos\omega-2r\cos\theta](1-r^2)\sin\omega}. \tag{33.6}$$

The magnitude squared function can be written as

$$|H_1(e^{j\omega})|^2=\frac{4K^2}{\left[\frac{(1+r^2)\cos\omega-2r\cos\theta}{\sin\omega}\right]^2+(1-r^2)^2}. \tag{33.7}$$

Maximum value of $|H_1(e^{j\omega})|^2$ is reached when the first term in the denominator vanishes, i.e. at $\omega = \omega_0$, where

$$\cos\omega_0 = 2r\cos\theta/(1+r^2). \tag{33.8}$$

Comparing with Eq. 33.2, we, therefore, have

$$2r\cos\theta/(1+r^2)=\beta. \tag{33.9}$$

Also, combining Eqs. 33.7 and 33.8, we note that the maximum magnitude will be unity if

$$K=(1-r^2)/2, \tag{33.10}$$

and, combining Eq. 33.7 with Eq. 33.10, we can write

$$|H_1(e^{j\omega})|^2=\frac{1}{1+\left[\frac{(1+r^2)\cos\omega-2r\cos\theta}{(1-r^2)\sin\omega}\right]^2}. \tag{33.11}$$

Now, suppose, instead of the range $1/\sqrt{2} < |H(e^{j\omega})| < 1$, the pass-band is defined by $1/\sqrt{1+\varepsilon^2} < |H(e^{j\omega})| < 1$, where $\varepsilon < 1$ is arbitrary. Then from Eq. 33.11, the pass-band edge frequencies ω_2 and ω_1 will satisfy the equation

$$(1+r^2)\cos\omega-2r\cos\theta=\pm\epsilon(1-r^2)\sin\omega. \tag{33.12}$$

We let (at the moment arbitrarily, but justified by later results)

$$(1+r^2)\cos\omega_1-2r\cos\theta=\epsilon(1-r^2)\sin\omega_1, \tag{33.13}$$

and

$$(1+r^2)\cos\omega_2-2r\cos\theta=-\epsilon(1-r^2)\sin\omega_2. \tag{33.14}$$

Subtracting Eq. 33.14 from Eq. 33.13 gives

$$\begin{aligned}&(1+r^2)(\cos\omega_1-\cos\omega_2)\\&\quad=\epsilon(1-r^2)(\sin\omega_1+\sin\omega_2).\end{aligned} \tag{33.15}$$

Or,

$$(1+r^2)\sin\frac{\omega_2+\omega_1}{2}\sin\frac{\omega_2-\omega_1}{2} = \varepsilon(1-r^2)\sin\frac{\omega_2+\omega_1}{2}\cos\frac{\omega_2-\omega_1}{2}. \quad (33.16)$$

This gives

$$\tan\frac{\omega_2-\omega_1}{2} = \frac{\varepsilon(1-r^2)}{1+r^2}. \quad (33.17)$$

Using the relationships

$$\cos 2\,\theta = 2\cos^2\theta - 1 \text{ and } \sec^2\theta = 1+\tan^2\theta, \quad (33.18)$$

we get from Eq. 33.17, after simplification,

$$\cos(\omega_2-\omega_1) = \frac{(1+r^2)^2-\varepsilon^2(1-r^2)^2}{(1+r^2)^2+\varepsilon^2(1-r^2)^2}. \quad (33.19)$$

The 3-dB bandwidth B is obtained as $\omega_2 - \omega_1$ with $\varepsilon = 1$. Hence,

$$\cos\beta = \frac{2r^2}{1+r^4}. \quad (33.20)$$

Comparing Eq. 33.20 with Eq. 33.2, we get

$$r^2 = \alpha. \quad (33.21)$$

Substituting Eqs. 33.9, 33.10 and 33.21 in Eq. 33.5, it becomes identical with Eq. 33.1.

Note, in passing, that for small pass-band tolerance ($\varepsilon \to 0$) or small bandwidth ($\omega_2 - \omega_1 \to 0$) or both addition of Eqs. 33.13 and 33.14 gives

$$(1+r^2)(\cos\omega_1+\cos\omega_2) \approx 4r\cos\theta. \quad (33.22)$$

Or,

$$\cos\omega_1+\cos\omega_2 \approx 4r\cos\theta/(1+r^2) = 2\beta = 2\cos\omega_0. \quad (33.23)$$

Thus, $\cos\omega_1$ and $\cos\omega_2$ are approximately arithmetically symmetrical about $\cos\omega_0$.

Design for Arbitrary Pass-band Tolerance

Combining Eqs. 33.17 and 33.21, and letting β_ε denote the bandwidth for an arbitrary pass-band tolerance specified by ε, we get

$$\tan\frac{B_\varepsilon}{2} = \frac{\varepsilon(1-\alpha)}{1+\alpha}. \quad (33.24)$$

Thus, given ε and B_ε, one can choose α from

$$\alpha = \frac{\varepsilon - \tan\frac{B_\varepsilon}{2}}{\varepsilon + \tan\frac{B_\varepsilon}{2}}. \quad (33.25)$$

Equation 33.25 puts a constraint on the specifications of ε and B_ε. Since $0 < B_\varepsilon < \pi$, $\tan(B_\varepsilon/2) > 0$; also $\alpha = r^2$ is a real positive quantity. Hence, ε and B_ε must satisfy

$$\varepsilon > \tan\frac{B_\varepsilon}{2}. \quad (33.26)$$

This is quite logical because, with two parameters ε and β, one cannot satisfy three specifications, viz. ε, B_ε and ω_0. The constraint Eq. 33.26 is shown graphically in Fig. 33.1 in the form of a plot of $\varepsilon = \tan\frac{B_\varepsilon}{2}$. No specification point which lies below the curve can be met by the second-order BPF characterized by Eq. 33.1.

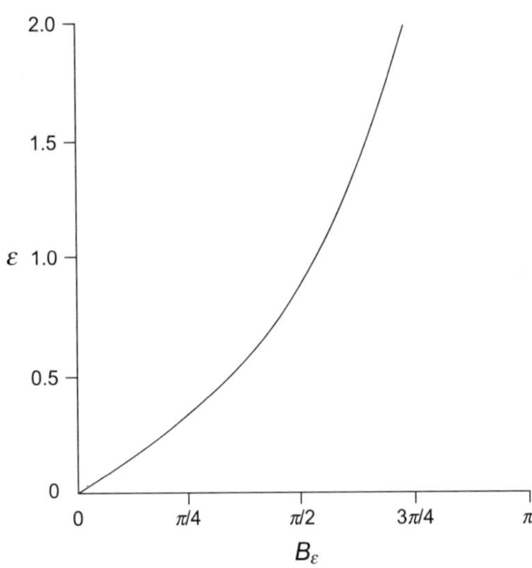

Fig. 33.1 Plot of $\varepsilon = \tan\frac{B_\varepsilon}{2}$. Specification points below the curve cannot be met by the second-order filter

Realization

In [1], the realizations of Eq. 33.1 and 33.3 have been derived from that of the all-pass filter

$$A_2(z) = \frac{\alpha - \beta(1+\alpha)z^{-1} + z^{-2}}{1 - \beta(1+\alpha)z^{-1} + \alpha z^{-2}}. \tag{33.27}$$

by noting that

$$H_{1,2}(z) = (1/2)[1 \pm A_2(z)]. \tag{33.28}$$

Figure 33.2 shows the implementation of Eq. 33.28, while Fig. 33.3 shows the realization of A_2(z) with a lattice structure using the two multipliers α and β.

We give here another realization of $A_2(z)$, starting from that of the transfer function

$$A_2(z) = \frac{d_2 + d_1 z^{-1} + z^{-2}}{1 + d_1 z^{-1} + d_2 z^{-2}}. \tag{33.29}$$

as given in [1] and reproduced in Fig. 33.4. For Eq. 33.28,

$$d_1 = -\beta(1+\alpha) \quad \text{and} \quad d_2 = \alpha. \tag{33.30}$$

The part marked by nodes A, B and C in Fig. 33.4 can be modified to the equivalent configuration shown in Fig. 33.5, where the two

Fig. 33.2 Implementation of Eq. 33.28

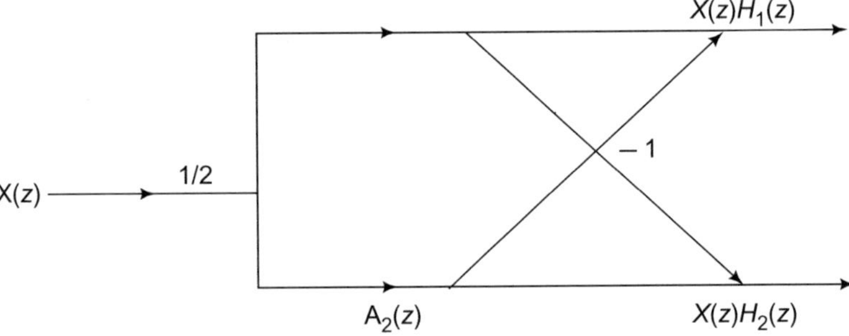

Fig. 33.3 Realization of A_2(z) of Eq. 33.27, as given in [1]

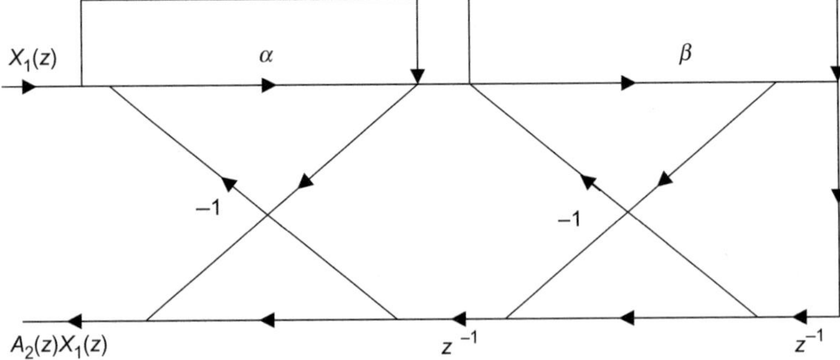

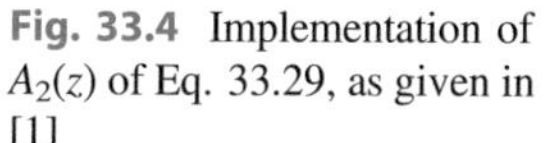

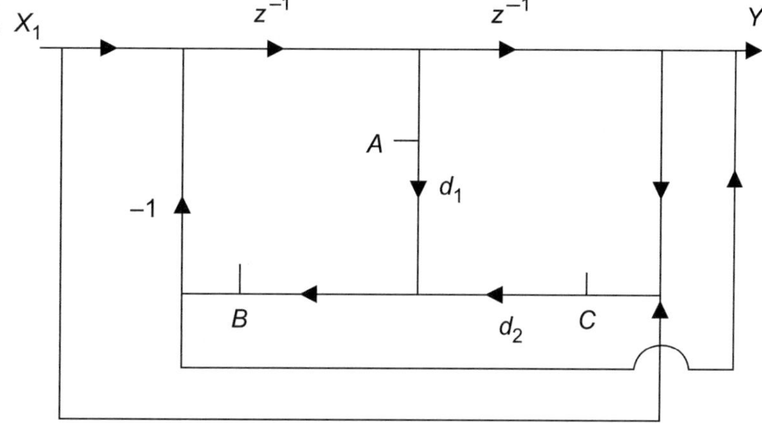

Fig. 33.4 Implementation of $A_2(z)$ of Eq. 33.29, as given in [1]

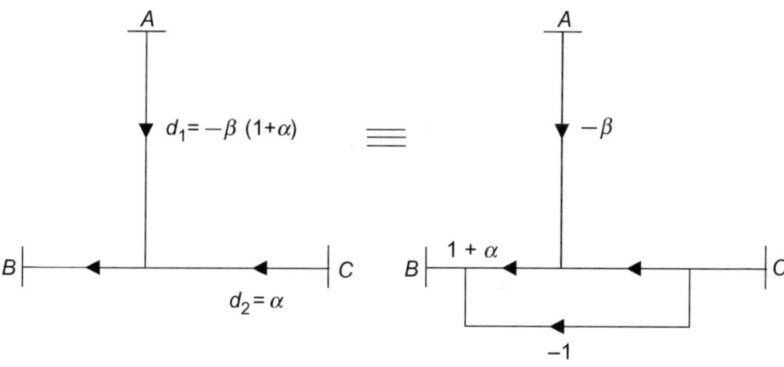

Fig. 33.5 Equivalent representation of the part ABC of Fig. 33.4

multipliers α and β have been separated out. Replacing the part ABC of Fig. 33.4 by part (b) of Fig. 33.5 gives an alternative to the lattice structure of Fig. 33.3. Whether other alternative structures are possible or not is left as an open problem for the reader.

Conclusion

A derivation has been given of the elegant second-order band-pass/band-stop filter transfer function, in which the two parameters α and β control the centre frequency and the difference between the pass-band edges, independently of each other. Design equations have been derived and the limitations of the design have been pointed out for arbitrary pass-band tolerance. An alternative canonic realization structure has also been presented, in which α and β are the only two multipliers.

Problems

P.1. Suppose $\beta = 0$ in Eq. 33.3. What kind of filter do you get?

P.2. In Eq. 33.3, investigate what happens when (i) $\beta = +1$ and (ii) $\beta = -1$.

P.3. Look at Eq. 33.19. Find cos $(\omega_2 + \omega_1)$ and find the product cos $(\omega_2 + 1)$ cos $(\omega_2 - \omega_1)$. Interpret the result.

P.4. What happens when $\beta = 0$ in Eq. 33.27.

P.5. What happens when (i) $\beta = +1$ and (ii) $\beta = -1$ in Eq. 33.27?

Reference

1. S.K. Mitra, *Digital Signal Processing—A Computer Based Approach, Second Edition* (McGraw-Hill, New York, 2000)

Derivation of Second-Order Canonic All-Pass Digital Filter Realizations

34

This chapter deals with the derivation of two canonic all-pass digital filter realizations, first proposed by Mitra and Hirano. In contrast to their derivation, which uses a three-pair approach, our derivation is much simpler because we use a two-pair approach, in which only four, instead of nine parameters have to be chosen.

Keywords
Canonical · All-pass · Digital filter Realizations

Introduction

All-pass digital filters have been recognized as basic building blocks of many digital signal processors [1]. Any arbitrary order all-pass filter can be realized by cascading first- and second-order ones only. Mitra and Hirano [2] proposed the canonic first- and second-order configurations shown in Figs. 34.1, 34.2 and 34.3, which realize, respectively, the following transfer functions:

$$A_1(z) = \frac{d_1 + z^{-1}}{1 + d_1 z^{-1}}, \tag{34.1}$$

$$A_2(z) = \frac{d_1 d_2 + d_1 z^{-1} + z^{-2}}{1 + d_1 z^{-1} + d_1 d_2 z^{2}}, \tag{34.2}$$

and

$$B_2(z) = \frac{d_2 + d_1 z^{-1} + z^{-2}}{1 + d_1 z^{-1} + d_2 z^{-2}}. \tag{34.3}$$

Note that the transformed forms of these structures [3] will also give canonic realizations of the same transfer functions; these are not being considered in this chapter.

The derivation of the first-order structure was given in [2] by using the two-pair approach, i.e. by assuming a multiplier-less two-pair with the single multiplier d_1 as its termination, as shown in Fig. 34.4. Using the two-pair relationship

$$\begin{bmatrix} Y_1 \\ Y_2 \end{bmatrix} = \begin{bmatrix} t_{11} & t_{12} \\ t_{21} & t_{22} \end{bmatrix} \begin{bmatrix} X_1 \\ X_2 \end{bmatrix}, \tag{34.4}$$

and the terminating constraint:

$$X_2 = d_1 Y_2, \tag{34.5}$$

Source: S. C. Dutta Roy, P. Uday Kiran, Bhargav R. Vyas,Tarun Aggarwal and D. G. Senthil Kumar, "Derivation of Second-Order Canonic All-Pass Digital Filter Realizations", *IETE Journal of Education*, vol. 47, pp. 153–157, October–December 2006.

S. C. Dutta Roy, *Circuits, Systems and Signal Processing*,
https://doi.org/10.1007/978-981-10-6919-2_34

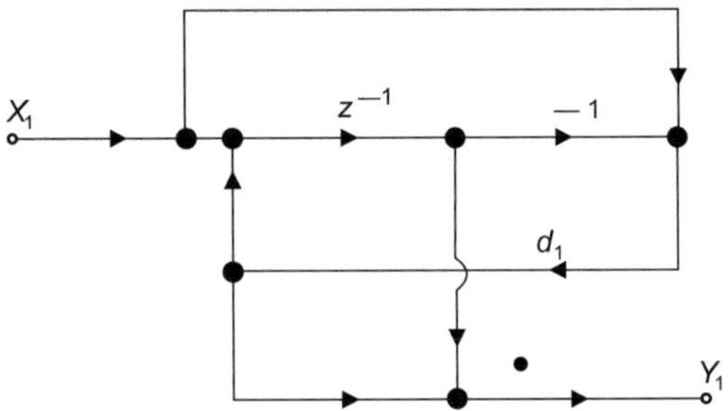

Fig. 34.1 Canonical realization of Eq. 34.1

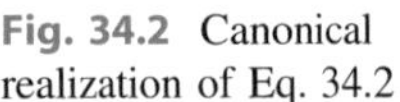

Fig. 34.2 Canonical realization of Eq. 34.2

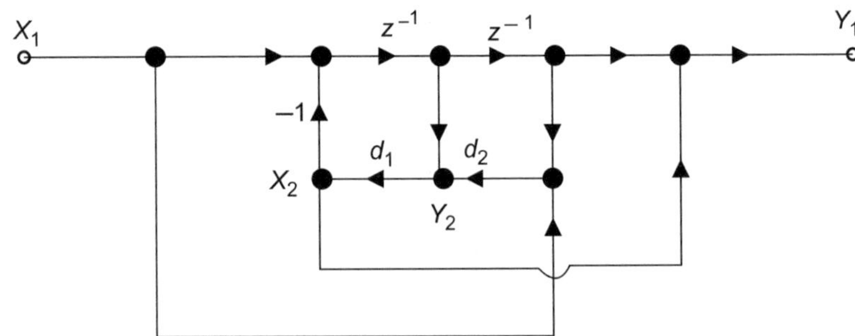

Fig. 34.3 Canonical realization of Eq. 34.3

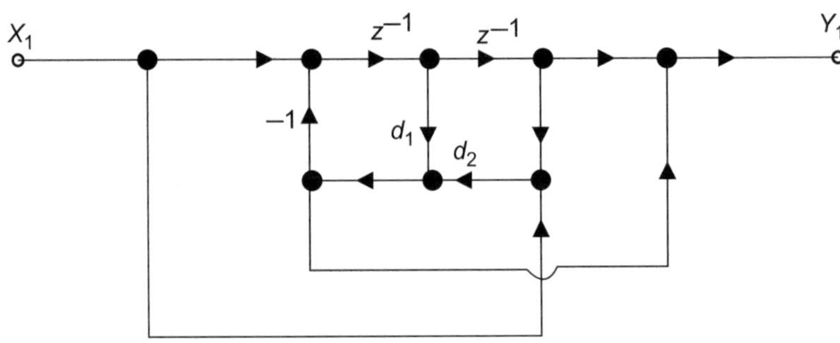

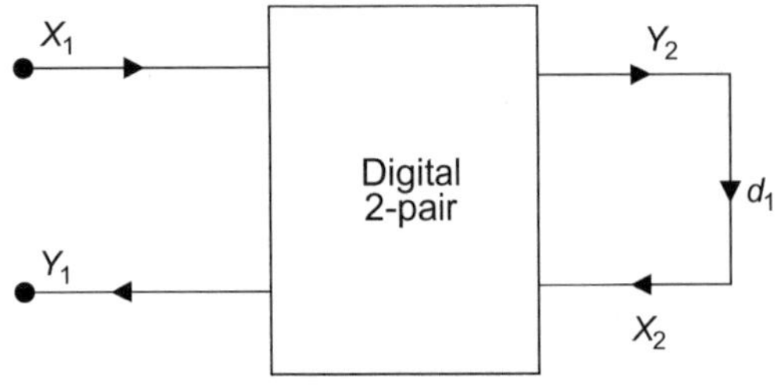

Fig. 34.4 Digital two-pair terminated in multiplier d_1

one obtains

$$A_1(z) = \frac{Y_1}{X_1} = \frac{t_{11} - d_1(t_{11}t_{12} - t_{12}t_{21})}{1 - d_1 t_{22}}. \quad (34.6)$$

Comparing Eq. 34.6 with Eq. 34.1, we see that various choices are possible, of which the set

$$t_{11} = z^{-1},\ t_{22} = -z^{-1},\ t_{12} = 1 + z^{-1},\ \text{and}$$

$$t_{21} = 1 - z^{-1} \quad (34.7)$$

gives the structure of Fig. 34.1, while interchanging the expressions of t_{12} and t_{21} in Eq. 34.7 gives the transposed form of Fig. 34.1.

If we follow the same procedure for deriving the structures for Eqs. 34.2 and 34.3, we have to start with a multiplier-less three-pair, two of whose pairs will be terminated in d_1 and d_2. As given in [2], analysis for the required t-parameters of the 3×3 t-matrix becomes quite involved. The purpose of this chapter is to present much simpler derivations of the structures of Figs. 34.2 and 34.3, by using the two-pair approach only.

Derivation of the Structure of Fig. 34.2

With the aim of deriving the structure of Fig. 34.2, we start with same constrained two-pair shown in Fig. 34.4, where the two-pair is no longer multiplier-less. Instead, it contains one multiplier (d_2) and two delays. Following the steps of Eqs. 34.4, 34.5 and 34.6, we now have to match the right-hand sides of Eqs. 34.6 and 34.2, i.e.

$$\frac{t_{11} - d_1(t_{11}t_{22} - t_{12}t_{21})}{1 - d_1 t_{22}} = \frac{d_1 d_2 + d_1 z^{-1} + z^{-2}}{1 + d_1 z^{-1} + d_1 d_2 z^{-2}} \tag{34.8}$$

An obvious set of simple choices is the following:

$$t_{11} = z^{-2}, t_{22} = -(z^{-1} + d_2 z^{-2}), \tag{34.9}$$

$$t_{11}t_{22} - t_{12}t_{21} = -(d_2 + z^{-1}).$$

From Eq. 34.9, we get

$$\begin{aligned} t_{12}t_{21} &= t_{11}t_{22} + d_2 + z^{-1} \\ &= -z^{-2}(z^{-1} + d_2 z^{-2}) + d_2 + z^{-1}, \end{aligned} \tag{34.10}$$

which can be simplified to the following:

$$t_{12}t_{21} = (1 - z^{-2})(d_2 + z^{-1} + d_2 z^{-2}) \tag{34.11}$$

Again, obvious choices of t_{12} and t_{21} are as follows:

$$t_{12} = 1 - z^{-2} \tag{34.12}$$

and

$$t_{21} = d_2 + z^{-1} + d_2 z^{-2}. \tag{34.13}$$

From Eqs. 34.4, 34.9, 34.12 and 34.13, we get

$$Y_1 = z^{-2}X_1 + (1 - z^{-2})X_2, \tag{34.14}$$

and

$$Y_2 = (d_2 + z^{-1} + d_2 z^{-2})X_1 - (z^{-1} + d_2 z^{-2})X_2. \tag{34.15}$$

Equations 34.14 and 34.15 can be rewritten in the following forms

$$Y_1 = z^{-2}(X_1 - X_2) + X_2, \tag{34.16}$$

and

$$Y_2 = z^{-1}(X_1 - X_2) + d_2\left[X_1 + z^{-2}(X_1 - X_2)\right] \tag{34.17}$$

It is easily verified that Fig. 34.2 is a realization of these two equations, where, for convenience, the locations of the signals Y_2 and X_2 are also indicated.

Derivation of the Structure of Fig. 34.3

In contrast to the derivation of [2], which again uses the three-pair approach, ab initio, for Eq. 34.3, we derive the structure of Fig. 34.3 from that of Fig. 34.2 by an elementary manipulation. Let, in Eq. 34.3, $d_2 = d_1\ p_2$. Then Eq. 34.3 becomes identical in form to Eq. 34.2 with d_2 replaced by p_2. In the resulting diagram, the part with p_2, Y_2 and d_1 is reproduced in

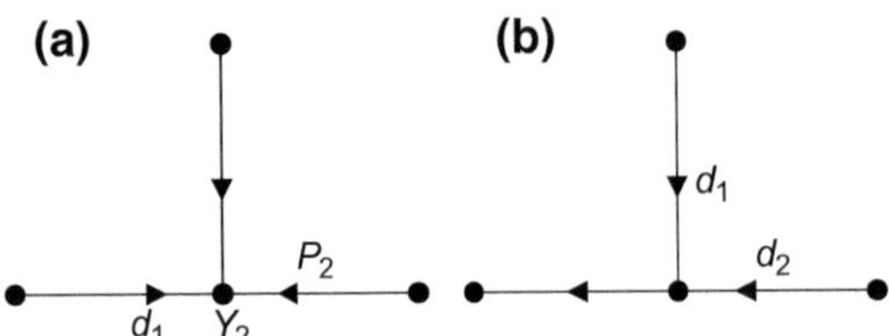

Fig. 34.5 **a** Part of modified Fig. 34.2 and **b** its equivalent

Fig. 34.5a. Shifting d_1 to the two inputs of summing point gives the equivalent diagram of Fig. 34.5b. Replacing the latter in the original diagram gives the configuration of Fig. 34.3.

Alternative Derivation of the Structure of Fig. 34.2

We now ask the question: If, in Fig. 34.4, we replace d_1 by d_2, can we get another canonic realization of Eq. 34.2? In effect, then, we should have

$$\frac{z^{-2}+d_1z^{-1}+d_1d_2}{d_1d_2z^{-2}+d_1z^{-1}+1}=\frac{t_{11}-d_2(t_{11}t_{22}-t_{12}t_{21})}{1-d_2t_{22}}. \tag{34.18}$$

In order to have the two denominators in Eq. 34.18 of the same form, we now divide both the numerator and denominator of the left-hand side of Eq. 34.18 by $(1+d_1z^{-1})$. Then, we identify

$$t_{22}=-\frac{d_1z^{-2}}{1+d_1z^{-1}}. \tag{34.19}$$

Correspondingly, for the modified numerator, we get

$$t_{11}t_{22}-t_{12}t_{21}=-\frac{d_1}{1+d_1z^{-1}}, \tag{34.20}$$

and

$$t_{11}=\frac{z^{-2}+d_1z^{-1}}{1+d_1z^{-1}}, \tag{34.21}$$

Then from Eqs. 34.19–34.21, we get

$$t_{12}t_{21}=\frac{d_1}{1+d_1z^{-1}}-\frac{z^{-2}+d_1z^{-1}}{1+d_1z^{-1}}\cdot\frac{d_1z^{-2}}{1+d_1z^{-1}}, \tag{34.22}$$

which, on simplification, gives

$$t_{12}t_{21}=\frac{d_1}{1+d_1z^{-1}}\cdot\frac{(1-z^{-2})(1+d_1z^{-1}+z^{-2})}{1+d_1z^{-1}}. \tag{34.23}$$

Subject to the constraint of Eq. 34.23, various choices for t_{12} and t_{21} are possible. By some trial and error, we have found the following choices to yield a canonic solution:

$$t_{12}=\frac{d_1(1-z^{-2})}{1+d_1z^{-1}}, \tag{34.24}$$

and

$$t_{21}=\frac{(1+d_1z^{-1}+z^{-2})}{1+d_1z^{-1}}=1+\frac{z^{-2}}{1+d_1z^{-1}}. \tag{34.25}$$

We now have the following basic equations:

$$Y_1=\frac{z^{-2}+d_1z^{-1}}{1+d_1z^{-1}}X_1+\frac{d_1(1-z^{-2})}{1+d_1z^{-1}}X_2, \tag{34.26}$$

and

$$Y_2=\left(1+\frac{z^{-2}}{1+d_1z^{-1}}\right)X_1-\frac{d_1z^{-2}}{1+d_1z^{-1}}X_2. \tag{34.27}$$

A systematic procedure for obtaining the realization diagram is depicted in Fig. 34.6. Part (a) of the figure shows how $t_{11}X_1$ is obtained with $X_2=0$, while part (b) of the same figure shows how the same hardware can realize $t_{12}X_2$ with $X_1=0$. Superimposing parts (a) and (b), we get the part (c) with the solid lines, which give the output Y_1. To obtain the output Y_2, note the value of the signal at node A, as indicated, and that just adding X_1 to it gives Y_2 according to Eq. 34.26.

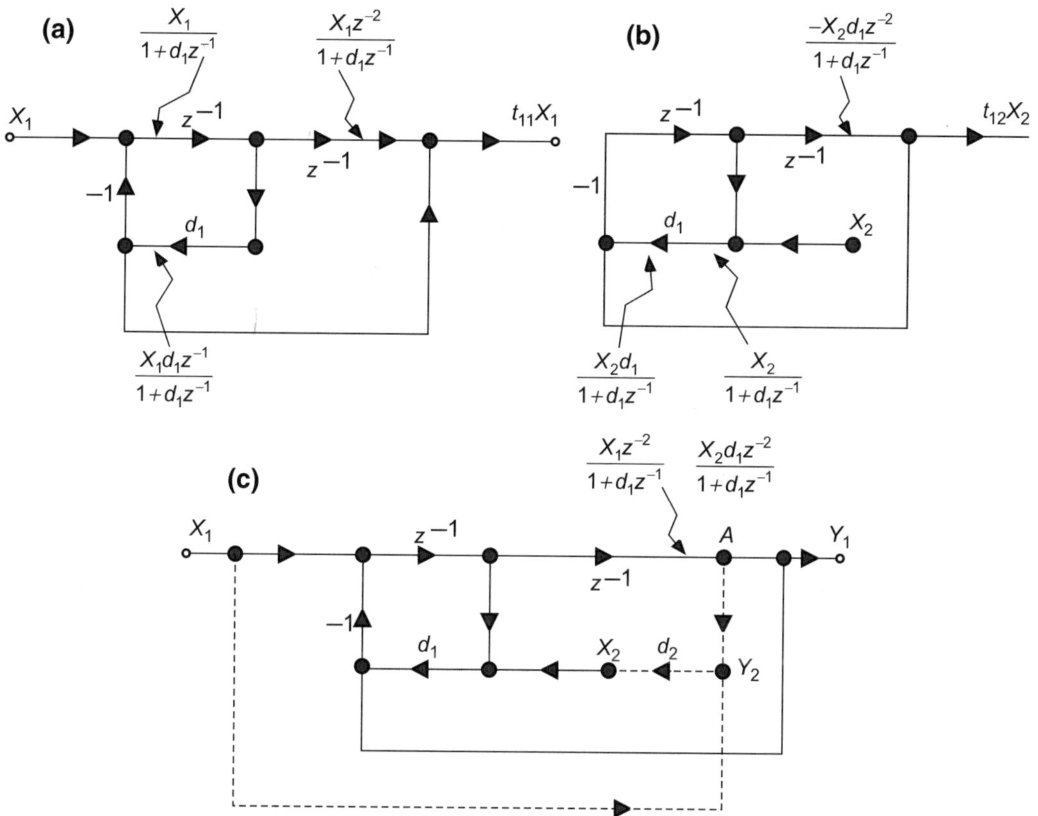

Fig. 34.6 Steps in the realization of Eqs. 34.25 and 34.26

Finally, multiplying Y_2 by d_2 gives X_2, as indicated by the broken lines in Fig. 34.6c. This configuration is indeed identical with that of Fig. 34.2.

Besides the transposed structures of Figs. 34.2 and 34.3, are there other canonical possibilities? We leave this as an open question to the reader.

Yet Another Derivation of the Structure of Fig. 34.2

Another question that arises at this point is the following: If, in Fig. 34.4, we replace d_1 by z^{-1}, and aim for a two-pair containing one delay and the multipliers d_1 and d_2, do we get a new structure? It is left to the readers to verify that under this constraint, one possible set of choices for the t-parameters is the following:

$$t_{11} = d_1d_2,\ t_{22} = -\left(d_1 + d_1d_2z^{-1}\right),\ t_{12} = (1-d_1d_2) \text{ and } t_{21} = d_1 + (1+d_1d_2)z^{-1}, \tag{34.28}$$

and that the resulting realization is the transpose of that of Fig. 34.2.

Conclusion

In this chapter, we have derived canonic realizations of second-order all-pass digital filters by a procedure, which is much simpler than that of the original three-pair approach of [2]. We have also derived the realization of Eq. 34.3 from that for Eq. 34.2 by an elementary manipulation of Fig. 34.2; this is drastically simpler as compared to the repetition of the three-pair approach, as done in [2].

Problems

P.1. Write down the transfer function of a third-order all-pass filter and draw its structure.

P.2. Do the same for a fourth-order transfer function.

P.3. When two digital 2-points are cascaded, how do you find the overall parameters. Do this in terms of t-parameter.

P.4. Two first-order all-pass transfer functions are cascaded. They have different d_1s (refer to Eq. 34.1). Find the overall transfer function and find its characteristics.

P.5. Do the same for two second-order ones, having different d_1 and d_2 (refer to Eq. 34.2).

References

1. P.A. Regalia, S.K. Mitra, P.P. Vaidyanathan, The digital all-pass network: a versatile signal processing building block. Proc. IEEE **76**, 19–37 (1988)
2. S.K Mitra, K. Hirano, Digital all-pass networks. IEEE Trans. Circ. Sys. **21**, 688–700 (September 1974)
3. A.V. Oppenheim, R.W. Schafer, J.R. Buck, *Discrete-Time Signal Processing* (Prentice Hall, New Jersey, 2000), p. 363

Derivation of the FIR Lattice Structure

35

A simple derivation is presented for the FIR lattice structure, based on the digital two-pair concept. Go ahead, read it and judge for yourself whether it is simple or not!

Keywords
Lattice structure · Realization

Introduction

In discussing the FIR lattice structure, it is usual in textbooks on digital signal processing (see, e.g. [1–3]) to assume the configuration of Fig. 35.1a, where each section is of the form shown in Fig. 35.1b, for realizing the transfer function

$$H_N(z) = \frac{X_N(z)}{X_0(z)} = 1 + \sum_{n=1}^{N} a_n^{(N)} z^{-n} \qquad (35.1)$$

and then derive the recursion formula for the coefficients of the lower order transfer functions

$$\begin{aligned} H_i(z) &= \frac{X_i(z)}{X_0(z)} \\ &= 1 + \sum_{n=1}^{i} a_n^{(i)} z^{-n}, \quad i = N-1 \text{ to } 1. \end{aligned} \qquad (35.2)$$

In the process, one finds the multipliers as

$$k_i = a_i^{(i)} \qquad (35.3)$$

and also the relationship

$$H_i'(z) = \frac{X_i'(z)}{X_0(z)} = z^{-i} H_i(z^{-1}). \qquad (35.4)$$

i.e. the two transfer functions $H_i(z)$ and $H_i'(z)$ are a pair of mirror image polynomials. Specifically, with $H_i(z)$ given by Eq. 35.2.

$$H_i'(z) = z^{-i} + \sum_{n=0}^{i-1} a_{i-n}^{(i)} z^{-n}. \qquad (35.5)$$

Even Mitra [1], who introduced the concept of digital two-pair [4] and used the same to derive IIR lattice structures, did not use it to derive FIR lattice structure. However, Vaidyanathan [5], a former student of Mitra, used this approach to derive a variety of FIR lattice structures for the

Source: S. C. Dutta Roy, "Derivation of the FIR Lattice Structure", *IETE Journal of Education*, vol. 45, pp. 211–212, October–December 2004.

S. C. Dutta Roy, *Circuits, Systems and Signal Processing*,
https://doi.org/10.1007/978-981-10-6919-2_35

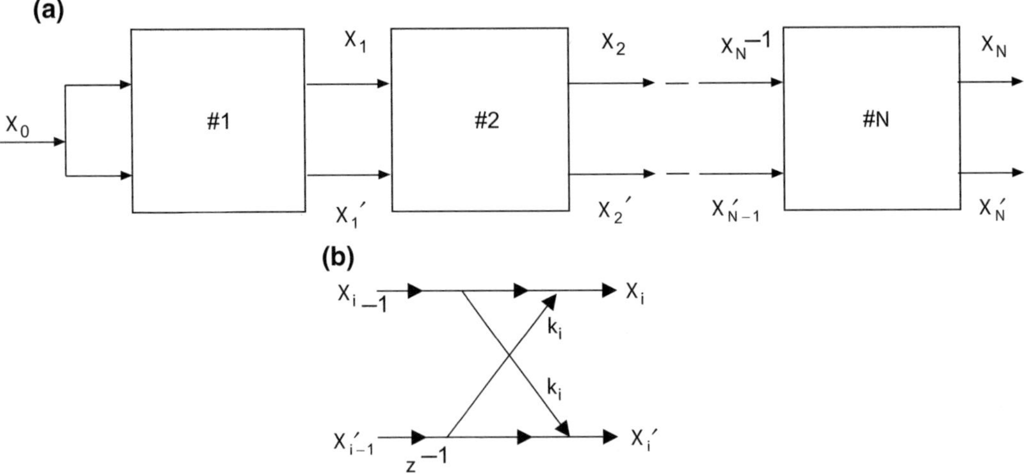

Fig. 35.1 **a** The general cascaded FIR lattice structure. **b** The *i*th section of (***a***)

so-called 'lossless bounded real (LBR)' transfer functions.

We present here a simple derivation of the FIR lattice structure of Fig. 35.1 using the two-pair approach, for a general transfer function of the form Eq. 35.1 with the only constraints of Eqs. 35.2 and 35.4, and no others.

Derivation

Consider the ith stage of the FIR lattice, shown in Fig. 35.1b, and let it be characterized by the transmission matrix

$$T^{(i)} = \begin{bmatrix} t_{11}^{(i)} & t_{12}^{(i)} \\ t_{21}^{(i)} & t_{22}^{(i)} \end{bmatrix}. \qquad (35.6)$$

For simplicity, we shall drop the superscript (i) in the following discussion. Equation 35.6 implies that

$$X_i(z) = t_{11}X_{i-1}(z) + t_{12}X'_{i-1}(z) \qquad (35.7)$$

$$X'_i(z) = t_{21}X_{i-1}(z) + t_{22}X'_{i-1}(z). \qquad (35.8)$$

In terms of the transfer functions $H_i(z)$ and $H'_i(z)$, Eqs. 35.7 and 35.8 translate to the following:

$$H_i(z) = t_{11}H_{i-1}(z) + t_{12}H'_{i-1}(z), \qquad (35.9)$$

$$H'_i(z) = t_{21}H_{i-1}(z) + t_{22}H'_{i-1}(z). \qquad (35.10)$$

Observe that in Eq. 35.9, t_{11} has to be unity in order to satisfy the requirement of Eq. 35.2 that the constant term in $H_i(z)$ should be unity. Also, for satisfying the requirement that $H_i(z)$ should be a polynomial of order i in z^{-1}, t_{12} must be of the form $k_i z^{-1}$. Thus, we have

$$t_{11} = 1 \text{ and } t_{12} = k_i z^{-1}. \qquad (35.11)$$

We next put the constraint of Eq. 35.4 and get, from Eqs. 35.9 and 35.11,

$$\begin{aligned} H'_i(z) &= z^{-i}H_i(z^{-1}) \\ &= z^{-i}\left[H_{i-1}(z^{-1}) + k_i z H'_{i-1}(z^{-1})\right] \\ &= z^{-i}H_{i-1}(z^{-1}) + k_i z^{-(i-1)}H'_{i-1}(z^{-1}) \\ &= z^{-1}z^{-(i-1)}H_{i-1}(z^{-1}) + k_i z^{-(i-1)}H'_{i-1}(z^{-1}) \\ &= k_i H_{i-1}(z) + z^{-1}H'_{i-1}(z) \end{aligned} \qquad (35.12)$$

Comparing the last line of Eq. 35.12 with Eq. 35.10, we observe that

$$t_{21} = k_i \text{ and } t_{22} = z^{-1} \qquad (35.13)$$

The structure resulting from Eqs. 35.11 and 35.13 is precisely that of Fig. 35.1b.

To obtain the coefficients of $H_{i-1}(z)$ from those of $H_i(z)$, one follows the same procedure as in [1]. The result is

$$a_n^{(i-1)} = \frac{a_n^{(i)} - k_i a_{i-n}^{(i)}}{1 - k_i^2}, \ n = i - 1 \text{ to } 1,$$
$$i = N - 1 \text{ to } 1 \tag{35.14}$$

Concluding Comments

A simple derivation has been presented for the FIR lattice structure on the basis of the constraint that the two transfer functions obtained at the output of any section bear a mirror image relationship to each other.

It is not difficult to appreciate that other lattice structures are possible to derive by assuming some other relationship between $H_i(z)$ and $H_i'(z)$, e.g.

$$H_i'(z) = \pm z^{-i} H_i(-z^{-1}). \tag{35.15}$$

These structures, when carefully derived, differ from those of Fig. 35.1b in one or more of the following aspects: the position of the delay branch; positions of the multipliers; relative signs of the two multipliers, i.e. one multiplier may be the negative of the other; and nonuniformity of the signs of the multipliers from one section to thc next. Examples of such structures can be found in [5]. Detailed derivation of these structures for the general transfer function of Eq. 35.1 and their recurrence relations will be presented in a later chapter.

Problems

P.1. What happens if the lower bound slanting arrows in Fig. 35.1b points in the opposite direction? What if the upper bound slanting arrow points in the opposite direction? What if both do the same?

P.2. Suppose all arrows in Fig. 35.1a are reversed. What kind of transfer function do we get?

P.3. Suppose x_1, x_1' arrows point in the opposite direction. What kind of transfer function would you get?

P.4. Re-derive the equations in terms of transmission parameters.

P.5. Besides *t*- and transmission parameters, what other parameter are meaningful in the context of digital two-pairs? How are they related *t*- and transmission parameters?

References

1. S.K. Mitra, *Digital Signal Processing—A Computer-Based Approach* (McGrawHill, New York, 2001)
2. A.V. Oppenheim, R.W. Schafer, *Discrete-Time Signal Processing, Englewood Cliffs* (Prentice Hall, NJ, 1989)
3. J.G. Proakis, D.G. Manolakis, *Introduction to Digital Signal Processing* (Macmillan, New York, 1989)
4. S.K. Mitra, R.J. Sherwood, Digital ladder networks. IEEE Trans. Audio Electroacoust. AU-**21**, 30–36 (February 1973)
5. P.P. Vaidyanathan, Passive cascaded lattice structures for low-sensitivity FIR design, with applications to filter banks. IEEE Trans. Circ. Syst. CAS-**33**, 1045–1064 (November 1986)

36 Solution to a Problem in FIR Lattice Synthesis

In FIR lattice synthesis, if at any but the last stage, a lattice parameter becomes ± 1, then the synthesis fails. A linear phase transfer function is an example of this situation. This chapter, written in a tutorial style, is concerned with a simple solution to this problem, demonstrated through simple examples, rather than detailed mathematical analysis, some of which is available in (Dutta Roy in IEE Proc-Vis Image Signal Process 147:549–552, 2000 [1]).

Keywords
FIR filters · Lattice synthesis

Introduction

Consider the FIR transfer function

$$H_N(z) = 1 + \sum_{n=1}^{N} h_N(n) z^{-n} \tag{36.1}$$

which is known to be realizable by the lattice structure of Fig. 36.1a, where each k_i block has the structure shown in Fig. 36.1b, provided that at no stage in the synthesis procedure except the last one, one encounters a parameter $k_i = \pm 1$. For example, if $h_N(N) = \pm 1$, one cannot obtain a lattice by the usual procedure. This problem was not adequately addressed to in the literature (see e.g. [3–5]), and was solved in [1] for the linear phase case with rigorous mathematical analysis and proofs. However, the solution for the non-linear phase case with $h_N(N) = \pm 1$ was given in [1] in terms of parallel lattices, which, in general, is neither delay canonic nor multiplier canonic. In this chapter, written in a tutorial style, we present the essence of [1] through several simple examples for easy comprehension by the students. We also give a canonic solution to the nonlinear phase case through a tapped lattice structure.

Note that although two multipliers have been shown in Fig. 36.1b, each lattice section should also be realized by a single multiplier structure. Unfortunately, however, such a structure does not exist as yet. However, we shall refer to the structure of Fig. 36.1 as multiplier canonic, even if we show two multipliers in each lattice section.

Source: S. C. Dutta Roy, 'Solution to a Problem in FIR Lattice Synthesis', *IETE Journal of Education*, vol. 43, pp. 33–36, January–March 2002 (Corrections on p. 219, October–December 2002).

S. C. Dutta Roy, *Circuits, Systems and Signal Processing*,
https://doi.org/10.1007/978-981-10-6919-2_36

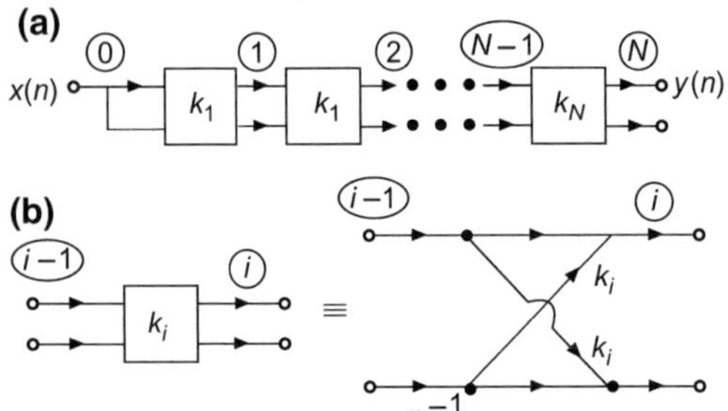

Fig. 36.1 **a** General FIR lattice structure **b** Composition of the *i*th block in (***a***)

Conventional Synthesis Procedure

We first review the conventional FIR lattice synthesis procedure [2]. By analysis of Fig. 36.1, it is easily shown that if $H_i(z)$ is the transfer function from node 0 to node i, then

$$H_i(z) = H_{i-1}(z) + k_i z^{-i} H_{i-1}(z^{-1}), \; i = 1 \to N \tag{36.2}$$

Also, $H_i(z)$ is of the form

$$H_i(z) = 1 + \sum_{n=1}^{i} h_i(n) z^{-n} \tag{36.3}$$

The lattice parameters (k_i) of Fig. 36.1 are given by

$$k_i = h_i(i), \; i = 1 \to N \tag{36.4}$$

To obtain $H_{i-1}(z)$ from $H_i(z)$, $i = N \to 2$, one uses the following recursion formula:

$$h_{i-1}(n) = [h_i(n) - k_i h_i(i-n)]/(1-k_i^2) \quad i = N \to 2 \tag{36.5}$$

Obviously, if $k_i = \pm 1$, then the synthesis fails!

We now illustrate the conventional procedure by an example.

Fig. 36.2 Lattice structure for the transfer function given by Eq. 36.6

Example 1

Let

$$H_3(z) = 1 + 0.5z^{-1} + 0.3z^{-2} - 0.2z^{-3} \tag{36.6}$$

Here, $h_3(1) = 0.5$, $h_3(2) = 0.3$ and $h_3(3) = -0.2$. Hence $k_3 = h_3(3) = -0.2$. From Eq. 36.5,

$$h_2(1) = [h_3(1) - k_3 h_3(2)]/(1-k_3^2) = 0.53846 \tag{36.7}$$

and

$$h_2(2) = [h_3(2) - k_3 h_3(1)]/(1-k_3^2) = 0.38462 \tag{36.8}$$

Hence,

$$H_2(z) = 1 + 0.53846z^{-1} + 0.38462z^{-2} \tag{36.9}$$

so that $k_2 = 0.38462$. Apply Eq. 36.5 again to get

$$\begin{aligned} h_1(1) &= [h_2(1) - k_2 h_2(1)]/(1-k_2^2) \\ &= h_2(1)/(1+k_2) = 0.38889 \end{aligned} \tag{36.10}$$

Thus

$$H_1(z) = 1 + 0.38889z^{-1} \tag{36.11}$$

giving $k_1 = 0.38889$. The synthesis is thereby complete and the resulting structure is shown in Fig. 36.2.

Linear Phase Transfer Function

As is well known, there are four different types of linear phase transfer functions, viz. (1) symmetrical impulse response of even length;

(2) symmetrical impulse response of odd length; (3) asymmetrical impulse response of even length; and (4) asymmetrical impulse response of odd length. We shall consider each of these cases through simple examples.

Example 2: Illustrating Case 1
Let

$$H_5(z) = 1 + h_5(1)z^{-1} + h_5(2)z^{-2} + h_5(2)z^{-3} + h_5(1)z^{-4} + z^{-5} \tag{36.12}$$

Here $k_5 = 1$. Note that Eq. 36.12 can be rewritten as

$$H_5(z) = \left[1 + h_5(1)z^{-1} + h_5(2)z^{-2}\right] + (1)z^{-5}\left[1 + h_5(1)z + h_5(2)z^2\right] \tag{36.13}$$

Comparing this with Eq. 36.2 with $i = 5$, we note that

$$H_4(z) = 1 + h_5(1)z^{-1} + h_5(2)z^{-2} = H_2(z) \tag{36.14}$$

because the order of the polynomial is 2. This order reduction means that $k_4 = k_3 = 0$. Also $k_2 = h_5(2)$, and by the formula in Eq. 36.10, we get

$$\begin{aligned} k_1 &= h_1(1) = h_2(1)/(1 + k_2) \\ &= h_5(1)/[1 + h_5(2)] \end{aligned} \tag{36.15}$$

The synthesis is now complete and the resulting structure is shown in Fig. 36.3. Note that only two non-trivial lattice parameters are needed for the synthesis of a fifth-order transfer function; this is what it should be, because there are only two independent parameters in the transfer function 36.12. This can be easily generalized to the Nth-order case, N odd, which will require $(N-1)/2$ non-trivial lattice sections to begin with, followed by the same number of simple delays, and ending in one unity parameter lattice section.

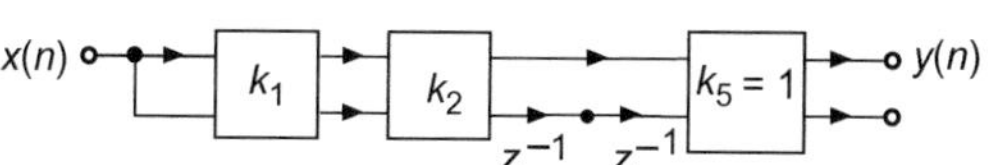

Fig. 36.3 Synthesis of Eq. 36.12

Example 3: Illustrating case 2
Let

$$_4(z) = 1 + h_4(1)z^{-1} + h_4(2)z^{-2} + h_4(1)z^{-3} + z^{-4} \tag{36.16}$$

Here we have $k_4 = 1$; also Eq. 36.16 can be rewritten as

$$H_4(z) = \left[1 + h_4(1)z^{-1} + (1/2)h_4(2)z^{-2}\right] + (1)z^{-4}\left[1 + h_4(1)z + (1/2)h_4(2)z^2\right] \tag{36.17}$$

Combining this with Eq. 36.2 with $i = 4$, we get

$$H_3(z) = 1 + h_4(1)z^{-1} + (1/2)h_4(2)z^{-2} = H_2(z) \tag{36.18}$$

which implies that $k_3 = 0$ and $k_2 = (1/2)h_4(2)$. Finally, as in Eq. 36.15, we have

$$\begin{aligned} k_1 &= h_1(1) = h_2(1)/(1 + k_2) \\ &= h_4(1)/[1 + (1/2)h_4(2)] \end{aligned} \tag{36.19}$$

The resulting lattice is shown in Fig. 36.4. In general, for an Nth-order transfer function with N even, we shall require $N/2$ non-trivial lattice sections, $(N/2)-1$ simple delays, and a unity parameter lattice section.

Example 4: Illustrating case 3
Let

$$H_5(z) = 1 + h_5(1)z^{-1} + h_5(2)z^{-2} - h_5(2)z^{-3} - h_5(1)z^{-4} - z^{-5} \tag{36.20}$$

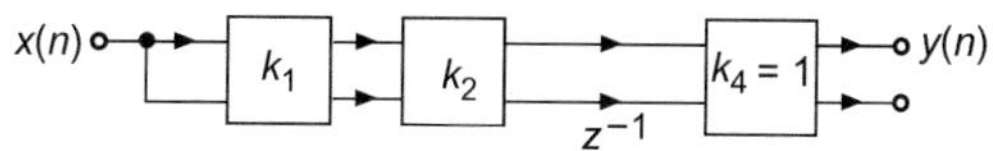

Fig. 36.4 Synthesis of Eq. 36.16

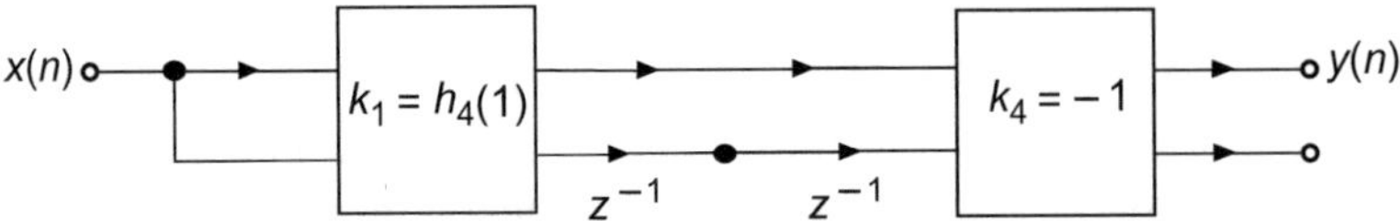

Fig. 36.5 Synthesis of Eq. 36.22

Here $k_5 = -1$ and we can rewrite Eq. 36.20 as

$$H_5(z) = \left[1 + h_5(1)z^{-1} + h_5(2)z^{-2}\right] + (-1)z^{-5}\left[1 + h_5(1)z + h_5(2)z^2\right] \quad (36.21)$$

Comparing with Example 2, we see that the lower order polynomial is the same in this case also. Hence, the realization of Fig. 36.2 is valid for this case also, except that the last lattice section will have $k_5 = -1$.

Example 5: Illustrating case 4
Let

$$H_4(z) = 1 + h_4(1)z^{-1} - h_4(1)z^{-3} - z^{-4} \quad (36.22)$$

Note that because of asymmetry, $h_4(2)$ is identically zero. Here, we have $k_4 = -1$ and we can write

$$H_4(z) = \left[1 + h_4(1)z^{-1}\right] + (-1)z^{-4}[1 + h_4(1)z] \quad (36.23)$$

Thus,

$$H_3(z) = 1 + h_4(1)z^{-1} = H_1(z) \quad (36.24)$$

i.e. $k_3 = k_2 = 0$ and $k_1 = h_4(1)$. The synthesis is, therefore, complete and the resulting structure is shown in Fig. 36.5. In general, for N even and asymmetric impulse response, there will be $(N/2) - 1$ non-trivial lattice sections, $N/2$ simple delays and a last lattice section with the parameter -1.

Nonlinear Phase FIR Function with $h_N(N) = \pm 1$

Let

$$H_N(z) = 1 + \sum_{n=1}^{N-1} h_N(n)z^{-n} \pm z^{-N} \quad (36.25)$$

where N may be even or odd. We first consider the case of even N and illustrate our new procedure with examples.

Example 6
Let

$$H_4(z) = 1 + h_4(1)z^{-1} + h_4(2)z^{-2} + h_4(3)z^{-3} + z^{-4} \quad (36.26)$$

This can be decomposed as follows:

$$H_4(z) = \left[1 + h_4(3)z^{-1} + h_4(2)z^{-2} + h_4(3)z^{-3} + z^{-4}\right] + \left\{[h_4(1) - h_4(3)]z^{-1}\right\} \quad (36.27)$$

The first transfer function (within square brackets) is linear phase and is the same as Eq. 36.16 with $h_4(1)$ replaced by $h_4(3)$. The second transfer function in Eq. 36.27 (within curly brackets) can be realized by tapping the k_1 block after the delay z^{-1}, multiplying it by $[h_4(1) - h_4(3)]$, and adding it to the main output, as shown in Fig. 36.6.

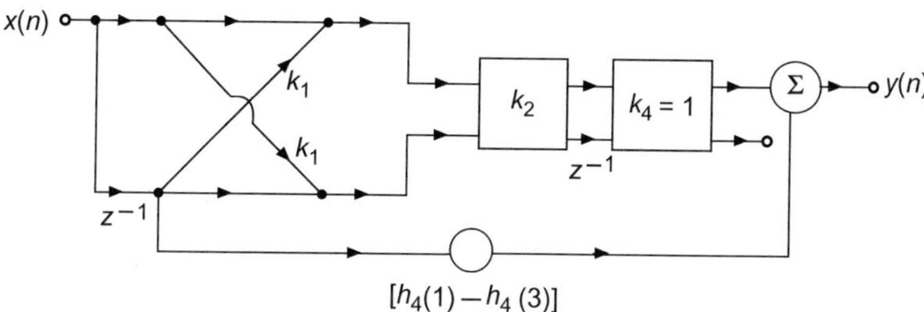

Fig. 36.6 Synthesis of Eq. 36.26: $k_1 = h_4(3)/[1 + (1/2)h_4(2)]$ and $k_2 = (1/2)h_4(2)$

Note that if each lattice can be realized by a single multiplier structure; then the realization of Fig. 36.6 will be a delay, as well as multiplier canonic. Unfortunately, however, such a structure does not exist; this is still an unsolved problem. For higher order transfer functions, one would require more tappings at the end of delays and the multiplier coefficients have to be appropriately chosen. The next example illustrates both of these points, although the order of the transfer function is the same.

Example 7
Let

$$H_4(z) = 1 + h_4(1)z^{-1} + h_4(2)z^{-2} + h_4(3)z^{-3} - z^{-4} \quad (36.28)$$

Here $k_4 = -1$ and the necessary decomposition is as follows:

$$\begin{aligned} H_4(z) = &\left[1 - h_4(3)z^{-1} + h_4(3)z^{-3} - z^{-4}\right] \\ &+ \left\{[h_4(3) + h_4(1)]z^{-1} + h_4(2)z^{-2}\right\} \end{aligned} \quad (36.29)$$

The first of these transfer functions is linear phase with $k_4 = -1$, $k_3 = k_2 = 0$ and $k_1 = -h_4(3)$. The second transfer function is realized by taking tappings after the first and second delays, as shown in Fig. 36.7. The three outputs are then combined. The multipliers α and β are found from the following equation:

$$z^{-1}\alpha + \left(k_1 + z^{-1}\right)z^{-1}\beta = [h_4(3) + h_4(1)]z^{-1} + h_4(2)z^{-2} \quad (36.30)$$

This gives

$$\begin{aligned} \beta = -h_4(2) \text{ and } \alpha \\ = h_4(3) + h_4(1) - h_4(2)h_4(3) \end{aligned} \quad (36.31)$$

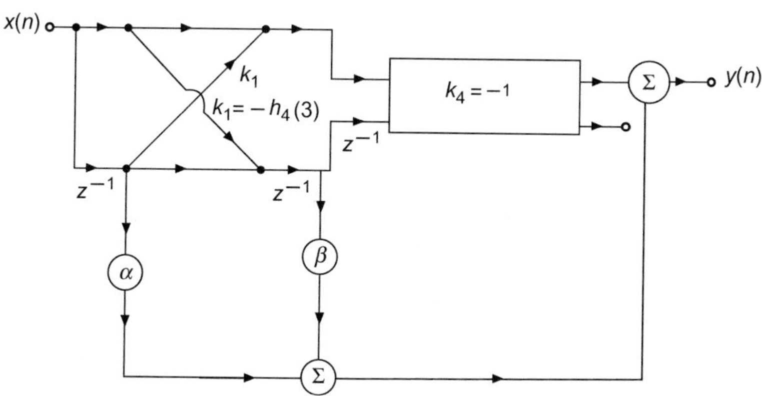

Fig. 36.7 Synthesis of Eq. 36.28: α and β are given by Eq. 36.31

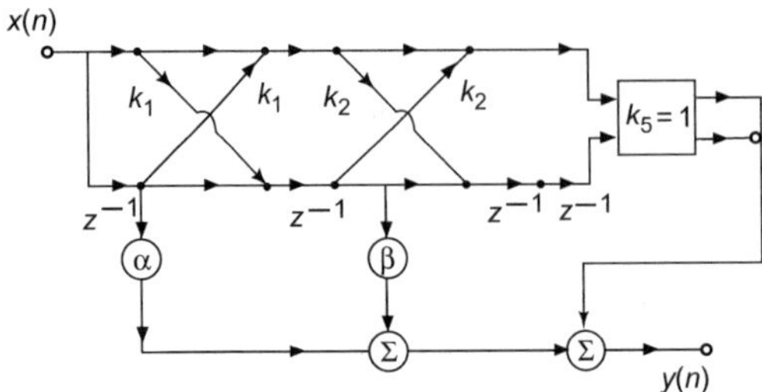

Fig. 36.8 Synthesis of Eq. 36.33: $k_1 = h_5(4)/[1 + h_5(3)]$, $k_2 = h_5(3)$, $\alpha = h_5(1) - h_5(4) - \beta k_1$ and $\beta = h_5(2) + h_5(3)$

Example 8

Let

$$H_5(z) = 1 + h_5(1)z^{-1} + h_5(2)z^{-2} + h_5(3)z^{-3} + h_5(4)z^{-4} + z^{-5} \quad (36.32)$$

Here $k_5 = 1$ and we can rewrite Eq. 36.32 as

$$H_5(z) = [1 + h_5(4)z^{-1} + h_5(3)z^{-2} + h_5(3)z^{-3} + h_5(4)z^{-4} + z^{-5}] + \{[h_5(1) - h_5(4)]z^{-1} + [h_5(2) - h_5(3)]z^{-2}\} \quad (36.33)$$

The first transfer function has already been realized in Example 2, except that here $h_5(4)$ takes the place of $h_5(1)$ and $h_5(3)$ takes the place of $h_5(2)$. The second transfer function in Eq. 36.33 can be realized by tapping the signals after the first and the second delays. The procedure is the same as in Example 7 and is not repeated here. The final result is shown in Fig. 36.8.

Example 9

Let

$$H_5(z) = 1 + h_5(1)z^{-1} + h_5(2)z^{-2} + h_5(3)z^{-3} + h_5(4)z^{-4} - z^{-5} \quad (36.34)$$

The necessary decomposition is as follows:

$$H_5(z) = [1 - h_5(4)z^{-1} - h_5(3)z^{-2} + h_5(3)z^{-3} + h_5(4)z^{-4} - z^{-5}] + \{[h_5(1) + h_5(4)]z^{-1} + [h_5(2) + h_5(3)]z^{-2}\} \quad (36.35)$$

The first transfer function is linear phase, antisymmetrical and can be realized by the procedure already illustrated. We shall have

$$k_5 = -1,\ k_4 = k_3 = 0,\ k_2 = -h_5(3) \text{ and } k_1 = -h_5(4)/[1 - h_5(3)] \quad (36.36)$$

The second transfer function in Eq. 36.35 is realized by tappings after the first and second delays. The final realization is the same as that shown in Fig. 36.8 with k values given by Eq. 36.36 and

$$\alpha = h_5(1) + h_5(4) - \beta k_1 \text{ and } \beta = h_5(2) + h_5(3) \quad (36.37)$$

Conclusion

We have demonstrated, through simple examples, how an FIR lattice transfer function $H_N(z) = 1 + \sum_{n=1}^{N-1} h_N(n)z^{-n} \pm z^{-N}$ can be realized for both linear and nonlinear phase cases with canonic delays and multipliers. The procedure is expected to be useful to students and teachers, as well as designers concerned with digital signal processing.

Problems

P.1. What happens in Fig. 36.1b if one k_1 is in the reverse direction?

P.2. Obtain a lattice structure for

$$H_0(z) = 1 + 0.5\,z^{-1} + 0.3\,z^{-2} + 0.5\,z^{-3} + z^{-4}$$

P.3. Obtain a lattice if z^{-5} in Eq. 36.12 has a negative sign.

P.4. Obtain a lattice for Eq. 36.16 with $h_4(2) = 0$.

P.5. Can you decompose Eq. 36.26 in any fashion other than Eq. 36.27? Give as many as possible if you can.

References

1. S.C. Dutta Roy, Synthesis of FIR lattice structures. *IEE Proc-Vis Image Signal Process* **147**, 549–552 (2000)
2. A.V. Oppenheim, R.W. Schafer, *Discrete Time Signal Processing* (Prentice Hall, New Jersey, 1989)
3. M. Bellanger, *Digital Processing of Signals* (Wiley, Hoboken, 1984)
4. J.G. Proakis, D.G. Manolakis, *Digital Signal Processing* (McMillan, Basingstoke, 1992)
5. L.B. Jackson, *Digital Filters and Signal Processing* (Kluwer, Alphen aan den Rijn, 1989)

37 FIR Lattice Structures with Single-Multiplier Sections

An alternative derivation is given for the linear prediction FIR lattice structures with single-multiplier sections. As compared to the previous approaches, this method is believed to be conceptually simpler and more straightforward.

Keywords
FIR lattice · Single-multiplier realization

Introduction

The linear prediction FIR lattice filter, shown in Fig. 37.1, realizes the transfer function

$$H_N(z) = \frac{Y(z)}{X(z)} = 1 + \sum_{n=1}^{N} h_n z^{-n} \qquad (37.1)$$

and its mirror image transfer function

$$H'_N(z) = \frac{Y'(z)}{X(z)} = z^{-N} H_N(z^{-1}). \qquad (37.2)$$

This structure uses two identical multipliers in each section, and is attributed to Itakura and Saito [1]. Most textbooks on Digital Signal Processing (DSP) refer to this structure and its variations in details. They assume, rather than derive, the structure and then analyze it for finding the transfer function as well as some recurrence formulas. A simple derivation of the structure has recently been given in [2], and an alternative lattice structure in which the two transfer functions $H_N(z)$ and $H_N'(z)$ are complementary to each other, i.e. $H_N'(z) = z^{-N} H_N(-z^{-1})$, has been given in [3].

There exists a corresponding lattice structure for all-pass IIR transfer functions, which also uses two multipliers per section. This structure has been derived by Mitra by using his two-pair or multiplier extraction approach [4]. Using the same approach, Mitra also derived a modified IIR lattice structure which uses a single multiplier per section, and is therefore canonic in multipliers. As it is well known, the basic all-pass IIR structure can be used with additional multipliers and summers to realize any arbitrary IIR transfer function [5].

A question arises as to whether a single-multiplier structure is possible for the FIR lattice also. The answer was given by Makhoul, as early as 1978 [6]. He derived two structures, called LF2 and LF3, each section of which contains three multipliers, and then converted them to single-multiplier ones by a clever

Source: S. C. Dutta Roy, "FIR Lattice Structures with Single-Multiplier Sections," *IETE Journal of Education*, vol. 47, pp. 119–122, July–September 2006.

S. C. Dutta Roy, *Circuits, Systems and Signal Processing*,
https://doi.org/10.1007/978-981-10-6919-2_37

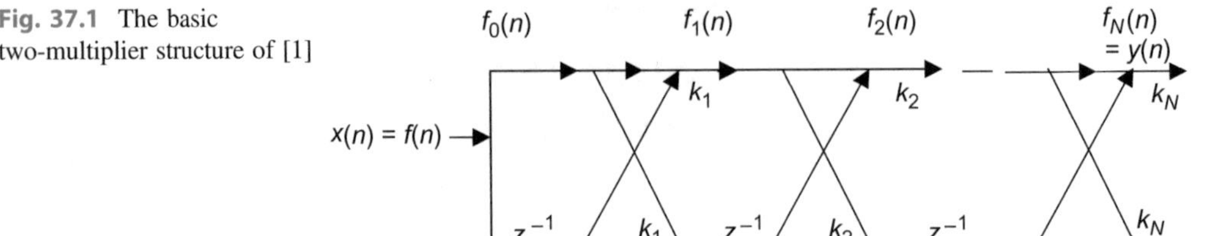

Fig. 37.1 The basic two-multiplier structure of [1]

manipulation. Nine years after the appearance of [6], Dognata and Vaidyanathan [7] gave an alternative derivation of the single-multiplier FIR lattice by the multiplier extraction approach, and another 2 years later, Krishna [8] indicated that the same structure could also be arrived at by the eigen-decomposition approach.

Surprisingly, no mention could be found of Makhoul's work in any of the large number of textbooks scanned by the author. The purpose of this chapter is, first, to bring this fine piece of work to the attention of teachers and students of DSP, and, second, to present an alternative, class tested and conceptually simpler procedure for deriving Makhoul's single-multiplier structures.

Derivation

Consider the mth section of the two-multiplier structure, shown in Fig. 37.2 with signals transformed to the z-domain. Also, for simplicity, let

$$F_{m-1}(z) = X_1(z), G_{m-1}(z) = X_2(z), \quad (37.3a)$$

and

$$F_m(z) = Y_1(z), \text{ and } G_m(z) = Y_2(z), \quad (37.3b)$$

By inspection of Fig. 37.2, we get

$$Y_1 = X_1 + k_m z^{-1} X_2, \quad (37.4)$$

and

$$Y_2 = k_m X_1 + z^{-1} X_2. \quad (37.5)$$

The input variables occurring in Eqs. 37.4 and 37.5 are X_1 and z^{-1} X_2. Clearly, in order to obtain single-multiplier realizations, we require one or more additional variables involving a linear combination of X_1 and z^{-1} X_2 without any multiplication. The possibilities for a third variable are $X_1 + z^{-1}$ X_2; $X_1 - z^1$ X_2; $-X_1 + z^{-1}$ X_2; and $-X_1 - z^{-1}$ X_2. We shall now investigate some of these cases.

Realization 1

In terms of the variable set (X_1, z^{-1} X_2, $X_1 + z^{-1}$ X_2), Eqs. 37.4 and 37.5 can be rewritten as

$$Y_1 = (1-k_m)X_1 + k_m(X_1 + z^{-1}X_2) \quad (37.6)$$

and

$$Y_2 = (1-k_m)z^{-1}X_2 + k_m(X_1 + z^{-1}X_2). \quad (37.7)$$

Dividing both sides of Eqs. 37.6 and 37.7 by $(1 - k_m)$, we get

$$\frac{Y_1}{1-k_m} = X_1 + \frac{k_m}{1-k_m}(X_1 + z^{-1}X_2), \quad (37.8)$$

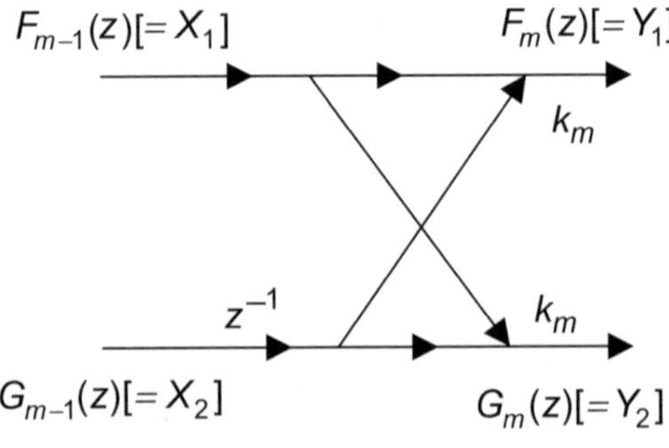

Fig. 37.2 The mth section of Fig. 37.1: Is $m = 1$?

and

$$\frac{Y_2}{1-k_m} = z^{-1}X_2 + \frac{k_m}{1-k_m}(X_1 + z^{-1}X_2) \quad (37.9)$$

The resulting structure involves only one multiplier, as shown in Fig. 37.3a. However, each output is now scaled by the factor $1/(1 - k_m)$. Y_1 and Y_2 can, of course, be recovered by multiplying the outputs by the factor $(1 - k_m)$. If each stage of Fig. 37.1 is thus converted into a single-multiplier one, then all output multipliers can be clubbed into a single multiplier of value $\prod_{m=1}^{N}(1 - k_m)$ at the input of the overall lattice, thus reducing the total number of multipliers from $2N$ to $N + 1$. Note that instead of lumping the multipliers at the input, one can also distribute them appropriately in order to prevent overflow and/or minimize quantization errors. The total number of multipliers may still be much less than $2N$, as required in the lattice of Fig. 37.1. One should also keep in mind that considerations of overflow and quantization errors may dictate the use of additional lumped/distributed scaling in the structure of Fig. 37.1 too. Also, observe that Fig. 37.3a is the same as the one-multiplier lattice form LF2/1(a) of Makhoul [6].

Since the procedure for other realizations is similar, we shall, for brevity, only give the main results for three other cases.

Realization 2

Taking the variable set as (X_1, z^{-1} X_2, $X_1 - z^{-1}$ X_2), Eqs. 37.4 and 37.5 can be rewritten as

$$Y_1 = (1 + k_m)X_1 - k_m(X_1 - z^{-1}X_2), \quad (37.10)$$

and

$$Y_2 = (1 + k_m)z^{-1}X_2 + k_m(X_1 - z^{-1}X_2). \quad (37.11)$$

Single-multiplier realization is obtained by dividing both sides of Eqs. 37.10 and 37.11 by $(1 + k_m)$, and is shown in Fig. 37.3b which is the same as LF2/1(b) of [6]. The multiplier needed at the input of the lattice is $\prod_{m=1}^{N}(1 + k_m)$ in this case.

Realization 3

For the same variable set as in realization 1, Eqs. 37.4 and 37.5 can also be rewritten as follows:

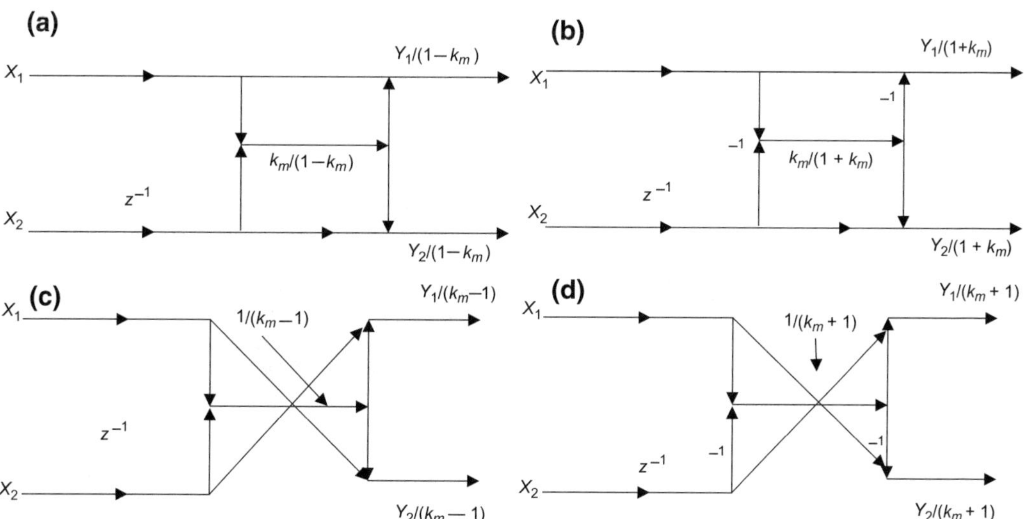

Fig. 37.3 Four single-multiplier realizations of the lattice section of Fig. 37.2

$$Y_1 = (X_1 + z^{-1}X_2) + (k_m - l)z^{-1}X_2, \quad (37.12)$$

and

$$Y_2 = (X_1 + z^{-1}X_2) + (k_m - 1)z^{-1}X_1. \quad (37.13)$$

The corresponding single-multiplier realization is obtained by dividing both sides of Eqs. 37.12 and 37.13 by $(k_m - 1)$ and is shown in Fig. 37.3c, which is the same as LF3/l(a) of [6]. The multiplier needed at the input N of the lattice is $\prod_{m=1}^{N}(k_m - 1)$.

Realization 4

Taking the same variable set as in realization 2, we can also modify Eqs. 37.4 and 37.5 as follows:

$$Y_1 = (X_1 - z^{-1}X_2) + (1 + k_m)z^{-1}X_2, \quad (37.14)$$

and

$$Y_2 = -(X_1 - z^{-1}X_2) + (1 + k_m)X_1. \quad (37.15)$$

Single-multiplier realization is obtained by dividing both sides of Eqs. 37.14 and 37.15 by $(1 + k_m)$, and is shown in Fig. 37.3d, which is the same as LF3/l(b) of [6]. The multiplier needed at the input of the lattice is now $\prod_{m=1}^{N}(1 + k_m)$.

As can be easily verified, the other input variable sets like (X_1, z^{-1} X_2, $-X_1 \pm z^{-1}$ X_2), ($-X_1$, z^{-1} X_2, $-X_1 \pm z^{-1}$ X_2), etc., give simple variations of the four structures shown in Fig. 37.3.

Conclusion

As indicated in the introduction, Makhoul's work [6] has not received adequate recognition in textbooks on DSP. It is hoped that this chapter will facilitate inclusion of this work in books to be written or in the new editions of the books which exist. It is also hoped that the conceptually simpler approach presented here for deriving the single-multiplier FIR lattice would appeal to students and teachers of DSP.

Problems

P.1. In Fig. 37.1, if all arrows are reversed, what kind of transfer function is obtained?
P.2. What if only the lower line arrows are reversed in Fig. 37.1?
P.3. Write the equations in Fig. 37.2 with only the lattice arrows reversed.
P.4. In Fig. 37.3a and b, again reverse all the arrows, and comment on the transfer function so obtained.
P.5. Do the same for Fig. 37.3c and d.

Acknowledgments The author thanks R. Vishwanath for helpful discussions.

References

1. F. Itakura, S. Saito, Digital Filtering Techniques for Speech Analysis and Synthesis, in *Proc 7th Int Cong Acoust*, (Budapest, Hungary, 1971) pp. 261–264
2. S.C. Dutta Roy, R. Vishwanath, Derivation of the FIR lattice. IETE J. Educ. **45**, 211–212 (October–December 2004)
3. S.C. Dutta Roy, R. Vishwanath, Another FIR lattice structure, Int. J. Circ. Theor. Appl. **33**, 347–351, (July–August 2005)
4. S.K. Mitra, *Digital Signal Processing: A Computer Based Approach* (McGrawHill, New York, 2001)
5. A.H. Gray, J.D. Markel, Digital lattice and ladder filter synthesis. IEEE Trans. Audio Electroacoust. AU-**21**, 491–500 (December 1973)
6. J. Makhoul, A class of all-zero lattice digital filters: properties and applications. IEEE Trans. Acoust. Speech Sig. Process. ASSP-**26**, 304–314, (August 1978)
7. Z. Doganata, P.P. Vaidyanathan, On one-multiplier implementations of FIR lattice structures. IEEE Trans. Circ. Syst. CAS-**34**, 1608–1609 (December 1987)
8. H. Krishna, An eigen-decomposition approach to one-multiplier realizations of FIR lattice structures. IEEE Trans. Circ. Syst. CAS-**36**, 145–146, (January 1989)

A Note on the FFT

38

This chapter gives a formula for the exact number of non-trival multipliers required in the basic N-point FFT algorithms, where N is an integral power of 2. Now proceed further, but not too far!

Keywords
FFT · Computation · Number of multipliers

Introduction

In the usual presentation of the Fast Fourier Transform (FFT) in textbooks, it is mentioned that the basic N-point FFT algorithms reduce the number of multipliers from N^2 to $N \log_2 N$ ([1], p 287), if N is a power of 2. If one looks at the actual FFT diagrams, either decimation-in-time (DIT) or decimation-in-frequency (DIF), of an 8-point sequence, as shown in Fig. 38.1a and b, respectively, one finds that the actual number of non-trivial multipliers is 5, instead of 24, as predicted by the $N \log_2 N$ formula. As another example, the FFT diagrams for a 32-point sequence, in both DIT and DIF forms show the number of non-trivial multipliers to be 49, instead of 160, as predicted by the $N \log_2 N$ formula. This reduction is effected by using the butterfly simplification and the facts that $W_N^0 = 1$ and $W_N^{N/2} = -1$, where $W_N = \exp(-j2\pi/N)$. It is, therefore, of interest to find out the actual number of non-trivial multipliers needed in a general N-point FFT, where $N = 2^q$, q being a positive integer. This note gives a formula for this purpose.

Derivation of the Formula

Consider the DIT algorithm of an N-point FFT, incorporating butterfly simplification. The last (qth) stage of the computation will have $N/2$ multipliers, of which W_N^0 is trivial and the others are $W_N^1, W_N^2, \ldots, W_N^{(N/2)-1}$. Thus the number of multipliers at this stage is $[(N/2) - 1]$. The preceding stage $[(q-1)\text{th}]$ will have two groups of multipliers, each having $(N/4)$ members. In each group, there will be a W_N^0 multiplier. Hence, the number of multipliers at the $[(q-1)\text{th}]$ stage is $2[(N/4)-1]$. Similarly, at the $(q-2)$th stage, the

Source: S. C. Dutta Roy, "A Note on the FFT," *IETE Journal of Education*, vol. 46, pp. 61–63, April–June 2005.

S. C. Dutta Roy, *Circuits, Systems and Signal Processing*,
https://doi.org/10.1007/978-981-10-6919-2_38

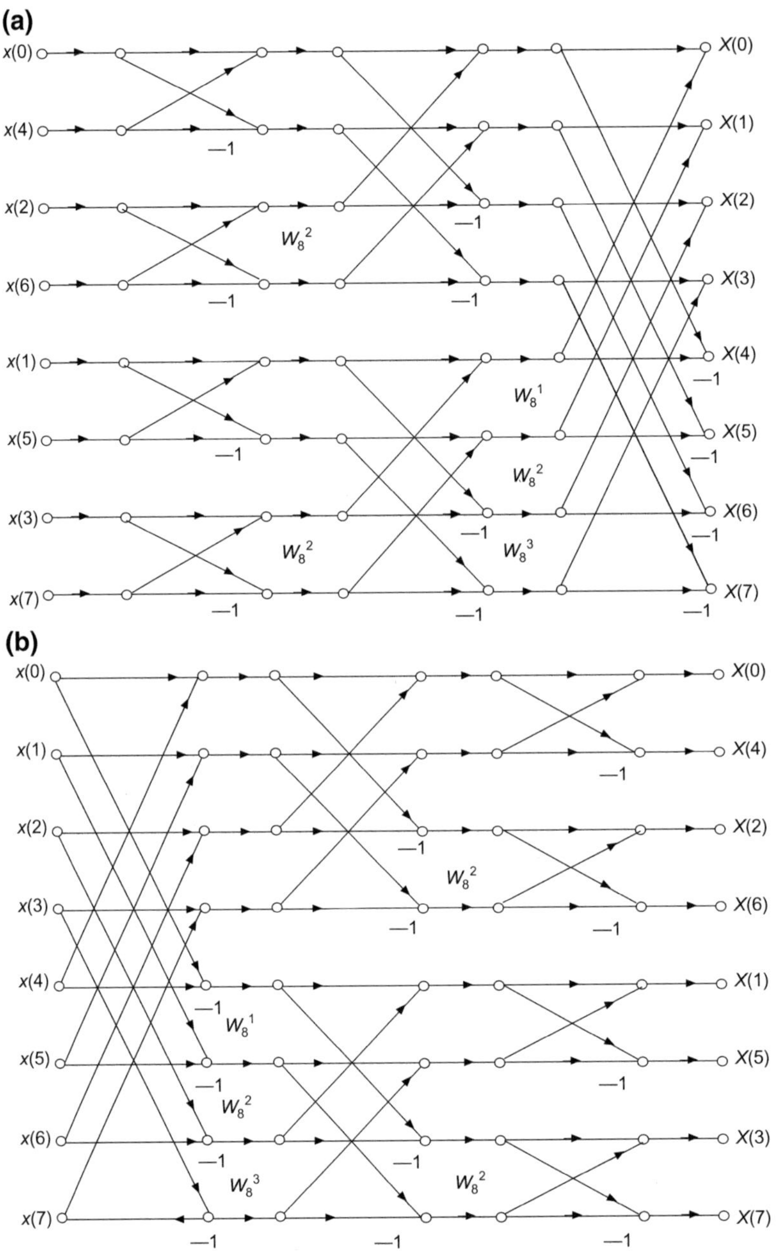

Fig. 38.1 **a** Decimation-in-time FFT flow diagram for a 8-point sequence, showing only the non-trivial multipliers; **b** Decimation-in-frequency FFT flow diagram for a 8-point sequence, again showing only the non-trivial multipliers

number of multipliers is $4[(N/8) - 1]$ and so on, till we reach the second stage where there are $(N/4)$ groups of multipliers, with two multipliers in each group, one of them being W_N^0. Hence the second stage contributes to $(N/4)$ multipliers. The first stage has $(N/2)$ multipliers, each of value W_N^0. Thus, the total number of non-trivial multipliers in an N-point FFT becomes

$$\begin{aligned}M(N) =& 2^0\left(\frac{N}{2}-1\right)+2^1\left(\frac{N}{4}-1\right)+2^2\left(\frac{N}{8}-1\right)\\ &+\cdots+\frac{N}{4}(2-1)\\ &\uparrow \qquad\qquad \uparrow \qquad\qquad \uparrow \qquad\qquad \uparrow\\ &q\text{th stage} \mid (q-1)\text{th stage} \mid (q-2)\text{th stage} \mid 2\text{nd stage}\\ &=\left(\frac{N}{2}-1\right)+\left(\frac{N}{2}-2\right)+\left(\frac{N}{2}-4\right)\\ &+\cdots+\left(\frac{N}{2}-\frac{N}{4}\right)\\ &=(q-1)\frac{N}{2}-\left(1+2+4+\cdots+\frac{N}{4}\right)\\ &=\frac{N}{2}(\log_2 N-1)-\left(1+2+2^2+\cdots+2^{q-2}\right)\\ &=\frac{N}{2}(\log_2 N-1)-\frac{1-2^{q-1}}{1-2}\\ &=\frac{N}{2}(\log_2 N-1)+\left(1-\frac{N}{2}\right).\end{aligned} \tag{38.1}$$

Finally, therefore,

$$M(N)=\frac{N}{2}(\log_2 N-2)+1. \tag{38.2}$$

The same number of multipliers arises in DIF also, by recognizing that the index of stages will be reversed in Eq. 38.1. Table 38.1 shows a comparison of the actual number of non-trivial multipliers $M(N)$ and the number predicted by the $N \log_2 N$ formula.

Table 38.1 Number of multipliers in FFT

N	$M(N)$	$N \log_2 N$
2	0	2
4	1	8
8	5	24
16	17	64
32	49	160
64	129	384
128	321	896
256	769	2048
512	1793	3584

Recurrence Relation

A recurrence formula for $M(N)$ can be derived as follows. For a $2N$-point FFT, Eq. 28.2 gives

$$\begin{aligned}M(2N) &= N[\log_2(2N)-2]+1\\ &= N[\log_2 2+\log_2 N-2]+1\\ &= N(I+\log_2 N-2)+1\\ &= N(\log_2 N-1)+1.\end{aligned} \tag{38.3}$$

Also,

$$2M(N)=N(\log_2 N-2)+2.$$

From Eqs. 38.3 and 38.4, we get

$$M(2N)=2M(N)+N-1,$$

which is the required recurrence formula.

Alternative Derivation for *M*(*N*)

An alternative derivation of the formula for $M(N)$ follows by noting that the actual number of multipliers after using the butterfly simplification is $(N/2)\log_2 N$, in which the number of W_N^0 multipliers is, using the DIT,

1 at the qth stage;
2 at the $(q-1)$th stage;
4 at the $(q-2)$th stage;
……………………,
$(N/4)$ at the 2nd stage; and
$(N/2)$ at the 1st stage.

Hence, the total number of W_N^0 multipliers is

$$\begin{aligned}&1+2+4+\cdots+\frac{N}{2}\\ &=1+2+2^2+\cdots 2^{q-l}\\ &=\frac{1-2^q}{1-2}\\ &=N-1.\end{aligned}$$

Thus,

$$M(N) = \frac{N}{2}\log_2 N - (N-1) = \frac{N}{2}(\log_2 N - 2) + 1, \quad (38.4)$$

which is the same as Eq. 38.2.

Problems

P.1. What looks simpler? DIT or DIF? Why?
P.2. Draw the DIF diagram for a 16-point FFT.
P.3. What if the number of points is 15? I mean is DIF.
P.4. What if the number of points is 6? DIF again.
P.5. Same as P.4 for DIT.

Reference

1. A.V. Oppenheim, R.W. Schafer, *Digital Signal Processing* (Prentice Hall, New Jersey, 1975)

Appendix: Some Mathematical Topics Simplified

In the Appendix, I give some simple, common sense methods for deriving mathematical formulas frequently used in CSSP. Appendix A.1 gives a semi-analytical method for finding the roots of a polynomial. Euler's relation forms the basis of complex numbers. A fresh look at it forms the content of Appendix A.2. The square root of the sum of two squares is required in finding the magnitude of complex quantity. An approximation appears in Appendix A.3.

It is well known that algebraic equations of order more than 2 are difficult to solve. For third and fourth orders, analytical solutions exist, but are difficult to implement. For still higher orders, numerical methods have to be resorted to. For cubic and quartic equations, simplified procedures are given in Appendix A.4. Appendix A.5 gives many ways of solving an ordinary linear second order differential equation. A simple method has been presented in Appendix A.5 for this purpose.

Chebyshev was out-and-out a mathematician. Little did he know that his polynomials would be found so useful by filter designers. They do appear to be complicated to students, but a reading of Appendices A.6 and A.7 would show that you can derive Chebyshev polynomial identities with ease and compute the coefficients of Chebyshev polynomials without difficulty.

As in Section I, all chapters in this section end up with five carefully designed problems. Each problem requires a thorough understanding of the contents of the corresponding chapter. Work them out carefully and the joy of finding the clue can perhaps be compared with the joy you derive when you get a piece of your most favourite food. Learning is, by all accounts, feeding yourself. You can never overeat, and if you think you have done so, it will cause no uncomfortable feeling. Learning is consuming food for your intellectual development. The more you learn, the more you would like to learn.

Happy learning, dear students!

S. C. Dutta Roy, *Circuits, Systems and Signal Processing*,
https://doi.org/10.1007/978-981-10-6919-2

A.1: A Semi-analytical Method for Finding the Roots of a Polynomial

A systematic method, which combines graphical, analytical, and numerical techniques, is presented for finding the roots of a polynomial $P_0(s)$ of any degree. Real roots are first found by a simple graphical method, and then the purely imaginary roots are found by the Hurwitz test. When all the real and purely imaginary roots are removed from $P_0(s)$, the remainder polynomial $P_2(s)$ will have only complex conjugate roots and hence will be of even degree. When this degree is 2, the roots are obvious. For $P_2(s)$ of degree 4, a variation of a previously published analytical method, combined with a graphical display, is presented which is easier to apply. When the degree of $P_2(s)$ is greater than 4, only numerical methods have to be used.

Keywords

Hurwitz test • Polynomial roots • Quartic polynomial • Solution of algebraic equations

Introduction

The problem of finding the roots of a given polynomial arises in all fields of science and engineering, particularly in electrical engineering, in connection with the determination of poles and zeros of transfer functions, and in testing for stability of a given system. It is well known that analytical solutions are possible only for polynomials of degree 4 or less, and for higher degrees, numerical methods have to be resorted to.

In this section, we present a semi-analytical method, consisting of a combination of graphical, analytical, and numerical techniques for finding the roots of a polynomial $P_0(s)$ of any degree. When $P_0(s)$ contains only one or two pairs of complex conjugate roots, besides those on the real and imaginary axes, it is shown that a combination of graphical and analytical methods suffices. When $P_0(s)$ contains three or more pairs of complex conjugate roots, numerical methods have to be resorted to, after extracting all the real and imaginary axis roots from $P_0(s)$.

Roots on the Real Axis

Let

$$P_0(s) = s^N + a_{N-1}s^{N-1} + \cdots + a_2s^2 + a_1s + a_0 \tag{A.1.1}$$

where the coefficients are real and $a_0 \neq 0$. (If $a_0 = 0$, then there is a root at $s = 0$, and the polynomial degree is reduced by one.) Let $s = \sigma + j\omega$, where σ and ω are real and can be positive or negative. If $P_0(s)$ contains real roots, all on the negative σ axis, then all a_i's will be positive, while the existence of one or more positive real roots will be indicated by one or more a_i's being

Source: S. C. Dutta Roy, "A Semi-Analytical Method for Finding the Roots of a Polynomial," IETE Journal of Education, vol. 55, pp. 90–93, July–December 2014.

S. C. Dutta Roy, *Circuits, Systems and Signal Processing*,
https://doi.org/10.1007/978-981-10-6919-2

negative or missing. Such real roots can be simply obtained by plotting $|P_0(\sigma)|$ versus σ for one or both sides of $\sigma = 0$, as appropriate. We plot the magnitude rather than the value because visually the zero crossings from negative to positive values, or vice versa, of $P_0(\sigma)$ are not as appealing (or perhaps not as accurate) as the position of the nulls, similar to those occurring in the magnitude response of null networks.

As an example, consider the 10th-degree polynomial

$$P_0(s) = s^{10} + 8s^9 + 31s^8 + 87s^7 + 188s^6 + 317s^5 + 428s^4 + 452s^3 + 372s^2 + 204s + 72. \quad \text{(A.1.2)}$$

Since there are no negative coefficients, we need to plot $|P_0(\sigma)|$ only for negative values of σ. This plot is shown in Fig. A.1.1 which clearly indicates that there are real roots at $s = -2$ and $s = -3$.

In general, if $P_0(s)$ of Eq. A.1.1 contains real roots at $s = \sigma_i, i = 1, 2, \ldots, M, M \leq N$, where σ_i can be positive as well as negative, then $P_0(s)$ can be written as

$$P_0(s) = \prod_{i=1^M} (s - \sigma_i) P_1(s) \quad \text{(A.1.3)}$$

where $P_1(s)$ does not have any real roots. In the case of Eq. A.1.2, the continued product term simplifies to $(s^2 + 5s + 6)$ and $P_1(s)$, obtained by dividing Eq. A.1.2 by this quadratic becomes

$$P_1(s) = s^8 + 3s^7 + 10s^6 + 19s^5 + 33s^4 + 38s^3 + 40s^2 + 24s + 12. \quad \text{(A.1.4)}$$

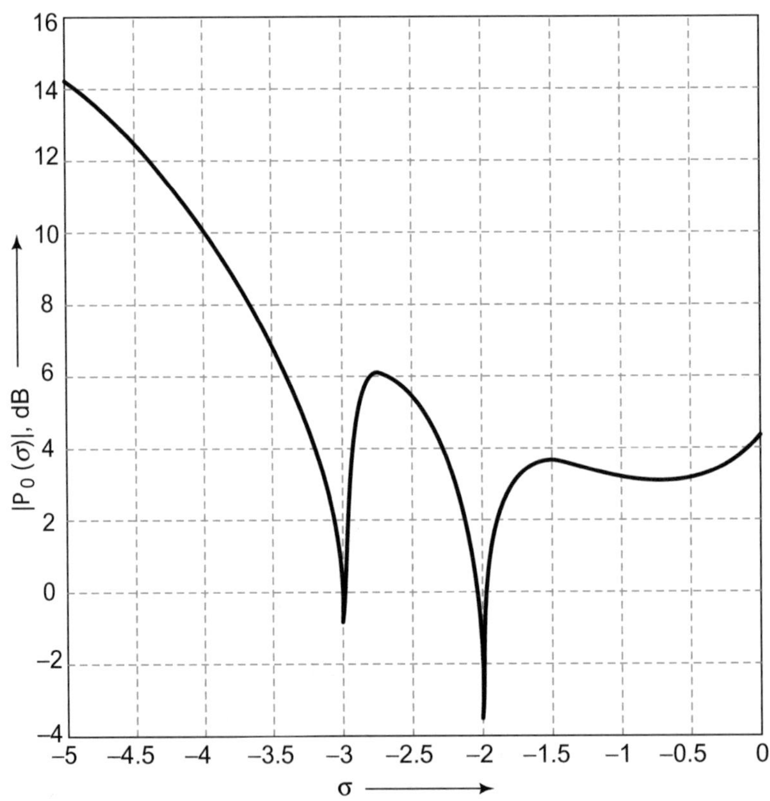

Fig. A.1.1 Plot of $|P_0(\sigma)|$, in dB, versus σ for the example of Eq. A.1.2

Roots on the Imaginary Axis

As is well known [1], roots on the imaginary axis are revealed by performing the Hurwitz test. It consists of performing a continued fraction expansion (CFE), starting with the highest powers, of the odd rational function (even part of the polynomial)/(odd part of the polynomial) or its reciprocal, depending on which one has a pole at infinity. The existence of $j\omega$-axis roots makes the CFE end prematurely, and the last divisor contains all these roots. (Note, in passing, that if the coefficient of any quotient in the CFE is negative, then the polynomial has roots in the right half plane; this is important in stability testing.)

In the present case of $P_1(s)$, if the CFE mentioned above does end prematurely, and if the last divisor is $D(s)$, then we can write

$$P_1(s) = D(s)P_2(s) \tag{A.1.5}$$

where $D(s)$ is of the form

$$D(s) = \prod_{k=1}^{Q} (s^2 + \omega_k^2). \tag{A.1.6}$$

Note that we have taken $D(s)$ to be an even polynomial because a possible root at $s = 0$ can be taken out either at the beginning or while finding the real roots. $P_2(s)$ is of degree $N - M - 2Q$ and contains only complex conjugate roots. If the degree of $D(s)$ is high, it may not be possible to find its roots analytically. In such a case, put $s^2 = S$. The resulting polynomial in S will have roots only on the negative real axis of the complex variable S, and hence the graphical procedure used in Sect. “Roots on the Real Axis” can be used.

Clearly, Hurwitz test could also be avoided by plotting $|P_1(j\omega)|^2$ versus ω^2 which will show nulls at $\omega^2 = \omega_k^2$. For the example of Eq. A.1.4, CFE of the even part/odd part ends prematurely at the fourth step, and the last divisor is

$$D(s) = s^4 + 5s^2 + 6 = (s^2 + 3)(s^2 + 2). \tag{A.1.7}$$

Also, the plot of $|P_1(j\omega)|^2$ versus ω^2, as given in Fig. A.1.2, shows nulls at $\omega^2 = 2$ and 3, thus confirming Eq. A.1.7.

$P_2(s)$ of Eq. A.1.5 can be obtained by long division of $P_1(s)$ by $D(s)$. For the example case, this process gives

$$P_2(s) = s^4 + 3s^3 + 5s^2 + 9s + 2 \tag{A.1.8}$$

This will have two pairs of complex conjugate roots. How to find them will be discussed in the next section.

Complex Conjugate Roots

In general, $P_2(s)$ will have $(N - M - 2Q)/2$ pairs of complex conjugate roots. If this number is 1, then $P_2(s)$ is a quadratic and its roots are easily found. If $P_2(s)$ is of degree 4, as in Eq. A.1.8, the method given in [2] or [3] may be followed. However, a confusion is likely to arise about signs in following this procedure. A variation of this procedure will now be given, which avoids this confusion and also does not require the analytical solution of the ‘resolvent’ cubic equation. Let

$$P_2(s) = s^4 + a_3s^3 + a_2s^2 + a_1s + a_0. \tag{A.1.9}$$

We express the right-hand side of Eq. A.1.9 as the difference of two squares, rather than the product of two quadratics, as in [2] and [3], as follows:

$$\begin{aligned} &s^4 + a_3s^3 + a_2s^2 + a_1s + a_0 \\ &= (s^2 + as + b)^2 - (cs + d)^2 \end{aligned} \tag{A.1.10}$$

where a, b, c, and d are constants to be determined. Equating the coefficients of powers of s on both sides of Eq. A.1.10, we get the following set of four equations:

$$2a = a_3 \tag{A.1.11}$$

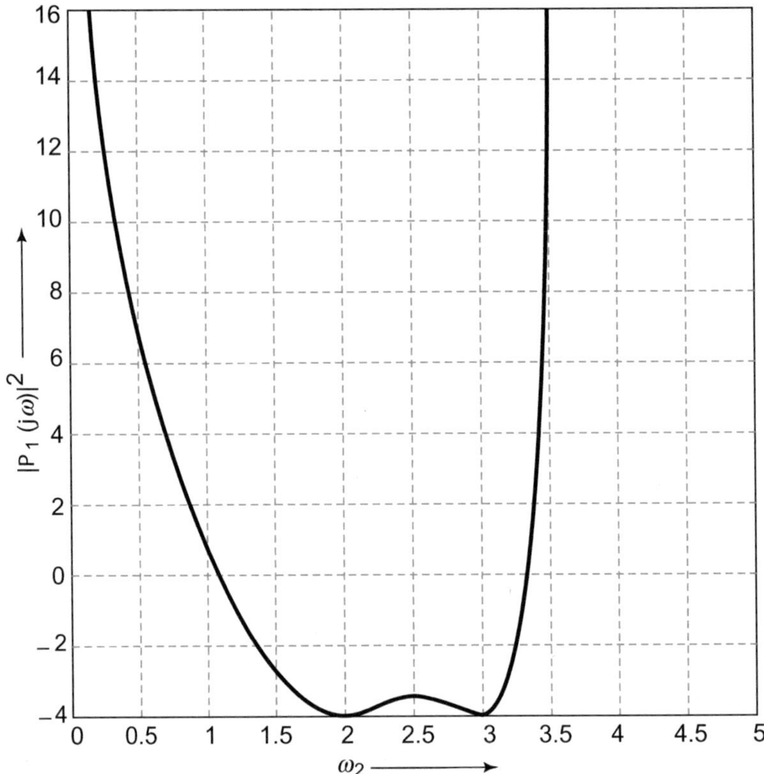

Fig. A.1.2 Plot of $|P_1(j\omega)|^2$ versus ω^2 for Eq. A.1.4

$$a^2 + 2b - c^2 = a_2 \tag{A.1.12}$$

$$2(ab - cd) = a_1 \tag{A.1.13}$$

$$b^2 - d^2 = a_0. \tag{A.1.14}$$

Equation A.1.11 gives

$$a = a_3/2. \tag{A.1.15}$$

To solve for b, c, and d, we have found it convenient to express c and d in terms of $2b$, which, for reasons to be made clear a little later, will be denoted by y. Then from Eqs. A.1.12 and A.1.15, we get

$$c^2 = -a_2 + y + (a_3^2/4) \tag{A.1.16}$$

while Eqs. A.1.13 and A.1.15 give

$$d = [(a_3 y/2) - a_1]/(2c). \tag{A.1.17}$$

Finally, Eqs. A.1.14 and A.1.15 give

$$d^2 = (y^2/4) - a_0. \tag{A.1.18}$$

Now combine Eqs. A.1.16–A.1.18; after simplification, we get the following cubic equation in y:

$$F(y) = y^3 - a_2 y^2 + (a_1 a_3 - 4a_0)y + (4a_0 a_2 - a_3 - a_0 a_3^2 - a_1) = 0. \tag{A.1.19}$$

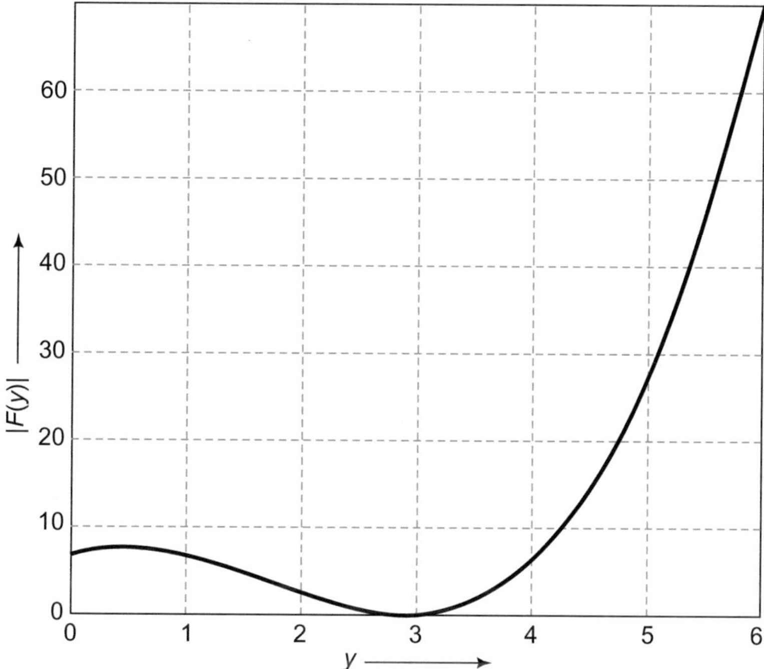

Fig. A.1.3 Plot of $|F(y)|$ versus y

It is interesting to note that although the approaches are slightly different, Eq. A.1.19 is the same as the 'resolvent' cubic Eq. A.1.12 of [3]. This is not unexpected though, because $B + b$ of [3] is the same as $2b$ in the approach adopted here. This is why $2b$ was denoted by y earlier in this section.

$F(y)$, being a cubic polynomial, must have at least one real root. Instead of following the analytical procedure of [3], we can plot $|F(y)|$ versus y to get the real root(s) from the null location(s). If y_1 is a real root, then our final solution will be as follows:

$$a = a_3/2 \tag{A.1.20}$$

$$b = y_1/2 \tag{A.1.21}$$

$$c = \pm\sqrt{\left[-a_2 + y_1 + (a_3^2/4)\right]} \tag{A.1.22}$$

$$d = [(a_3 y_1/2) - a_1]/(2c) \tag{A.1.23}$$

For the example of Eq. A.1.8, Eq. A.1.19 becomes

$$F(y) = y^3 - 5y^2 + 4y + 6 = 0. \tag{A.1.24}$$

The plot of $|F(y)|$ versus y, as shown in Fig. A.1.3, reveals only one real root at $y_1 = 3$. Substituting this value in Eqs. A.1.20–A.1.23, along with the values of a_i's, we get $a = b = \pm 1.5$ and $c = d = \pm 0.5$ (these are coincidences and not true in general). Substituting these values in Eq. A.1.10 and factorizing, we get

$$P_2(s) = (s^2 + 2s + 2)(s^2 + s + 1). \tag{A.1.25}$$

If $P_2(s)$ is of degree 6 or more, there exists no graphical or analytical method for finding the roots, and one has to take help of numerical methods.

Conclusions

In this section, a semi-analytical method has been presented for finding the real and purely imaginary roots of an arbitrary polynomial. After removing the factors corresponding to these two types of roots, the remaining polynomial will only have complex conjugate roots. If there is only one pair of such roots, then the roots are obtained by solving a quadratic equation. For two pairs of complex conjugate roots, a variation of an earlier published method is given, which is easier to apply. If there are more than two pairs of complex conjugate roots, then there is no alternative but to use numerical methods.

Problems

P.1. Read Ref. 3, and try solving a cubic equation $s^3 + as^2 + bs + c = 0$ by decomposing the LHS as $(s^2 + ds + e)\,(s + f) = 0$ and eliminating all constants except one. What do you get in this remaining constant?

P.2. Solve $s^3 + 2s^2 + 4s + 1 = 0$ by any method.

P.3. Solve $s^4 + 2s^2 + 3s + 4 = 0$ by any method.

P.4. Write $P(s) = s^5 + as^4 + bs^3 + cs^2 + d = (s + e)(s^2 + fs + g)(s^2 + hs + j)$ and by solving for $P(s) = 0$.

P.5. Find the roots of $(s + 1)^4 = 0$.

Any resemblance to Butterworth polynomial roots? Show the roots graphically.

Acknowledgements This work was supported by the Indian National Science Academy through the Honorary Scientist programme. The author thanks Professor Y. V. Joshi for his help in the preparation of this manuscript.

References

1. E.A. Guillemin, *Synthesis of Passive Networks* (Wiley, Hoboken, NJ, 1964)
2. M. Abramowitz, I.A. Stegun (eds.), *Handbook of Mathematical Functions* (Dover, New York, 1965)
3. S.C. Dutta Roy, On the solution of quadratic and cubic equations. IETE J. Educ. **47**(2), 91–95 (2006)

A.2: A Fresh Look at the Euler's Relation

A direct proof of the Euler's relation $e^{j} = \cos\theta +$ $j\sin\theta$*, is presented. It is direct in the sense that unlike the existing proofs, it does not presume any connection between* $e^{j\theta}$ and the trigonometric functions $\cos\theta$ and $\sin\theta$.

Keywords
Euler's formula • Proof

Euler's relation

$$e^{j\theta} = \cos\theta + j\sin\theta \tag{A.2.1}$$

is usually proved in mathematics and circuit theory texts [1, 2] by appealing to the infinite series expansions of $e^{j\theta}$, $\cos\theta$ and $\sin\theta$. Another way [3, 4] of showing the truth of the formula is based on the observation that if

$$y(\theta) = \cos\theta + j\sin\theta \tag{A.2.2}$$

then

$$dy(\theta)/y(\theta) = j d\theta. \tag{A.2.3}$$

Integrating Eq. A.2.3 and using the initial condition $y(0) = 1$, it follows that $y(\theta) = e^{j\theta}$.

Both of these proofs *presume* that there is a connection between $e^{j\theta}$ and the trigonometric $\cos\theta$ and $\sin\theta$. Presented here is a proof which does not do so, and in this sense it can be considered as a direct proof.

The Proof

Let the real and imaginary parts of the complex quantity $e^{j\theta}$ be denoted by $f(\theta)$ and $g(\theta)$, respectively; then

$$e^{j\theta} = f(\theta) + jg(\theta). \tag{A.2.4}$$

Differentiating Eq. A.2.4 and denoting $d()/d\theta$ by $()'$, one obtains

$$f'(\theta) + jg'(\theta) = e^{j\theta} = jf(\theta) - g(\theta) \tag{A.2.5}$$

Equating the real and imaginary parts on the two sides of Eq. A.2.5 gives

$$f'(\theta) - g'(\theta) \text{ and } g'(\theta) = f(\theta) \tag{A.2.6}$$

Differentiating one equation in Eq. A.2.6 and combining with the other yields the following differential equations for $f(\theta)$ and $g(\theta)$:

$$f''(\theta) + f(\theta) = 0 \text{ and } g''(\theta) + g(\theta) = 0 \tag{A.2.7}$$

Source: S. C. Dutta Roy, "A Fresh Look at the Euler's Relation," Students' Journal of the IETE, vol. 22, pp. 1–2, January 1981.

S. C. Dutta Roy, *Circuits, Systems and Signal Processing*,
https://doi.org/10.1007/978-981-10-6919-2

Each of these equations describes a simple harmonic motion; the solutions for $f(\theta)$ and $g(\theta)$ are, therefore

$$f(\theta) = K_f \cos(\theta + \theta_f) \text{ and } g(\theta) = K_g \cos(\theta + \theta_g) \quad \text{(A.2.8)}$$

where K_f, θ_f, K_g and θ_g are constants. Putting in Eq. A.2.4 gives the initial conditions $f(0) = 1$ and $g(0) = 0$. Substituting these in Eq. A.2.8, there results

$$K_f = 1/\cos\theta_f \text{ and } \theta_g = (2r+1)\pi/2, \quad r = 0, \pm 1, \pm 2, \ldots \quad \text{(A.2.9)}$$

From Eqs. A.2.8 and A.2.9, one obtains

$$f(\theta) = \cos(\theta + \theta_f)/\cos\theta_f \text{ and } g(\theta) = K \sin\theta \quad \text{(A.2.10)}$$

where $K = \pm K_g$. Substituting Eq. A.2.10 in Eq. A.2.6 results in the following two equations:

$$\sin(\theta + \theta_f)/\cos\theta_f = K \sin\theta \text{ and}$$

$$\cos(\theta + \theta_f)/\cos\theta_f = K \cos\theta. \quad \text{(A.2.11)}$$

Dividing the first equation in Eq. A.2.11 by the second gives

$$\tan(\theta + \theta_f) = \tan\theta, \quad \text{(A.2.12)}$$

which is satisfied only if

$$\theta_f = 2p\pi, p = 0, \pm 1, \pm 2, \ldots \quad \text{(A.2.13)}$$

Substituting for θ_f in either of the equations in Eq. A.2.11 gives

$$K = 1 \quad \text{(A.2.14)}$$

Thus, finally,

$$f(\theta) = \cos\theta \text{ and } g(\theta) = \sin\theta \quad \text{(A.2.15)}$$

and from equations A.2.4 and A.2.15, Eq. A.2.1 follows.

Q.E.D.

Problems

P.1. Solve for θ: $(\cos\theta)^4 = 1$.

P.2. Repeat for $(\sin\theta)^4 = 1$.

P.3. Solve for θ: $e^{jN\theta} = 1$, $N > 1$. Any relation with the solutions of P.1 and P.2? Any relation with Chebyshev? Maybe, maybe not.

P.4. Solve for $\theta: e^{j\theta} + e^{jN\theta} = 1$, again $N > 1$.

P.5. Solve for θ: $\left(e^{j\theta}\right)^{N_1} + \left(e^{j\theta}\right)^{N_2} = 1, N_1 \neq N_2, N_{1,2} > 1$.

References

1. H. Sohon, *Engineering Mathematics* (Van Nostrand, New York, 1953), p. 65
2. W.H. Hayt Jr., J.E. Kemmerly, *Engineering Circuit Analysis* (McGraw-Hill, New York, 1978), pp. 747–748
3. W.H. Hayt Jr., J.E. Kemmerly, *Engineering Circuit Analysis* (McGraw-Hill, New York, 1962), p. 283
4. A.G. Beged-Dov, Another look at Euler's relation. IEEE Trans. Educ. **E-9**, 44 (1966)

A.3: Approximating the Square Root of the Sum of Two Squares

It is shown that $\sqrt{(x^2+y^2)}$, with $x > y$, can be approximated by $x + y^2/(2x)$ for $0 \le y/x \le 1/2$, and by $0.816\,(x + 0.722\,y)$ for $1/2 \le y/x \le 1$ to within a relative error of 0.64%. This should be useful in computations, but more so in analytical developments involving such expressions.

Keywords
Square root • Sum of two squares

When dealing with complex numbers, as in circuit analysis or FFT computation, it is often required to calculate the value of

$$S = \sqrt{(x^2+y^2)} \qquad \text{(A.3.1)}$$

where, without loss of generality, it may be assumed that

$$0 < y < x \qquad \text{(A.3.2)}$$

As is well known, evaluating the square root is somewhat tedious and time consuming. To speed up the processing time, without significant loss of accuracy, we propose, in this section, an approximation for S for $\widetilde{S}$, consisting of two expressions valid for the ranges $0 \le y/x \le 0.5$ and $0.5 \le y/x \le 1$, and show that the maximum percentage relative error ε, defined by

$$\varepsilon = \left|(S - \widetilde{S})/S\right| \times 100 \qquad \text{(A.3.3)}$$

is thereby reduced to a value of 0.64 only $\widetilde{S}$ being the approximated value.

Derivation

In view of Eq. A.3.2, we can write

$$S = x\sqrt{(1+t^2)} \qquad \text{(A.3.4)}$$

where

$$0 < t = y/x < 1 \qquad \text{(A.3.5)}$$

The problem is therefore to approximate $\sqrt{(1+t^2)}$, a plot of which is shown in Fig. A.3.1 (not to scale). Note that the latter part of the graph (for $t > t_1$ say) can be approximated by a straight line, a possible candidate for which is indicated as $a + bt$. Also, when t is small, one can approximate $\sqrt{(1+t^2)}$ by $1 + t^2/2$, a plot of which is also shown in Fig. A.3.1 (in a slightly exaggerated form). In order to obtain a uniformly

Source: S. C. Dutta Roy, "Approximating the Square Root of the Sum of Two Squares," Students' Journal of the IETE, vol. 32, pp. 11–13, April–June 1991.

S. C. Dutta Roy, *Circuits, Systems and Signal Processing*,
https://doi.org/10.1007/978-981-10-6919-2

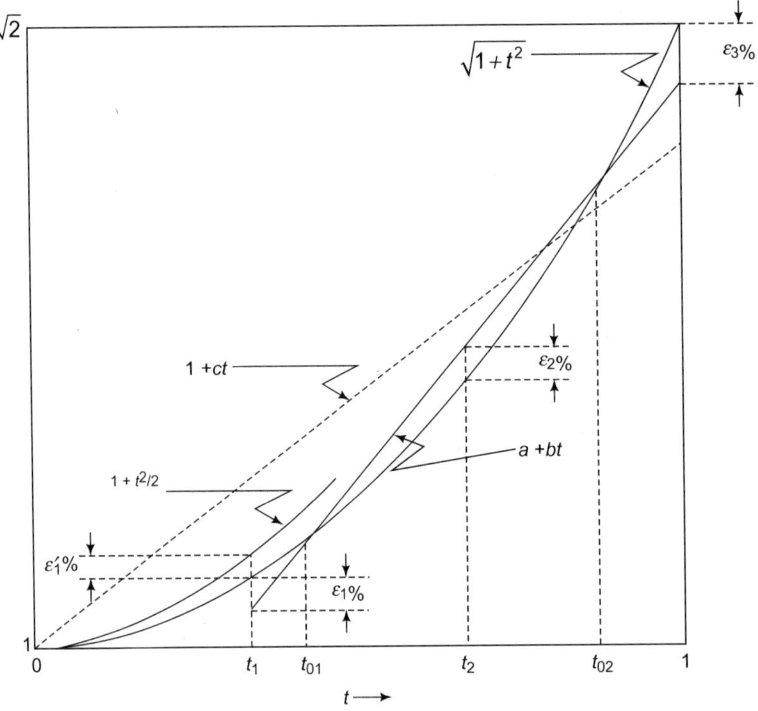

Fig. A.3.1 Showing the variation (not to scale) of $\sqrt{(1+t^2)}$, $(1+t^2/2)$ and $a+bt$

'good' approximation over the entire range of t, we assume that

$$\sqrt{(1+t^2)} \cong \begin{cases} 1+t^2/2 & 0 \le t \le t_1 \\ a+bt & t_1 \le t \le 1 \end{cases} \quad \text{(A.3.6a, b)}$$

and that the following relative errors are equal:

(i) ε_1' at $t = t_1$ computed by using Eq. A.3.6a,b
(ii) ε_1 at $t = t_1$ computed by using Eq. A.3.6a,b
(iii) ε_3 at $t = 1$, and
(iv) ε_2 at $t = t_2$, which is the maximum value of ε in the range $t_{01} \le t \le t_{02}$ (see Fig. A.3.1).

We therefore have the three equations

$$\varepsilon_t' = \varepsilon_2 \quad \text{(A.3.7a)}$$

$$\varepsilon_1 = \varepsilon_3 \quad \text{(A.3.7b)}$$

$$\varepsilon_2 = \varepsilon_3 \quad \text{(A.3.7c)}$$

for determining the unknown quantities t_1, a and b. Once these are known, the maximum relative error $\varepsilon_m (= \varepsilon_1 = \varepsilon_1' = \varepsilon_2 = \varepsilon_3)$ will also be known.

Now, according to Eq. A.3.7b, we have

$$\left[\sqrt{(1+t_1^2)} - (a+bt_1)\right]\sqrt{(1+t_1^2)} = \left[\sqrt{2} - (a+b)\right]/\sqrt{2} \quad \text{(A.3.8)}$$

which can be simplified to the following:

$$b = a\beta \quad \text{(A.3.9)}$$

where

$$\beta = \left[\sqrt{2} - \sqrt{(1+t_1^2)}\right]/\sqrt{(1+t_1^2)} - \sqrt{2}t_1] \quad \text{(A.3.10)}$$

To determine ε_2, note that in the range $t_{01} \le t \le t_{02}$.

$$\varepsilon = 100\left[a(1+\beta t) - \sqrt{(1+t^2)}\right]/\sqrt{(1+t^2)} \quad \text{(A.3.11)}$$

The maximum of ε occurs when $d\varepsilon/dt = 0$; carrying out the differentiation and simplifying gives the rather simple result

$$t_2 = \beta \quad \text{(A.3.12)}$$

Putting this in Eq. A.3.11, we get

$$\varepsilon_2 = 100\left[a\sqrt{(1+\beta^2)} - 1\right] \quad \text{(A.3.13)}$$

Now using Eq. A.3.7c, we get

$$a\sqrt{/(1+\beta^2)} - 1 = 1 - a(1+\beta)/\sqrt{2} \quad \text{(A.3.14)}$$

This can be simplified to the following;

$$a = 2/\left[\sqrt{(1+\beta^2)} + (1+\beta)/\sqrt{2}\right] \quad \text{(A.3.15)}$$

Thus if β is known, which in turn requires that t_1 is known, we can compute a from Eq. A.3.15 and b from Eq. A.3.9. To find t_1, we use Eq. A.3.7a, which dictates that

$$\left[(1+t_1^2/2) - \sqrt{(1+t_1^2)}\right] \Big/ \sqrt{(1+t_1^2)} = \left[a + b\,t_2 - \sqrt{(1+t_2^2)}\right] \Big/ \sqrt{(1+t_2^2)} \quad \text{(A.3.16)}$$

Combining this with Eqs. A.3.9 and A.3.12, and simplifying, we get

$$(1+t_1^2/2)/\sqrt{(1+t_1^2)} = a\sqrt{(1+\beta^2)} \quad \text{(A.3.17)}$$

Now substituting the values of a and β from Eqs. A.3.15 and A.3.10 respectively, and simplifying, we get

$$\frac{1+t_1^2/2}{\sqrt{(1+t_1^2)}} = \frac{2}{1+(1-t_1)/\sqrt{[4(1+t_1^2) - 2(1+t_1)\sqrt{\{2(1+t_1^2)\}}]}} \quad \text{(A.3.18)}$$

Numerical experimentation with this seemingly hopeless equation for t_1 reveals that the solution is, surprisingly, $t_1 \cong 0.5$. Further refinement shows that $t_1 \cong 0.5035$, and a simple calculation shows that under this condition, ε_m = 0.64. Finally, we calculate β and a from Eqs. A.3.10 and A.3.15 as

$$\beta = 0.722 \text{ and } a = 0.816 \quad \text{(A.3.19)}$$

Concluding Comments

We have shown that

$$\sqrt{(x^2+y^2)} \cong \begin{cases} x + y^2/(2x) & 0 \le y/x \le 0.5 \\ 0.816(x+0.722y) & 0.5 \le y/x \le 1 \end{cases} \quad \text{(A.3.20a, b)}$$

to within a relative error of 0.64%. This represents a uniformly 'good' approximation, and should be useful in speeding up the computation of $\sqrt{(x^2+y^2)}$, but more so in analytical developments involving such expressions. As an example, let it be required to find out if the equation

$$\sqrt{(1+x^2)} + x^2 + x = 2 \quad \text{(A.3.21)}$$

has a real root in the range $0.5 \le x \le 1$ and if so, what is its approximate value. Using Eq. A.3.20a,b, Eq. A.3.21 becomes

$$0.816\,(1+0.722x) + x^2 + x - 2 = 0 \quad \text{(A.3.22)}$$

Solving this quadratic gives one value of x as 0.5528. Putting this value in Eq. A.3.21, the left-hand side becomes 2.001, which differs from the right-hand side by 0.05% only. Note that the exact solution of Eq. A.3.21 will require the solution of a quadratic equation.

Problems

P.1. Can you approximate $\sqrt{(a+bxy)(x^2+y^2)}$? Take $b \ll a$.

P.2. How about approximating $\sqrt{(x^2+y^2)(f^2+g^2)}$?

P.3. Will the answer to P.2. be $0.816(x + 0.722y)(f + 0.722g)$?

P.4. What will be the approximation of $\sqrt{(x^2+y^2)}$ if $y \ll x$?

P.5. Repeat this for P.1.

A.4: On the Solution of Quartic and Cubic Equations

A method of solving a quartic equation, which does not require extracting the roots of complex numbers is explained in detail. In the process, the solution of a cubic equation has also been presented, with the same degree of simplicity.

Keywords

Quartic equation • Cubic equation • Solution

Introduction

When faced with a problem in mathematics, electrical engineers–students, faculty, researchers and practitioners alike–usually consult mathematics handbooks and encyclopaedias for a quick solution. While trying to design a dualband band-pass filter by using frequency transformation of a normalized low-pass filter, the author was confronted with the problem of solving a quartic equation of the form

$$z^4 + a_3 z^3 + a_2 z^2 + a_1 z + a_0 = 0, \qquad \text{(A.4.1)}$$

to find the pass-band edges. Abramowitz and Stegun's Handbook [1], which has been considered as 'The Bible' by scientists and engineers for ages, was consulted, but to the author's surprise, the correct solution could not be obtained. On deeper examination, it was found that there is a typographical mistake in signs in the last equation on page 17 and that the opening statement on page 18 is ambiguous. Also, the method requires handling square roots of complex numbers. A number of other such references [2–7] and internet sources [8, 9] were also consulted and it was found that they were either sketchy or had typographical mistakes or required finding the square and cube roots of complex numbers. We present here a solution to the problem which does not require messy calculations with complex numbers. In the process, we also deal with the solution of a general cubic equation of the form

$$y^3 + b_2 y^2 + b_1 y + b_0 = 0 \qquad \text{(A.4.2)}$$

with the same kind of simplicity, as compared to the solutions given in [1–9] and also in [10] and [11]. The treatment is based on simplification and consolidation of a monograph [12] by S Neumark, a British aeronautical engineer, whose work does not appear to have been appreciated or even referred to in the literature. We illustrate the procedures by examples whose solutions are known beforehand.

Source: S. C. Dutta Roy, "On the Solution of Quartic and Cubic Equations," IETE Journal of Education, vol. 47, pp. 91–95, April–June 2006.

S. C. Dutta Roy, *Circuits, Systems and Signal Processing*,
https://doi.org/10.1007/978-981-10-6919-2

Solution to the Quartic

Equation A.4.1 can be written as

$$(z^2+Az+B)(z^2+a\,z+b)=0, \quad \text{(A.4.3)}$$

where

$$A+a=a_3, \quad \text{(A.4.4)}$$

$$B+b+Aa=a_2, \quad \text{(A.4.5)}$$

$$Ab+aB=a_1 \quad \text{(A.4.6)}$$

and

$$Bb=a_0. \quad \text{(A.4.7)}$$

Let

$$B+b=y. \quad \text{(A.4.8)}$$

Then from Eqs. A.4.7 and A.4.8, we get a quadratic equation in *B* or *b,* the solution of which gives

$$B=\frac{y\pm\sqrt{y^2-4a_0}}{2} \text{ and } b=\frac{y\mp\sqrt{y^2-4a_0}}{2} \quad \text{(A.4.9)}$$

From Eqs. A.4.4, A.4.6 and A.4.9 (henceforth, we take the positive sign for *B* and the negative sign for *b,* without any loss of generality), we get

$$A,a=\frac{a_3}{2}\pm\frac{a_3y-2a_1}{2\sqrt{y^2-4a_0}}. \quad \text{(A.4.10)}$$

Finally, substituting for *A, a, B* and *b* in Eq. A.4.5, and simplifying gives

$$y+\frac{a_3^2}{4}-\frac{(a_3y-2a_1)^2}{4\sqrt{y^2-4a_0}}=a_2. \quad \text{(A.4.11)}$$

Simplification of Eq. A.4.11 results in the following cubic equation in *y*:

$$y^3-a_2y^2+(a_1a_3-4a_0)y-$$

$$(a_1^2+a_0a_3^2-4a_0a_2)=0. \quad \text{(A.4.12)}$$

This is the so-called 'resolvent' cubic and checks with the equation given in [1], page 17. If $y=y_1$ satisfies Eq. A.4.12, then from Eqs. A.4.3, A.4.9 and A.4.10, the roots of the quartic Eq. A.4.1 are obtained by solving the following two quadratic equations:

$$z^2+\left(\frac{a_3}{2}\pm\frac{a_3y_1-2a_1}{2\sqrt{y_1^2-4a_0}}\right)z+\frac{y_1}{2}\pm\sqrt{\frac{y_1^2}{4}-a_0}=0. \quad \text{(A.4.13)}$$

From Eq. A.4.11, the second term in the coefficient of z in Eq. A.4.13 can be written as $\pm\sqrt{\frac{a_3^2}{4}+y_1-a_2}$. Choosing the positive sign here leads to the mistake in [1] as pointed out in the Introduction. Choosing the negative sign gives the correct results, as demonstrated in the Example worked out later. Hence we get the simplified form of Eq. A.4.13 as:

$$z^2+\left(\frac{a_3}{2}\mp\sqrt{\frac{a_3^2}{4}+y_1-a_2}\right)z+\frac{y_1}{2}\pm\sqrt{\frac{y_1^2}{4}-a_0}=0. \quad \text{(A.4.14)}$$

We next consider the solution of the cubic equation A.4.12, and shall do so with general coefficients, as in Eq. A.4.2.

Solution of the Cubic Equation

Consider Eq. A.4.2. In the literature, it is the usual practice to derive a 'depressed' cubic i.e. another cubic equation in which the y^2 term is missing. To this end, we include the first two terms of Eq. A.4.2 in $\left(y+\frac{b_2}{3}\right)^3$. Then Eq. A.4.2 can be written as

$$\left(y+\frac{b_2}{3}\right)^3-\left(\frac{b_2^2}{3}-b_1\right)y+b_0-\frac{b_2^3}{27}=0. \quad \text{(A.4.15)}$$

We next supplement y in the second term by $+b_2/3$; then Eq. A.4.15 becomes

$$\left(y+\frac{b_2}{3}\right)^3-\left(\frac{b_2^2}{3}-b_1\right)\left(y+\frac{b_2}{3}\right)+b_0-\frac{b_2^3}{27}+\frac{b_2}{3}\left(\frac{b_2^2}{3}-b_1\right)=0. \tag{A.4.16}$$

Let

$$y+\frac{b_2}{3}=kx. \tag{A.4.17}$$

Usually, k is taken as unity in the literature. With a general k, Eq. A.4.16 can be simplified to the following:

$$27k^3x^3-9k(b_2^2-3b_1)x+(27b_0+2b_2^3-9b_1b_2)=0. \tag{A.4.18}$$

Dividing both sides by $(27k^3/4)$, we get

$$4x^3-\frac{4(b_2^2-3b_1)}{3k^2}x+\frac{4(27b_0+2b_2^3-9b_1b_2)}{27k^3}=0. \tag{A.4.19}$$

Now comes the brilliant idea of forcing the coefficient of x as -3 by choosing

$$k=\pm\frac{2\sqrt{b_2^2-3b_1}}{3} \tag{A.4.20}$$

Using the negative sign in Eq. A.4.20, Eq. A.4.19 becomes

$$4x^3-3x=R, \tag{A.4.21}$$

where

$$R=\frac{27b_0+2b_2^3-9b_1b_2}{2(b_2^2-3b_1)^{3/2}}. \tag{A.4.22}$$

Since a cubic equation is required to have one real root, Eq. A.4.22 will be applicable only when $b_2^2-3b_1>0$. We shall consider the other case, *i.e.* $b_2^2-3b_1<0$ shortly.

Notice that the left-hand side of Eq. A.4.21 is the third order Chebyshev polynomial in x which oscillates between -1 and $+1$ for $-1\le x\le +1$, and is monotonically increasing or decreasing for $|x|>1$, as shown in Fig. A.4.1. This figure will only give real root(s), and clearly, it suffices to consider $R\ge 0$, because changing the sign of R, simply leads to the roots changing their signs. It is also obvious from the figure that for $R<1$, there will be three real roots, of which one is positive and the other two are negative. When $R>1$, there will be only one real (positive) root. When $R=1$, there will be a double root at $-1/2$ and a single root at $+1$.

For $R<1$, the real positive root $x=x_1$ occurs between the point A, where $x=\sqrt{3}/2$ and $x=1$. Since $x<1$, we can write

$$x_1=\cos\theta, \tag{A.4.23}$$

where $0<\theta<\pi/6$. Equation A.4.21 then becomes

$$\cos 3\theta=R. \tag{A.4.24}$$

Hence

$$y_1=\cos\left[(\cos^{-1}R)/3\right]. \tag{A.4.25}$$

Equation A.4.21 can now be rewritten in the form

$$4x^3-3x-R=(x-x_1)(4x^2+c_1x+c_0)=0. \tag{A.4.26}$$

Comparing coefficients, we get $c_1=4x_1$ and $c_0=R/x_1=4x_1^2-3$. Thus the other two real roots of Eq. A.4.21 are the solutions of the quadratic equation

$$4x^2+4x_1x+4x_1^2-3=0 \tag{A.4.27}$$

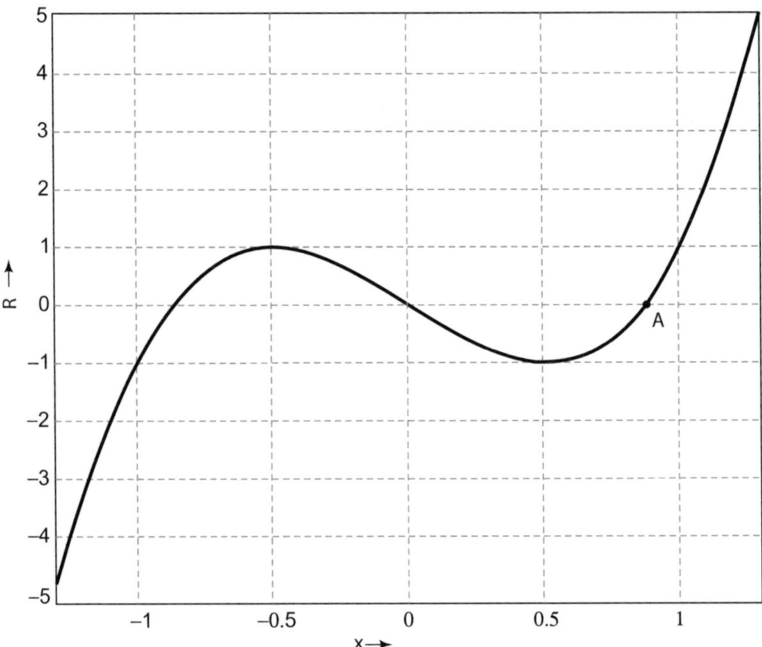

Fig. A.4.1 Plot of $R = 4x^3 - 3x$

so that

$$\begin{aligned} x_{2,3} &= -(x_1/2) \pm \left(\sqrt{3}/2\right)\sqrt{1 - x_1^2} \\ &= -[(\cos\theta)/2)] \pm \left(\sqrt{3}/2\right)\sin\theta \\ &= -\cos\left(\frac{\pi}{3} \pm \theta\right). \end{aligned} \qquad \text{(A.4.28)}$$

Now consider the case $R > 1$. Figure A.4.1 shows that there is only one real (positive) root at $x = x_1 > 1$. We can therefore write

$$x_1 = \cosh\alpha. \qquad \text{(A.4.29)}$$

Substituting this in Eq. A.4.21 and simplifying, we get

$$x_1 = \cosh\left[(\cosh^{-1} R)/3\right]. \qquad \text{(A.4.30)}$$

Equations A.4.26 and A.4.27 are applicable here also, so that the two complex roots are given by

$$x_{2,3} = -\left[(\cosh\alpha)/2)\right] \pm j(\sqrt{3}/2)\sinh\alpha \quad \text{(A.4.31)}$$

Finally, there remains the case $b_2^2 - 3b_1 < 0$. In this case, choose k such that the coefficient of x in Eq. A.4.19 becomes +3, i.e. let

$$k = -\frac{2\sqrt{3b_1 - b_2^2}}{3} \qquad \text{(A.4.32)}$$

Then Eq. A.4.21 becomes

$$4x^3 + 3x = R = \frac{27b_0 + 2b_2^3 - 9b_1b_2}{2|b_2^2 - 3b_1|^{3/2}} \qquad \text{(A.4.33)}$$

Notice that for uniformity with the previous case, we have modified the denominator of R. Here also, as in the previous case, we need to solve only for positive R; for negative R, all roots will change sign. The plot of the left-hand side of Eq. A.4.33 is shown in Fig. A.4.2, from which it is clear that there is only one real root x_1 which is

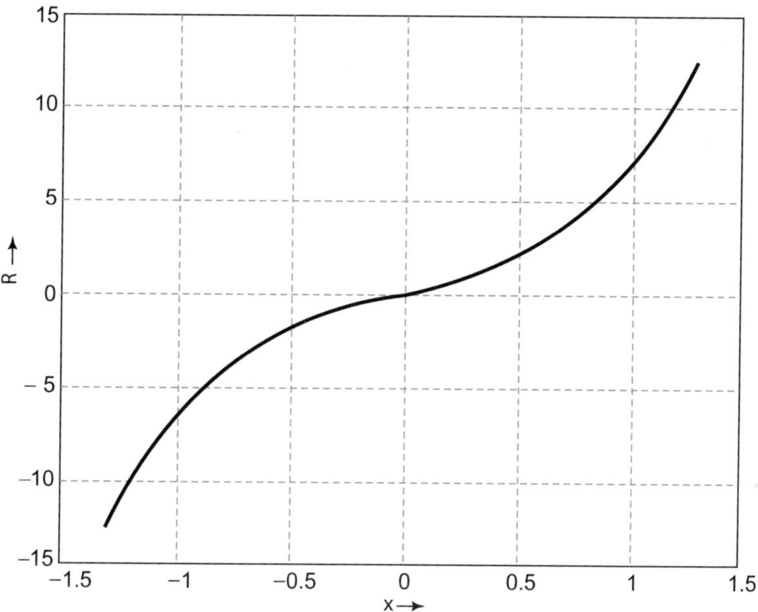

Fig. A.4.2 Plot of $R = 4x^3 + 3x$

positive for positive R. Since x_1 can have any value, we let

$$x_1 = \sinh\beta. \tag{A.4.34}$$

Then Eq. A.4.33 gives

$$x_1 = \sinh\left[\left(\sinh^{-1} R\right)/3\right]. \tag{A.4.35}$$

As in the earlier case, the cubic equation Eq. A.4.33 can be factorized as

$$(x - x_1)(4x^2 + 4x_1 x + 4x_1^2 + 3) = 0 \tag{A.4.36}$$

so that the other two roots which are complex conjugates, are given by

$$x_{2,3} = -[(\sinh\beta)/2)] \pm j\left(\sqrt{3}/2\right)\cosh\beta. \tag{A.4.37}$$

Table A.4.1 gives a summary of the procedure for solving a general cubic equation.

Example

Let the equation to be solved be

$$z^4 - 5z^3 + 5z^2 + 5z - 6 = 0. \tag{A.4.38}$$

As can be easily verified, Eq. A.4.38 has the roots −1, +1 and +5. Let us see what our procedure gives. From Eq. A.4.12, the resolvent cubic becomes

$$y^3 - 5y^2 - y + 5 = 0. \tag{A.4.39}$$

As can be easily verified, the roots of Eq. A.4.39 are −1, +1 and 5, but let us follow the procedure as given here. From Eqs. A.4.22 and A.4.39, R is calculated as $-10/(7\sqrt{7})$. Since $|R| < 1$, we get, by applying Table A.4.1,

$$x_{1,2,3} = 0.756, 0.189, -0.945. \tag{A.4.40}$$

Table A.4.1 Procedure for solving a cubic equation

Equation to be solved: $y^3 + b_2y^2 + b_1y + b_0 = 0$		
Compute $R = \frac{27b_0 + 2b_2^3 - 9b_1b_2}{2\lvert b_2^2 - 3b_1\rvert^{3/2}}$		
$b_2^2 - 3b_1 > 0$		$b_2^2 - 3b_1 < 0$
$0 < R < 1$	$R > 1$	Any R
$\theta = (\cos^{-1} R)/3$	$\alpha = (\cosh^{-1} R)/3$	$\beta = (\sinh^{-1} R)/3$
$x_1 = \cos\theta$	$x_1 = \cosh\theta$	$x_1 = \sinh\beta$
$x_{2,3} = -\cos(\frac{\pi}{3} \pm R)$	$x_{2,3} = -(\cosh\alpha \pm j\sqrt{3}\sinh\alpha)/2$	$x_{2,3} = -(\sinh\beta \pm j\sqrt{3}\cosh\beta)/2$
$y_i = -\left(b_2 + 2\lvert b_2^2 - 3b_1\rvert^{1/2} x_i\right)/3, i = 1, 2, 3$		

Note The solutions are valid for positive R. For negative R, compute the solutions for $|R|$, and then negate the values of x before substitution in the equation for Y_i

Correspondingly,

$$y_{1,2,3} = \left(5 - 4\sqrt{7}x_{1,2,3}\right)/3 = -1, 1, 5. \quad (A.4.41)$$

These values check with the solutions obtained by inspection of Eq. A.4.39. Selecting $y_1 = 1$ and using Eq. A.4.14, the quadratics then become

$$z^2 - 4z + 3 = 0 \text{ and } z^2 - z - 2 = 0. \quad (A.4.42)$$

These give $z = -1$, 1, 2 and 3 as expected. The same results are obtained had we selected $y_1 = -1$ or 5. It can be verified that a reversal of signs in the constant term of Eq. A.4.14 gives wrong results. This example also demonstrates that the opening sentence on p. 18 of [1] is ambiguous, because it does not say what is to be done if all the three real roots give real coefficients in the quadratic equations. As the present example shows, any of them can be used. A question that arises at this point is the following: Would it give the correct answer if we choose a complex, instead of a real root for y_1? The answer is yes.

For example, the quartic equation

$$z^4 - 2z^2 + \sqrt{3}z - 0.5 = 0 \quad (A.4.43)$$

has the following resolvent cubic:

$$y^3 + 2y^2 + 2y + 1 = 0 \quad (A.4.44)$$

Equation A.4.44 has a real root at $y = -1$ and a pair of complex roots at $y = \left(-1 + j\sqrt{3}\right)/2$. Using either of the complex roots, it can show that we get the correct roots as obtained by using $y_1 = -1$. The real root is preferred because it reduces the computational effort, considerably.

Problems

P.1. Can you solve a sixth order equation by decomposing as in Eq. A.4.3?
P.2. How about an eight order?
P.3. Could you solve a cubic equation by starting with trigonometry right from the start?
P.4. Solve $x^4 + x^2 + 1 = 0$ by any method.

P.5. Will the trigonometric approach work for P.4? Yes or no answer will not do. You must have the necessary mathematical support to justify your answer.

Acknowledgement The author thanks his former student and current colleague, Professor Jayadeva, for many helpful discussions on this topic during their evening walks in the corridors of IIT Delhi.

References

1. M. Abramowitz, I.A. Stegun, *Handbook of Mathematical Functions* (Dover, 1965)
2. G.A. Korn, J.M. Korn, *Mathematical Handbook for Scientists and Engineers* (McGraw-Hill, 1968)
3. R.S. Burington, *Handbook of Mathematical Tables and Formulas* (McGraw-Hill, 1973)
4. C.E. Pearson, *Handbook of Applicable Mathematics* (Van Nostrand, 1974)
5. W. Gellet et al., *The VNR Concise Encyclopaedia of Mathematics* (Van Nostrand, 1975)
6. E.W. Weisstein, *CRC Concise Encyclopaedia of Mathematics* (Chapman and Hall, 1999)
7. I.N. Bronshtein et al., *Handbook of Mathematics* (Springer, Berlin, 2000)
8. http://www.sosmath.com/algebra/factor/fac12/fac12.html: The quartic formula
9. http://mathforum.org/dr.math/faq/cubic.equations.html: Cubic and quartic equations
10. http://www.sosmath.com/algebra/factor/fac1/fac11.html: The cubic formula
11. http://mathforum.org/dr.math/faq/cubic.equations.html: Cubic equations—another solution
12. S. Neumark, *Solution of Cubic and Quartic Equations* (Pergamon, 1965)

A.5: Many Ways of Solving an Ordinary Linear Second Order Differential Equation with Constant Coefficients

There are many different ways of solving an ordinary linear second order differential equation with constant coefficients. Some of them are available in textbooks while some others are scattered in journal publications. A comprehensive survey of these methods is presented in this section, along with the essential steps in each method and the relevant references.

'As many faiths, as many ways'—Shri Ramakrishna Paramhansa

Keywords

ODE • Solution

Introduction

There are many problems in electrical engineering and physics where one is required to solve the following ordinary linear second order differential equation:

$$y'' + 2\alpha y' + \omega_0^2 y = 0, \tag{A.5.1}$$

where, in the usual situation, the prime denotes differentiation with respect to time *t*, and *a* and ω_0 are constants, subject to the initial conditions

$$y(0) = y_0 \text{ and } y'(0) = p_0. \tag{A.5.2}$$

For example, when a capacitor *C* is charged to a voltage *V* and discharged through an inductance *L* in series with a resistance *R*, the current *y* in the circuit obeys Eq. A.5.1 with $\alpha = R/(2L)$ and $\omega_0^2 = 1/(LC)$ [1]. Many techniques exist in the literature for solving Eq. A.5.1, of which the following are commonly available in one textbook or the other; (1) Laplace transform method; (2) assuming an exponential solution; and (3) operator method. Several other methods have appeared in the literature, mostly in journals, some of which are quite simple, innovative, and/or of pedagogical interest. We present here a survey of all these techniques, along with the essential steps in each method and the relevant reference(s).

Source: S. C. Dutta Roy, "Many Ways of Solving an Ordinary Linear Second Order Differential Equation with Constant Coefficients," IETE Journal of Education, vol. 48, pp. 73–76, April–June 2007.

S. C. Dutta Roy, *Circuits, Systems and Signal Processing*,
https://doi.org/10.1007/978-981-10-6919-2

Laplace Transform Method

Taking the Laplace transform of Eq. A.5.1 and denoting the Laplace transform of $y(t)$ by $Y(s)$, we get

$$s^2Y(s)-sy_0-p_0+2\alpha[sY(s)-y_0]+\omega_0^2Y(s)=0. \quad (A.5.3)$$

On simplification and factorization, this gives

$$Y(s)=\frac{(s+2\alpha)y_0+p_0}{(s-s_1)(s-s_2)}, \quad (A.5.4)$$

where

$$s_{1,2}=-\alpha\pm\beta \text{ and } \beta=\sqrt{\alpha^2-\omega_0^2}. \quad (A.5.5)$$

Expanding Eq. A.5.4 in partial fractions and taking the inverse Laplace transform, we get

$$y(t)=A_1e^{s_1t}+A_2e^{s_2t}, \quad (A.5.6)$$

where

$$A_{1,2}=\pm\frac{2\alpha y_0+p_0+s_{1,2}y_0}{2\beta} \quad (A.5.7)$$

Combining Eqs. A.5.5–A.5.7 and simplifying, one obtains

$$y(t)=(e^{-\alpha_t}/\beta)\,[(y_0\alpha+p_0)\sinh\beta t+y_0\beta\cosh\beta t]. \quad (A.5.8)$$

As shown in [1], this expression is adequate for considering all the three cases, *viz.* (*i*) overdamping: $\alpha > \omega_0$; (*ii*) critical damping: $\alpha = \omega_0$; and (*iii*) underdamping: $\alpha < \omega_0$. We shall not therefore pursue these cases separately, at this stage.

It is interesting to observe that expression Eq. A.5.8 is also adequate for considering the undamped case i.e. $a = 0$. Under this condition,

$$\beta=\pm j\omega_0, \sinh\beta t=\pm j\sin\omega_0 t \text{ and } \cosh\beta t = \cos\omega_0 t. \quad (A.5.9)$$

Putting these values in Eq. A.5.8 and simplifying, we get

$$y(t)=y_0\cos\omega_0 t+(p_0/\omega_0)\sin\omega_0 t, \quad (A.5.10)$$

which can be put in the form

$$y(t)=\sqrt{y_0^2+(p_0/\omega_0)^2}\cos\{\omega_0 t+\tan^{-1}[p_0/(y_0\omega_0)]\} \quad (A.5.11)$$

Assuming an Exponential Solution: Why Not Do It With Trigonometric Functions? Because of Approximation Errors

In this method, we assume a solution of the form Ae^{st} for $\alpha \neq \omega_0$ and $(A + Bt)e^{st}$ for $\alpha = \omega_0$. In the first case, putting the assumed solution in Eq. A.5.1 gives the so-called characteristic equation

$$s^2+2\alpha s+\omega_0^2+0 \quad (A.5.12)$$

which has the roots given by Eq. A.5.5. Thus both e^{s_1t} and e^{s_2t} are solutions of Eq. A.5.1 and the general solution is the same as that given by Eq. A.5.6. It is easily shown that for satisfying the initial conditions given in Eq. A.5.2, $A_{1,\,2}$ are the same as given in Eq. A.5.7 so that the required solution is given by Eq. A.5.8. As mentioned earlier, this expression can handle all the three cases of relative values of α and ω_0, but it is instructive to pursue the assumed solution $(A + Bt)e^{st}$ for the case $\alpha = \omega_0$ a bit further. As pointed out in [1], a heuristic justification for this assumed solution comes from the argument that in this case, one can try a general solution of the form $f(t)e^{st}$. instead of Ae^{st}, where $f(t)$ is to be determined. Substituting this trial solution in Eq. A.5.1 and simplifying, we get

$$f''(t)+(s^2+2\alpha s+\omega_0^2)f'(t)+2(s+\alpha)f(t)=0. \quad (A.5.13)$$

Since from Eq. A.5.12, $s = -\alpha = -\omega_0$ in this case, Eq. A.5.13 reduces to

$$f''(t) = 0, \qquad (A.5.14)$$

i.e.

$$f(t) = A + Bt \text{ and } y(t) = (A + Bt)e^{st}. \qquad (A.5.15)$$

Evaluating A and B from the initial conditions, we finally get, for this critically damped case,

$$y(t) = [y_0 + (p_0 + y_0\alpha)t]\, e^{-\alpha_t}. \qquad (A.5.16)$$

Operator Method

In this method, we define the operator $D = \frac{d}{d_t}$ and $D^2 = \frac{d^2}{d_{t^2}}$ so that Eq. A.5.1 becomes

$$(D^2 + 2\alpha D + \omega_0^2)y = 0. \qquad (A.5.17)$$

We then treat the quadratic operator $(D^2 + 2\alpha D + \omega_0^2)$ as an algebraic expression and factorize it to obtain the following changed form of Eq. A.5.17:

$$(D - s_1)(D - s_2)y = 0, \qquad (A.5.18)$$

where $s_{1,2}$ arc given by Eq. A.5.5. Now let

$$(D - s_2)y = z \qquad (A.5.19)$$

so that Eq. A.5.18 becomes the following first order homogeneous equation in z:

$$(D - s_1)z = 0. \qquad (A.5.20)$$

The solution of Eq. A.5.20 is

$$z(t) = K_1 e^{s_1 t}. \qquad (A.5.21)$$

where K_1 is a constant. Now combine equations A.5.19 and A.5.21 and solve the resulting non-homogeneous first order differential equation by the integrating factor method. The result is of the same form as Eq. A.5.6. The rest of the procedure is the same as in the previous section.

The critical damping case poses no problem with this method. Under critical damping, Eq. A.5.18 becomes

$$(D + \alpha)\,(D + \alpha)y = 0 \qquad (A.5.22)$$

which can be solved by following the same steps as in the case $\alpha \neq \omega_0$, and results in the same expression as Eq. A.5.16.

Solution by Change of Variable

As pointed out in [1], for the beginner student, who has not been exposed to Laplace transforms, the conventional approach is to use either an assumed solution or the operator method. In either case, the student has conceptual difficulty in accepting why a solution should be assumed, and that too of a particular type, or why with $D = \frac{d}{d_t}$, D^2y is $\frac{d^2y}{d_{t^2}}$ and not $\left(\frac{d_y}{d_t}\right)^2$, and how $(D^2 + 2\alpha D + \omega_0^2)$ can be treated as a polynomial and factorized. To obviate these difficulties, we proposed in [1] a change of variable from y to z with

$$z = y' - sy \qquad (A.5.23)$$

where s is an unknown constant. Obtain y'' from Eq. A.5.23 and substitute in Eq. A.5.1; the result is

$$z' + (s + 2\alpha)z + (s^2 + 2\alpha s + \omega_0^2)y = 0 \quad (A.5.24)$$

Now choose s such that the y term vanishes in Eq. A.5.24; this gives the same equation as Eq. A.5.12 with the possible values of s as given in Eq. A.5.5. Taking either value of s and solving the first order homogeneous equation in z, we get

$$z(t) = K_1 e^{-(s + 2\alpha)t}. \qquad (A.5.25)$$

Putting this value in Eq. A.5.23 and solving for y gives

$$y(t) = K_2 e^{st} + K_3 e^{-(s+2a)t}. \quad (A.5.26)$$

where $K_{2,3}$ are constants. Note that in Eq. A.5.25 taking either $s = s_1$ or $s = s_2$ makes no difference, because

$$\begin{aligned} -(s_{1,2} + 2\alpha) &= -(-\alpha \pm \beta + 2\alpha) \\ &= -\alpha \mp \beta = s_{2,1}. \end{aligned} \quad (A.5.27)$$

Thus the solution is of the same form as Eq. A.5.6.

It is to be recognized that the clue to the method is provided by the operator method but it has the advantage that the student has no difficulty in comprehending this solution. Also, note that when $s_1 = s_2 = -a$ (critical damping case), Eq. A.5.25 becomes $z(t) = K_1 e^{-\alpha_t}$. Putting this value in Eq. A.5.23, solving for y and evaluating the constants lead to the same result as Eq. A.5.16.

We next discuss some less known methods (LKM) for solving Eq. A.5.1, which do not appear to be included in textbooks, but deserve to be.

LKM 1: Modified Operator Method

This method is due to Garrison [2, 3] and starts with rewriting Eq. A.5.1 as

$$\left(\frac{d^2}{dt^2} + 2\alpha\frac{d}{dt}\right) y = -\omega_0^2 y \quad (A.5.28)$$

Add $a^2 y$ to both sides to get

$$\left(\frac{d^2}{dt^2} + 2\alpha\frac{d}{dt} + \alpha^2\right) y = \left(\alpha^2 - \omega_0^2\right) y. \quad (A.5.29)$$

As in the operator of Sect. 4, the operator on the left-hand side of Eq. A.5.29 can be factored as $\left(\frac{d}{d_t} + \alpha\right)\left(\frac{d}{d_t} + \alpha\right)$. Denoting either factor by D_m, Eq. A.5.29 combined with Eq. A.5.5 gives

$$D_m^2 y = \beta^2 y. \quad (A.5.30)$$

Since two operations by D_m is equivalent to multiplication by β^2, we see that

$$D_m y = \pm \beta y \quad (A.5.31)$$

If we take positive sign in Eq. A.5.31, and solve the resulting first order non-homogeneous equation, we get $y_1(t) = A_1 e^{s_1 t}$, while taking the negative sign gives $y_2(t) = A_2 e^{s_2 t}$.Thus the general solution is the same as Eq. A.5.6. For the critical damping case, let $z = D_m y$ [4]; then the equation to be solved is $D_m z = 0$; further procedure is similar to that in the operator method of Sect. 4.

LKM 2: Another Change of Variable Method

This method is due to Greenberg [5] and is based on a change of variable such that the first differential coefficient term is eliminated; simultaneously, it also removes the difficulty in analyzing the critical damping case. We let

$$y = z e^{-\alpha t}. \quad (A.5.32)$$

Substituting this in Eq. A.5.1 and simplifying, we get

$$z'' = \beta^2 z, \quad (A.5.33)$$

β being the same as in Eq. A.5.5. Now choose the trial solution $z = e^{\lambda t}$; putting this in Eq. A.5.33 given $\lambda^2 = \beta^2$ or $\lambda = \pm\beta$ so that the general solution for z becomes

$$z(t) = A_1 e^{\beta t} + A_2 e^{-\beta t}. \quad (A.5.34)$$

Combining Eq. A.5.34 with Eq. A.5.32, we get the same solution as Eq. A.5.6. For the critical damping case, $\beta = 0$, so that Eq. A.5.33 gives $z'' = 0$ or $z = A + Bt$, and combined with Eq. A.5.32, we get the same solution as Eq. A.5.15.

Instead of a trial solution, one could also make a change of variable from z to $x = z' - sz$ as in the method discussed in Sect. 4. Then $z'' = x' + sz' = x' + sx + s^2 z$. Substituting this in Eq. A.5.33 and choosing s such that the z-term is absent in the result leads to $s = \pm\beta$ and $x' + sx = 0$, which has the solution $x = K_1 e^{-st}$. Finally, solving the equation $z' - sz = K_1 e^{-st}$, we get $z(t) = K_2 e^{st} + K_3 e^{-st}$, which is of the same form as Eq. A.5.34, irrespective of whether $s = +\beta$ or $s = -\beta$.

State Variable Method

This method involves more efforts than in any other method discussed so far, but is of pedagogical interest, when the state variables are first introduced to undergraduate students [6]. We let

$$x_1 = y \text{ and } x_2 = y'. \qquad (A.5.35)$$

Then we can write

$$x_1' = x_2 \text{ and } x_2' = -\omega_0^2 x_1 - 2\alpha x_2. \qquad (A.5.36)$$

The two equations in Eq. A.5.36 can be written in the familiar matrix from

$$x' = Ax \qquad (A.5.37)$$

where

$$x = \begin{bmatrix} x_1 \\ x_2 \end{bmatrix} \text{ and } A = \begin{bmatrix} 0 & 1 \\ -\omega_0^2 & -2\alpha \end{bmatrix}. \qquad (A.5.38)$$

The solution for Eq. A.5.37, as is well known [7], is

$$x = e^{At} x_0, \qquad (A.5.39)$$

where

$$x_0 = \begin{bmatrix} y_0 \\ p_0 \end{bmatrix} \qquad (A.5.40)$$

and e^{At} is the so-called fundamental matrix. The latter can be calculated by using any of the well-known techniques [8]. The result, for this case, is

$$e^{At} = \frac{e^{-\alpha t}}{\beta} \begin{bmatrix} \alpha \sinh \beta t + \beta \cosh \beta t & \sinh \beta t \\ \omega_0^2 \sinh \beta t & \beta \cosh \beta t - \alpha \sinh \beta t \end{bmatrix} \qquad (A.5.41)$$

Combining Eq. A.5.41 with Eqs. A.5.39 and A.5.35 gives the desired $y(t)$, which is the same as that given by Eq. A.5.8.

Conclusion

A comprehensive survey has been presented here of the methods for solving Eq. A.5.1, some of which are well known and some are less known. Conceptually, the method based on change of variables given in Sect. 5 appears to be the simplest and easily comprehensible by the beginner. The less known methods given in Sects. 6 and 7 are quite instructive and should find a place in textbooks. The method based on state variables has a pedagogical value for introducing state variables to the beginner, rather than for solving Eq. A.5.1.

It should be mentioned here that except for the methods of Sects. 6 and 7, all other methods are applicable for solving higher order ordinary linear differential equations with constant coefficients [9].

Problems

P.1. Suppose, for Eq. A.5.1, the initial conditions Eq. A.5.2 are given at 0^-. How would you modify the solution?

P.2. Instead of 0 on the RHS of Eq. A.5.1, let there be a constant. How do you find a solution?

P.3. Repeat with RHS = $f(y)$.

P.4. Repeat with RHS = $f(x)$.

P.5. Suppose the middle term on the LHS of Eq. A.5.1 is $2ayy'$. Can you find a solution?

References

1. S.C. Dutta Roy, Transients in RLC networks revisited. IETE J. Educ. **44**, 207–211 (2003)
2. J.D. Garrison, On the solution of the equation for damped oscillation. Am. J. Phys. **42**, 694–695 (1974)
3. J.D. Garrison, Erratum: on the solution of the equation for damped oscillation. Am. J. Phys. **43**, 463 (1975)
4. S. Balasubramanian, R. Fatchally, Comment on the solution of the equation for damped oscillation. Am. J. Phys. **44**, 705 (1976)
5. H. Greenberg, Further remarks concerning the solution of the equation $\ddot{x}+2\alpha x+\omega^2 x=0$. Am. J. Phys. **44**, 1135–1136 (1976) (Note that in Eqs. (5) and (6) of this contribution, $(\omega^2 - a^2)$ should be replaced by $(a^2 - \omega^2)$)
6. D.S. Zrnic, Additional remarks on the equation $\ddot{x}+2\alpha x+\omega^2 x=0$. Am. J. Phys. **41**, 712 (1973) (Note that in Eq. (3) of this contribution, the sign of the (1, 2) element of the matrix should be positive)
7. S.C. Dutta Roy, An introduction to the state variable characterization of linear systems—part I. IETE J. Educ. **38**, 11–18 (1997)
8. S.C. Dutta Roy, An introduction to the state variable characterization of linear systems—part II. IETE J. Educ. **38**, 99–107 (1997)
9. S.C. Dutta Roy, *Solution of an Ordinary Linear Differential Equation With Constant Coefficients*, unpublished manuscript. That gives me an idea. I should try to publish this manuscript as soon as possible.

A.6: Proofs of Two Chebyshev Polynomial Identities Useful in Digital Filter Design

Alternate proofs of two Chebyshev polynomial identities, which are useful in the design of low-pass recursive digital filters, are presented. As compared to those provided by Yip [1], our proofs appear to be simpler and are direct, rather than inductive.

Keywords
Chebyshev polynomial • Identities • Application in DSP

In 1980, Yip provided proofs of the following two Chebyshev polynomial identities:

$$T_{2N}(x) + 1 = 2\left[T_N(x)\right]^2 \tag{A.6.1}$$

and

$$T_{2N+1}(x) + 1 = (1-x)\left[2\sum_{i=0}^{N}(-1)^i T_{N-1}(x) + (-1)^{N+1}\right]2, \tag{A.6.2}$$

where T_i is the i-th degree Chebyshev polynomial of the first kind. As shown by Shenoi and Agrawal [2], these identities are useful in the design of recursive low-pass digital filters. In proving Eq. A.6.1, Yip first proved the identity

$$T_n(x)T_m(x) = c\frac{1}{2}\left[T_{(m+n)}(x) + T_{|m-n|}(x)\right] \tag{A.6.3}$$

and then substituted $m = n = N$. In proving Eq. A.6.2, Yip used the method of induction. We present here simpler proofs of Eqs. A.6.1 and A.6.2, and in the latter case, we give a direct, rather than inductive proof, based solely on the properties of trigonometric functions.

Proof of the First Identity

Letting $x = \cos\theta$, we have

$$T_i(x) = \cos i\theta \tag{A.6.4}$$

Using Eq. A.6.4 and the trigonometric formula

$$\cos 2\phi = 2\cos^2\phi - 1 \tag{A.6.5}$$

Equation A.6.1 follows easily by putting $\phi = N\theta$.

Source: S. C. Dutta Roy, "Proofs of Two Chebyshev Polynomial Identities Useful in Digital Filter Design" Journal of the IETE, vol. 28, p. 605, November 1982. (Corrections in vol. 29, p. 132, March 1983).

S. C. Dutta Roy, *Circuits, Systems and Signal Processing*,
https://doi.org/10.1007/978-981-10-6919-2

Proof of the Second Identity

Using Eq. A.6.4, the right-hand side of Eq. A.6.2 becomes

$$RHS = (1 + \cos\theta)\left[2\sum_{i=0}^{N}(-1)^i\cos(N-i)\theta + (-1)^{N+1}\right]^2 \quad (A.6.6)$$

Putting

$$2\cos(N-i)\theta = e^{-j(N-i)\theta} + e^{+j(N-i)\theta} \quad (A.6.7)$$

in Eq. A.6.6, the first term within its square brackets, to be denoted by F for brevity, becomes

$$F = e^{jN\theta}\sum_{i=0}^{N}\left(-e^{-j\theta}\right)^i + e^{-jN\theta}\sum_{i=0}^{N}\left(-e^{\theta}\right)^i = \frac{e^{jN\theta}\left[1-\left(-e^{-j\theta}\right)^{N+1}\right]}{1+e^{-j\theta}} + \frac{e^{-jN\theta}\left[1-\left(-e^{j\theta}\right)^{N+1}\right]}{1+e^{j\theta}} \quad (A.6.8)$$

By routine simplification of Eq. A.6.8, we get

$$F = \frac{\cos N\theta + \cos(N+1)\theta - (-1)^{N+1}(1+\cos\theta)}{1+\cos\theta} \quad (A.6.9)$$

Combining Eq. A.6.6 with Eq. A.6.9, and simplifying gives

$$\mathrm{RHS} = [\cos N\theta + \cos(N+1)\theta]^2/(1+\cos\theta) \quad (A.6.10)$$

Now using the trigonometric identity

$$\cos C + \cos D = 2\cos\frac{C+D}{2}\cos\frac{|C-D|}{2} \quad (A.6.11)$$

and Eq. A.6.5 in Eq. A.6.10, the latter simplifies to

$$\mathrm{RHS} = 2\cos^2[(2N+1)\theta/2] \quad (A.6.12)$$

Using Eq. A.6.5 once again gives

$$\begin{aligned}\mathrm{RHS} &= \cos(2N+1)\theta + 1\\ &= T_{2N+1}(x) + 1 = \mathrm{LHS}\end{aligned} \quad (A.6.13)$$

which completes the proof.

Problems

P.1. Can these identities be proved with other kinds of polynomial. For example, a second order one? Try it and let me know.

P.2. Could the second identity be proved by bringing in Euler again? That is, replacing $\cos\theta$ by Re $e^{j\theta}$?

P.3. Prove Eq. A.6.5 with Euler's identity.

P.4. Write $T_{2N}(x)$ as a polynomial in $T_N(x)$.

P.5. Can you prove Eq. A.6.3 without invoking induction? That is, directly?

References

1. P.C. Yip, On a conjecture for the design of low-pass recursive filters. IEEE Trans. ASSP-28, **6**, 768 (1980)
2. K. Sbenoi, B.P. Agrawal, On the design of recursive low-pass digital filters. IEEE Trans. ASSP-28, **1**, 79–84 (1980)

A.7: Computation of the Coefficients of Chebyshev Polynomials

A simple derivation is given of a closed form formula for the computation of the coefficients of Chebyshev polynomials, which, as is well known, are required for the design of equal ripple filters. A modification of the formula is also given for facilitating fast computation.

Keywords
Chebyshev polynomials • Computation • Coefficients

Introduction

Chebyshev polynomials are required in the design of filters in which the pass-band or the stop-band is desired to have equal ripple characteristic [1, 2]. As is well known, elliptic filters, in which both pass- and stop-bands are equal ripple, are the optimum ones. However, their design is rather involved because of the computational complexity of the elliptic functions. Next to them in the category of optimum filters comes the Chebyshev filter. For a normalized Chebyshev low-pass filter with cutoff at 1 rad/s, the magnitude squared function is given by

$$|H(j\omega)|^2 = \frac{1}{1+\varepsilon^2 C_n^2(\omega)} \tag{A.7.1}$$

where $C_n(\omega)$ is the Chebyshev polynomial, defined by

$$C_n(\omega) = \begin{cases} \cos(n\cos^{-1}\omega), & \omega < 1, \\ \cosh(n\cosh^{-1}\omega), & \omega > 1. \end{cases} \tag{A.7.2}$$

$C_n(\omega)$ is usually computed by the recursion relation

$$C_{n+1}(\omega) = 2\omega C_n(\omega) - C_{n-1}(\omega) \tag{A.7.3}$$

with the initial conditions

$$C_0(\omega) = 1 \text{ and } C_1(\omega) = \omega. \tag{A.7.4}$$

Tables for low order $C_n(\omega)$ are available in textbooks. However when n is high, one starts from the two highest n for which entries exist in the Table and then uses Eq. A.7.3 recursively. Clearly, this computation is time consuming. Nguyen [3] derived a recursive formula for the coefficients and formulated some rules for cutting down on the computation time. Johnson and

Source: S. C. Dutta Roy, "Computation of the Coefficients of Chebyshev Polynomials", IETE Journal of Education, vol. 49, pp. 19–21, January–April 2008.

S. C. Dutta Roy, *Circuits, Systems and Signal Processing*,
https://doi.org/10.1007/978-981-10-6919-2

Johnson [4] derived the following closed form representation of $C_n(\omega)$:

$$C_n(\omega) = \sum_{k=0}^{\lfloor n/2 \rfloor} (-1)^k \binom{n}{2k} \omega^{n-2k}(1-\omega^2)^k \quad \text{(A.7.5)}$$

where $\lfloor n/2 \rfloor$ is the integer part of (n/2). They obtained this result by expressing $C_n(\omega)$ of Eq. A.7.2 as

$$C_n(\omega) = \text{Re}\left[\exp(jn\cos^{-1}\omega)\right]. \quad \text{(A.7.6)}$$

An alternative derivation of Eq. A.7.5 was given by Cole [5] by treating Eq. A.7.3 as a difference equation and applying z-transform to it.

In this section, we give a simple derivation of Eq. A.7.5 and a modification of this formula which directly gives the coefficients of ω^{n-2r}, $r =$ 0 to $\lfloor n/2 \rfloor$, and facilitates faster computation as compared to the existing methods.

Derivation

Let

$$\cos^{-1}\omega = \theta \quad \text{(A.7.7)}$$

so that

$$\begin{aligned} C_n(\omega) = \cos n\theta &= \frac{1}{2}\left(e^{jn\theta} + e^{-jn\theta}\right) = \frac{1}{2}\left[\left(e^{j\theta}\right)^n + \left(e^{-j\theta}\right)^n\right] \\ &= \frac{1}{2}\left[\left(\omega + j\sqrt{1-\omega^2}\right)^n + \left(\omega - j\sqrt{1-\omega^2}\right)^n\right] \\ &= \frac{1}{2}\left[\left(\omega - \sqrt{\omega^2-1}\right)^n + \left(\omega + \sqrt{\omega^2-1}\right)^n\right]. \end{aligned} \quad \text{(A.7.8)}$$

Note that this is the same as equation Eq. A.7.12 in [5], derived by using z-transforms. Using the Binomial theorem, we get

$$\left(\omega - \sqrt{\omega^2-1}\right)^n = \sum_{i=0}^{n} \binom{n}{i} (-1)^i \omega^{n-i}\left(\sqrt{\omega^2-1}\right)^i \quad \text{(A.7.9)}$$

If $(-1)^i$ is deleted from Eq. A.7.9, then we get the expansion for $\left(\omega + \sqrt{\omega^2-1}\right)^n$. Substituting these in Eq. A.7.8, we observe that the odd i-terms will cancel. Hence if we let $k = i/2$, then Eq. A.7.8 becomes

$$\begin{aligned} C_n(\omega) &= \sum_{k=0}^{\lfloor n/2 \rfloor} \binom{n}{2k} \omega^{n-2k}\left(\omega^2 - 1\right)^k \\ &= \sum_{k=0}^{\lfloor n/2 \rfloor} (-1)^k \binom{n}{2k} \omega^{n-2k}\left(1-\omega^2\right)^k \end{aligned} \quad \text{(A.7.10)}$$

The last form in Eq. A.7.10 is the same as Eq. A.7.5.

Simplification of Eq. A.7.10

We can write Eq. A.7.10 as

$$C_n(\omega) = \omega^n \sum_{k=0}^{\lfloor n/2 \rfloor} (-1)^k \binom{n}{2k} \left[\omega^{-2}\left(1-\omega^2\right)\right]^k \quad \text{(A.7.11)}$$

The term $[\omega^{-2}(1-\omega^2)]^k$ in Eq. A.7.11 can be expressed as

$$(\omega^{-2}-1)^k = (-1)^k(1-\omega^{-2})^k = (-1)^k \sum_{r=0}^{k} \binom{k}{r} (-\omega^{-2})^r = (-1)^k \sum_{r=0}^{k} \binom{k}{r} (-1)^r \omega^{-2r}. \quad \text{(A.7.12)}$$

Substituting the last form in Eq. A.7.12 for $[(\omega^{-2}(1-\omega^2)]^k$ in Eq. A.7.11, we get

$$C_n(\omega) = \sum_{k=0}^{\lfloor n/2 \rfloor} \binom{n}{2k} \sum_{r=0}^{k} (-1)^r \binom{k}{r} \omega^{n-2r}. \quad \text{(A.7.13)}$$

where (n/2) has the usual significance. Equation Eq. A.7.13 can also be written as

$$C_n(\omega) = \sum_{r=0}^{\lfloor n/2 \rfloor} a_{n-2r}\omega^{n-2r}, \qquad (A.7.14)$$

where

$$a_{n-2r} = \sum_{k=0}^{\lfloor n/2 \rfloor} \binom{n}{2k}(-1)^r\binom{k}{r}. \qquad (A.7.15)$$

Equations A.7.14 and A.7.15 constitute the simplified formula for computation. In using these, note the following additional simplifying features: (1) the quantities $\binom{n}{0}, \binom{n}{2}, \binom{n}{4}, \ldots \binom{n}{2\lfloor n/2 \rfloor}$ are required for each coefficient and can be pre-calculated and stored; (2) $\binom{k}{r} = 0$ for $k < r$; and (3) $\binom{k}{0} = \binom{k}{k} = 1$.

We now illustrate the computation with two examples.

Examples

Consider the case of n = 7. From Eqs. A.7.14 and A.7.15, we get

$$C_7(\omega) = \sum_{r=0}^{3} a_{7-2r}\omega^{7-2r}, \qquad (A.7.16)$$

where

$$a_{7-2r} = \sum_{k=0}^{3} \binom{7}{2k}(-1)^r\binom{k}{r}. \qquad (A.7.17)$$

For various values of r, the coefficients are calculated as follows:

$$r = 0 : a_7 = \sum_{k=0}^{3}\binom{7}{2k} = \binom{7}{0} + \binom{7}{2} + \binom{7}{4} + \binom{7}{6}$$
$$= 1 + 21 + 35 + 7 = 64,$$

$$r = 1 : a_5 = -\sum_{k=0}^{3}\binom{7}{2k}\binom{k}{1} = -\left[\binom{7}{2}\binom{1}{1} + \binom{7}{4}\binom{2}{1} + \binom{7}{6}\binom{3}{1}\right]$$
$$= -(21 \times 1 + 35 \times 2 + 7 \times 3) = -112,$$

$$r = 2 : a_3 = \sum_{k=0}^{3}\binom{7}{2k}\binom{k}{2} = \binom{7}{4}\binom{2}{2} + \binom{7}{6}\binom{3}{2}$$
$$= (35 \times 1 + 7 \times 3) = 56,$$

and

$$r = 3 : a_1 = -\sum_{k=0}^{3}\binom{7}{2k}\binom{k}{3}$$
$$= -\binom{7}{6}\binom{3}{3} = -7 \times 1 = -7.$$

Thus

$$C_7(\omega) = 64\omega^7 - 112\omega^5 + 56\omega^3 - 7\omega. \qquad (A.7.18)$$

Next, consider the example of n = 8. With the experience of the previous example, we can directly write

$$a_{8-2r} = \sum_{k=0}^{4}\binom{8}{2k}(-1)^r\binom{k}{r}$$
$$= (-1)^r\left[\binom{8}{0}\binom{0}{r} + \binom{8}{2}\binom{1}{r} + \binom{8}{4}\binom{2}{r} + \binom{8}{6}\binom{3}{r} + \binom{8}{8}\binom{4}{r}\right]$$
$$= (-1)^r\left[\binom{0}{r} + 28\binom{1}{r} + 70\binom{2}{r} + 28\binom{3}{r} + \binom{4}{r}\right].$$
$$(A.7.19)$$

For various values of r, Eq. A.7.19 gives

$$r = 0 : a_8 = 1 + 28 + 70 + 28 + 1 = 128,$$
$$r = 1 : a_6 = -(28 \times 1 + 70 \times 2 + 28 \times 3 + 4)$$
$$= -256,$$
$$r = 2 : a_4 = -(70 \times 1 + 28 \times 3 + 6) = 160,$$
$$r = 3 : a_2 = -(28 \times 1 + 4) = -32,$$

and

$$r = 4 : a_0 = 1.$$

Thus

$$C_8(\omega) = 128\omega^8 - 256\omega^6 + 160\omega^4 - 32\omega^2 + 1. \quad (A.7.20)$$

Equations A.7.18 and A.7.20 agree with those calculated by using any other method.

Conclusion

A method has been presented for rapid calculation of the coefficients of Chebyshev polynomials of high order. The method should be useful in designing high order equal ripple filters.

Problems

P.1. Any other method that you can find out for deriving Eq. A.7.6?

P.2. Compute $C_{15}(\omega)$.

P.3. Repeat for $C_{16}(\omega)$.

P.4. Compare Eq. A.7.18 with a Butterworth polynomial of the same order. What differences do you observe?

P.5. What about Legendre polynomials? Are you not familiar with Legendre, a cousin of Butterworth? Read Kuo and thy will come to know.

References

1. A. Budak, *Passive and Active Network Analysis and Synthesis* (Houghton Miffin, 1974)
2. H. Lam, *Analog and Digital Filters* (Prentice Hall, 1979)
3. T.V. Nguyen, A triangle of coefficients for Chebyshev polynomials, in *Proceedings of IEEE*, vol. 72 (July 1984), pp. 982–983
4. D.E. Johnson, J.R. Johnson, *Mathematical Methods in Engineering Physics* (Ronals Press, 1965)
5. J.D. Cole, A new derivation of a closed from expression for Chebyshev polynomials of any order. IEEE Trans. Educ. **32**, 390–392 (1989)

Printed in the United States
By Bookmasters